CHEMISTRY OF HIGH-TEMPERATURE MATERIALS

KHIMIYA VYSOKOTEMPERATURNYKH MATERIALOV

ХИМИЯ ВЫСОКОТЕМПЕРАТУРНЫХ МАТЕРИАЛОВ

CHEMISTRY OF HIGH-TEMPERATURE MATERIALS

Edited by N.A. Toropov
Director, I. V. Grebenshchikov Institute of Silicate Chemistry
Academy of Sciences of the USSR
Leningrad, USSR

Translated from Russian by
Cabell B. Finch
Metals and Ceramics Division
Oak Ridge National Laboratory
Oak Ridge, Tennessee

 Springer Science+Business Media, LLC 1969

ISBN 978-1-4899-4826-7   ISBN 978-1-4899-4824-3 (eBook)
DOI 10.1007/978-1-4899-4824-3

The original Russian text, comprising the Proceedings of the Second All-Union Conference on the High-Temperature Chemistry of Oxides, held in Leningrad, November 26-29, 1965, and published for the I. V. Grebenshchikov Institute of Silicate Chemistry by Nauka Press in Leningrad in 1967, has been corrected for the present edition.

Library of Congress Catalog Card Number 74-79891

© 1969 Springer Science+Business Media New York
Originally published by Consultants Bureau in 1969.
Softcover reprint of the hardcover 1st edition 1969
A Division of Plenum Publishing Corporation
227 West 17 Street, New York, N. Y. 10011

CONTENTS

III. SYSTEMS CONTAINING REFRACTORY OXIDE COMPOUNDS

A. Systems Containing Rare-Earth Oxides

B. Refractory Oxide Systems Not Containing Rare-Earth Elements

IV. SILICATE SYSTEMS WITH VOLATILE COMPONENTS

DEVELOPMENT OF HIGH-TEMPERATURE STUDIES
ON RARE-EARTH OXIDES

N. A. Toropov

Physicochemical studies on the simple and complex inorganic oxides which are important new materials of high-temperature technology have been intensified in recent years [1]. Considerable attention has been devoted to determining the physical (i.e., electrical, magnetic, optical) properties of specific crystalline and glassy phases in systems formed by these oxides. We now consider it important to investigate the relationships between the physicochemical diagram, the structure and properties of its various phases (regions of structural and chemical homogeneity apparent on physicochemical analysis), and the synthesis technique and physical characteristics of single crystals corresponding to these physicochemical phases. Single crystals, both natural minerals and synthetics, should serve as objects of this comparative study. Despite the complexity of such studies of diverse aspects of the physicochemical diagram (its several individual phases, natural and synthetic single crystals), we regard it as one of the most fruitful methods of further developing the physicochemical analysis of inorganic systems and of studying the crystalline compounds involved in natural and synthetic processes.

It is quite usual in the current stage of physicochemical research on high-temperature oxides to work on the rare-earth oxide systems. The fields of application for materials and components of high chemical purity are also expanding.

The intensive development of studies of the rare-earth oxide systems first began in 1957 [2] under a broad, general program at the Heterogeneous Equilibria Laboratory of the Institute of Silicate Chemistry.

The chief characteristic of our studies of rare-earth phase equilibria, in contrast to analogous work performed abroad, is the special emphasis placed on determination of the phase diagram of a given system, including a detailed structural and physical study of its phases as single crystals and glasses. Particular attention is devoted to crystallo-optical and x-ray studies on the condensed phases.

The first rare-earth oxide—silicon dioxide phase diagrams were published during 1959-1962 [3-5] by the Institute of Silicate Chemistry. After heterogeneous equilibria and phase diagram determinations on these systems, solid phase reactions were studied in the same systems at an appropriate laboratory of the latter Institute [6, 7]. Syntheses of several rare-earth silicates were performed during 1959 in laboratories abroad, but these studies do not treat the respective phase diagrams. These studies included systematic, analogous syntheses of many other nonsilicate rare-earth phases [10, 11].

The first rare-earth silicate diagram published in the USA (1964) pertains to the Nd_2O_3—SiO_2 system [12]. It basically agrees (in compound stoichiometry, location of solidus and liquidus, etc.) with the diagram we had earlier published in 1961 [13]. Besides the study of the trivalent rare-earth silicates, we ascribe no less importance to the investigation of systems containing the divalent [14] and occasionally tetravalent [15] rare earths. Four divalent europium silicates were prepared in [14], as well as two new divalent samarium silicates and four new divalent ytterbium silicates. The infrared spectra of the latter compounds were studied in [16]. It was shown still earlier, in studies at the Solid Phase Reactions Laboratory of the Institute of Silicate Chemistry, that trivalent cerium forms silicates of a different nature [15]. Cerium dioxide forms no chemical compounds with silicon dioxide.

The reactivity differences of the rare-earth oxides relative to silicon dioxide apparent from the above studies indicate the necessity of proposing new work on the rare-earth—oxygen systems, including determinations of the respective phase diagrams. Studies in this area are still quite scarce.

Two articles in this collection are devoted to a study of the stoichiometry in the rare-earth oxides. Work recently published in the USSR describing the effect of oxygen deficiency in quartz on the ease of its crystallization is likewise of interest [17].

Apart from studies on the rare-earth silicates of different valence states, there is currently a broad development of work on the physical chemistry of various analogs of these compounds. This is of great interest as applied to the problems of producing inorganic refractory materials and various components. For example, reference [18] includes data on gadolinium triphosphate, which are of considerable commercial interest. Systematic studies on the zirconium dioxide—rare-earth oxide phase diagrams are being conducted at the Chinese Institute of Silicates [19-21]. Solid phase reactions were earlier studied in the same system at the Institute of Silicate Chemistry [22-24].

A systematic investigation of the chromium sesquioxide—rare-earth oxide phase diagrams is being conducted in Kiev at the Institute of Material Fabrication Problems. Only recently a study was begun on single-crystal chromites in these systems [26].

Of considerable interest is the work by English researchers on the synthesis conditions, stoichiometry, structural types, and certain optical properties of compounds forming in the rare-earth oxide—niobium pentoxide, and rare-earth oxide—tantalum pentoxide systems [27, 28]. Worth mentioning are the many different structural types and unusual stoichiometries of these compounds.

Analogous studies are now being started at the Solid Phase Reaction Laboratory of the Institute of Silicate Chemistry [29].

X-ray studies are being intensified on variously synthesized single-crystal phases composed of diverse rare-earth silicates, aluminates, chromites, germanates, tantalates, and niobates.

These studies are essentially only in their beginning stage, although many interesting data have already been obtained through infrared spectroscopy [30] and unit cell parameter determinations [31, 32]. Work has started on fine-structure determinations.

LITERATURE CITED

1. N. A. Toropov, V. P. Barzakovskii, V. V. Lapin, and N. K. Kurtseva, Handbook: Phase Diagrams of Silicate Systems, No. 1 [in Russian], Izd. Nauka, Moscow (1965).
2. Bibliography of Works by Research Associates at the Institute of Silicate Chemistry, Academy of Sciences of the USSR, 1948-1961, Leningrad (1963).
3. N. A. Toropov and I. A. Bondar', Izv. Akad. Nauk SSSR, Otdel. Khim. Nauk, No. 3, p. 554 (1959).

4. N. A. Toropov and I. A. Bondar', Izv. Akad. Nauk SSSR, Otdel. Khim. Nauk, No. 2, p. 153 (1960).

5. N. A. Toropov, F. Ya. Galakhov, and S. F. Konovalova, Izv. Akad. Nauk SSSR, Otdel. Khim. Nauk, No. 8, p. 1365 (1961).

6. É. K. Keler, N. A. Godina, and E. P. Savchenko, Izv. Akad. Nauk SSSR, Otdel. Khim. Nauk, No. 10, p. 1728 (1961).

7. É. K. Keler, N. A. Godina, and E. P. Savchenko, Izv. Akad. Nauk SSSR, Otdel. Khim. Nauk, No. 10, p. 735 (1961).

8. J. Warshaw and R. Roy, Bull. Am. Ceram. Soc., Vol. 38, No. 4, p. 169 (1959).

9. J. Warshaw and R. Roy, Progress in Science and Technology of the Rare Earths, Vol. 1, Pergamon Press, New York (1964), p. 203.

10. S. J. Schneider and R. S. Roth, J. Res. Natl. Bur. Stds., Vol. 64A, No. 4, p. 317 (1960). USA.

11. S. J. Schneider, R. S. Roth, and J. L. Waring, J. Res. Natl. Bur. Stds, Vol. 65A, No. 4, p. 345 (1961). USA.

12. R. O. Miller and D. E. Rase, J. Am. Ceram. Soc., Vol. 47, No. 12, p. 653 (1964).

13. N. A. Toropov and T. P. Kiseleva, Tr. Leningr. Tekhnol. Inst. im. Lensoveta, No. 52, p. 76 (1961).

14. I. A. Bondar', N. A. Toropov, and L. M. Koroleva, Izv. Akad. Nauk SSSR, Neorg. Mat., Vol. 1, No. 4, p. 561 (1965).

15. A. I. Leonov, V. S. Rudenko, and É. K. Keler, Izv. Akad. Nauk SSSR, Otdel. Khim. Nauk, No. 11, p. 1925 (1961).

16. T. F. Tenisheva and A. N. Lazarev, Izv. Akad. Nauk SSSR, Neorg. Mat., Vol. 1, No. 4, p. 569 (1965).

17. A. G. Boganov, V. S. Rudenko, and G. L. Bashnina, Izv. Akad. Nauk SSSR, Neorg. Mat., Vol. 2, No. 2, p. 363 (1966).

18. I. V. Tananaev and N. A. Dzhatshivili, Izv. Akad. Nauk SSSR, Neorg. Mat., Vol. 1, No. 4, p. 514 (1965).

19. Lin Tsu-hsiang and Yü Huai-chii, Zh. Kitaisk. Silikatn. Obshch., Vol. 3, No. 3, p. 159 (1964).

20. Lin Tsu-hsiang and Yü Huai-chii, Zh. Kitaisk. Silikatn. Obshch., Vol. 3, No. 4, p. 229 (1964).

21. Lin Tsu-hsiang and Yü Huai-chii, Zh. Kitaisk. Silikatn. Obshch., Vol. 4, No. 1, p. 22 (1965).

22. N. A. Godina and É. K. Keler, Ogneupory, No. 9, p. 426 (1961).

23. É. K. Keler and N. A. Godina, Ogneupory, No. 9, p. 416 (1953).

24. É. K. Keler and N. A. Godina, Dokl. Akad. Nauk SSSR, Vol. 103, No. 2, p. 247 (1955).

25. A. V. Shevchenko, L. M. Lopato, and S. G. Tresvyatskii, Zh. Neorg. Khim., Vol. 1, No. 11, p. 1942 (1965).

26. V. N. Pavlikov, A. V. Shevchenko, L. M. Lopato, and S. G. Tresvyatskii, this volume, p. 57.

27. H. P. Rooksbury and E. A. D. White, Acta Cryst., Vol. 16, No. 9, p. 888 (1963).

28. A. J. Dyer and E. A. D. White, Trans. Brit. Ceram. Soc., Vol. 63, No. 6, p. 301 (1964).

29. E. P. Savchenko, N. A. Godina, and É. K. Keler, this volume, p. 108.

30. A. N. Lazarev, T. F. Tenisheva, I. A. Bondar', and N. A. Toropov, Izv. Akad. Nauk SSSR, Otdel. Khim. Nauk, No. 7, p. 1220 (1963).

31. N. V. Margolis, N. A. Toropov, and Yu. P. Udalov, Kristallografiya, Vol. 9, No. 3, p. 408 (1964); Yu. I. Smolin, Yu. F. Shepelev, I. A. Bondar', and N. A. Toropov, Izv. Akad. Nauk SSSR, Ser. Khim., No. 5, p. 925 (1965).

32. L. A. Harris and C. B. Finch, Am. Mineral., Vol. 50, No. 9, p. 1493 (1965).

PART I

CRYSTAL LATTICE ENERGETICS
AND GENERAL PROPERTY CHARACTERISTICS OF
OXIDE SYSTEMS

ENERGETICS OF MONOMERIC AND POLYMERIC
(RING AND CHAIN) METASILICATE
AND FLUOBERYLLATE RADICALS

R. G. Grebenshchikov

One of the chief problems in crystallochemical energetics is finding quantitative relationships linking the geometric and thermodynamic parameters of atoms, ions, molecules, and crystals. This includes further perfection of the computational methods for determining the energies of free radical formation and the crystal lattice energies.

The atomic, ionic, and metallic radius systems commonly used in crystallochemistry are approximations, and are based on the principle of additivity of atomic or ionic geometric characteristics. It is known that the atomic (ionic) sizes in condensed systems vary up to an average of 15%, depending on coordination and the types of chemical bond joining the surrounding atoms. An assumption of constant atomic (ionic) radii is usually justified for solution of many diverse structural and crystallochemical problems. The present article considers the development of a system of atomic geometric characteristics, the aim of which is to assign such to more complex, multiatomic groupings, and when possible, to determine all the factors influencing the effective radical sizes.

We calculated the thermochemical radii of several tetrahedral $[EO_4]^{2-}$ and $[BeF_4]^{2-}$ anions of cubic symmetry from their coefficients for incorporation into different crystal structures. The thermochemical radii of the $[Si_2O_7]^{6-}$ diortho group and the $[Si_3O_9]^{6-}$ ring radical were computed by an analogous method.

The present discussion, using the metasilicate $[SiO_3]^{2-}$ and fluoberyllate $[BeF_3]^{-}$ anions as examples, treats the energetics of monomeric and polymeric radicals of a single stoichiometric composition. This subject is of interest for determining the effect of polymerization on the stability of these anions under different conditions (vapor and condensed states).

The standard heats of formation of the free anionic radicals $[Si_mO_n]^{(2n-4m)-}$ and $[Be_mF_n]^{(n-2m)-}$ were calculated with an equation derived by K. B. Yatsimirskii, which utilizes the thermodynamic Born—Haber cycle:

$$U_c = m\Delta H_c^0 + n\Delta H_a^0 - \Delta H_{298}^0. \tag{1}$$

Here, ΔH_c^0, ΔH_a^0, and Δ_{298}^0 are the standard heats of formation, respectively, of a gaseous cation, an anionic radical, and a condensed compound; m and n are stoichiometric indices in the crystallochemical formula.

Let us examine the computation of the thermochemical radii of the monomeric SiO_3^{2-} and BeF_3^{-} anions. If the carbon and nitrogen atoms in CO_3^{2-} and NO_3^{-} anions in either the gaseous or crystalline state are at the center of a planar triangle, then (by analogy) the anionic triangle of the SiO_3^{2-} and BeF_3^- radicals should have a displaced central atom (cation), since the coordination ratios ($r_c/r_a \geq 0.3$) of the latter anions require the central atom to be tetrahedrally surrounded. Consequently, the gaseous SiO_3^{2-} and BeF_3^- anions should have a distorted triangular pyramidal configuration. These anions occur in a crystal lattice only as polymeric radicals with tetrahedrally coordinated central atoms. We computed the volume of a $Si_3O_9^{6-}$ ring radical as one third that of a SiO_3^{2-} radical (with $r_a = 2.16$ Å), using as data the volume $V = 128$ Å^3 and $r_a = 3.13$ Å for the $Si_3O_9^{6-}$ ring radical [1]. This computation was based on the identical character of the effective volume of a monomeric SiO_3^{2-} group and that of one link of an analogous tetrahedrally configurated composition in the polymeric (ring or chain) radical, $[n(SiO_3)]^{2n-}$. Taking the ratio $r_a(SiO_4^{4-})/r_a(BeF_4^{2-}) = 2.40/2.29 = 1.05$ and extending it to the SiO_3^{2-} and BeF_3^- monomeric anions, we find that the thermodynamic radius of the latter, $r_a(BeF_3^-)$, is 2.06Å. It is noteworthy that the crystallochemical classification of the fluoberyllates (the basic aspects of which are applicable to the silicates), while not applying to the pseudo-wollastonite-type structure, did fit the fluoberyllate analog $KZnBe_3F_9$, which has a benitoite structure and a ring-like $[Be_3F_9]^{3-}$ anion [12].

In [1] we computed the standard heat of formation of one link of the polymeric metasilicate radical, $\Delta H_a^0(SiO_3^{2-})$, as 120 kcal/g-ion on the basis of the monotonic change in the standard heat of formation for silicon-oxygen radicals of differing degrees of polymerization. Since the atomic SiO_3^{2-} groupings in the $Si_3O_9^{6-}$ rings and $[n(SiO_3)]^{2n-}$ chains are in the same structuro-chemical ratio as the number of closed Si—O—Si and free Si—O. . . (Me) bonds, the heat of formation of one link SiO_3^{2-} of a polymeric radical is independent of the structural type of the silicon-oxygen anion forming the complex. The high charge on the polymeric radicals in which the SiO_3^{2-} anion acts as hypothetical disruption may explain the endothermic reaction on formation of a polymeric SiO_3^{2-} anion, as opposed to the exothermic reactions on formation (under standard conditions) of the stoichiometrically analogous monomeric anions SiO_3^{2-}, CO_3^{2-}, SO_3^{2-}, NO_3^-, etc.

We earlier determined [3] the heats of formation of the Me_2BeF_4-type fluoberyllates on the basis of experimental data on heats of solution of alkali fluoberyllates relative to cyclical changes, using the comparative computational method of Karapet'yants [4]. A comparative evaluation of the heats of formation of the silicates and fluoberyllates using the C. A. Shchukareva isocomponent method allowed determination of the heats of formation. The values are presented in Table 1, which gives the complex lattice energies calculated by completion of cycles for Me_2BeF_4 [3], and by the method of isocomponent complex lattice energies for the $MeBeF_3$- and $MeBe_2F_5$-type compounds [1, 5].* The tabular data are used in Eq. (1), together with handbook values for the heats of formation of gaseous alkali cations, to compute the standard heats of formation of the free polymeric anion fluoberyllate radicals as $\Delta H_a^0(BeF_3^-)_{poly} = -286 \pm 3$ kcal/g-ion, and $\Delta H_a^0(Be_2F_5^-)_{compl\,poly} = -479 \pm 2$ kcal/g-ion.

As is known, the heats of formation of ions in the gaseous and dissolved states are related to their heats of hydration (solvation) by the strict thermodynamic relationship

$$\Delta H_i^0 = \Delta H_{aq}^0 - \Delta H_h^{\ddagger} \pm 101z. \tag{2}$$

The term ±101, the difference between the thermodynamic and electrochemical scales, is assumed to equal the heat of formation of a proton (H^+) in solution, and has the sign of the ionic charge z. Our work proves that with the fluoberyllates it is admissible to use the additive

---

*The heat of sublimation of BeF_2 is taken as 50 kcal/mol according to experimental determinations on the standard heats of formation of crystalline [6] and gaseous [17] BeF_2.

Table 1. Heats of Formation and Complex Lattice Energies for Alkali
Fluoberyllate Lattices (kcal/mole)

Compound	Alkali element				
	Li	Na	K	Rb	Cs
Me_2BeF_4	~8 542 498	18 531 450.2	28 538 414.4	30 534 401	41 536 386
$MeBeF_3$	~5 392_5 266	10 387 244	16 392 230	19 391_5 224	25 392 214
$MeBe_2F_5$	~5 634 315	11 629 294	18 635 282	21 635 276	29 633 264

method for computing the anionic heats of formation in solution. It also proves the justification for using the Born—Buerrum approximation [2] in calculating the heats of hydration of the monomeric silicate and fluoberyllate anions:

$$\Delta H_h^{\ddagger} = -\frac{165z^2}{r_a + 0.4}. \tag{3}$$

Here z is the charge, r_a is the crystallochemical anionic radius, and 0.4 is an empirical correction.

Knowing the radius of the BeF_3^- anion-monomer as r_a = 2.06 Å, we find with Eq. (3) a heat of hydration ΔH_n of -67 kcal/g-ion. For the hypothetical $Be_3F_9^{3-}$ ring anion with a r_a of 2.98 Å, we find a ΔH_n of -441 kcal/g-ion. In analogy to the $Si_3O_9^{6-}$ anion, the heat of formation of free $Be_3F_9^{3-}$ equals the value predicted for the heat of formation of one link of the polymeric BeF_3^- anion: $-286 \cdot 3 = -858$ kcal/g-ion. Solving Eq. (2) for the case of the hypothetical $Be_3F_9^{6-}$ anion, we obtain a value of $-858 - 441 + 101.3 = -996$ kcal/g-ion for ΔH_{aq}^0, or calculating for one link, we obtain ΔH_{aq}^0 (BeF_3^-) = $-996/3 = -332$ kcal/g-ion. The value of the heat of formation of BeF_3^- in solution computed by the additive method, ΔH_{aq}^0 (BeF_3^-) = ΔH_{aq}^0 (F^-) + ΔH_{aq}^0 (BeF_2) = $-78.66 - 251.4 = 330$ kcal/g-ion, practically coincides with the latter value. The energetics of $(BeF_3)^-$ anion formation in solution are therefore practically independent of polymerization for the radicals in question.

Taking the average value of the two ΔH_{aq}^0 of BeF_3^- as -331 kcal/g-ion and substituting this into Eq. (2), we obtain the heat of formation of the monomeric anion, ΔH_a^0 (BeF_3)$_{mon}$, as $-331 + 67 - 101 = -365$ kcal/g-ion. This value is visibly different from the heat of formation of one link of the polymeric anion, ΔH_a^0 (BeF_3^-)$_{pol}$ = -286 kcal/g-ion.

The heats of hydration of the monomeric SiO_3^{2-} anion and the ring-like $Si_3O_9^{6-}$ anion computed with formula (3) are -258 and -1690 kcal/g-ion, respectively. We computed the heat of formation of a SiO_3^{2-} anion in solution (ΔH_{aq}^0) by the method of comparative calculation, using the known value of the isobaric formation potential for this anion in solution: ΔG_{aq}^0(SiO_3^{2-}) = -212 kcal/g-ion [8]. Figure 1 depicts graphically the linear dependence between ΔG_{aq}^0 and ΔH_{aq}^0 for the different types of ions according to data from [8–10]. This gave an adequately reliable value for ΔH_{aq}^0 (SiO_3^{2-}) as -248 kcal/g-ion. A value approximating this is also obtained for the $Si_3O_9^{6-}$ anion through formula (2): ΔH_{aq}^0 = $120.3 - 1690 + 101.6 = -724$ kcal/g-ion, or for calculation of one link of the polymerical radical, ΔH_{aq}^0 (SiO_3^{2-}) = -241 kcal/g-ion. The heat of formation of the gaseous monomeric SiO_3^{2-} anion can be calculated from Eq. (2):

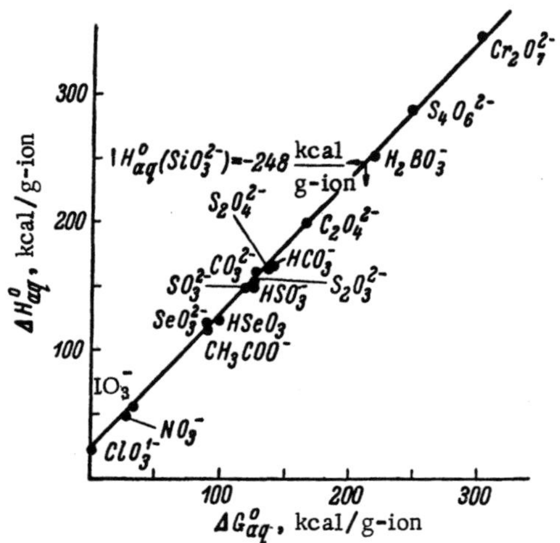

Fig. 1. Dependence between standard isobaric potentials (ΔG^0_{aq}) and heats of formation in solution (ΔH^0_{aq}) for several anions.

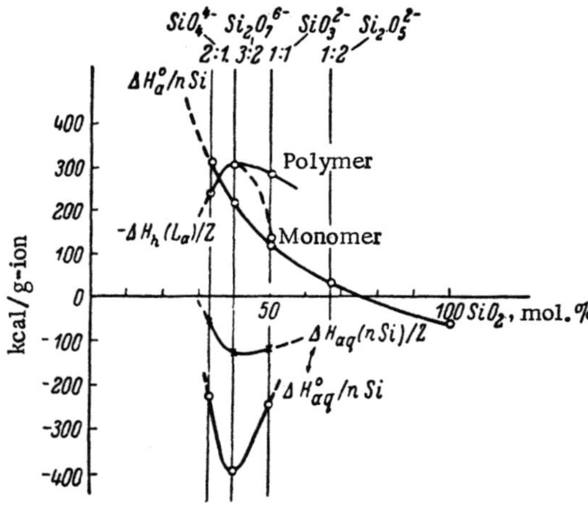

Fig. 2. Change in the basic thermodynamic parameters of silicon-oxygen anions as a function of their degree of polymerization.

$\Delta H^0_a = -248 + 258 - 101 \cdot 2 = 192$ kcal/g-ion. A similar value is also obtained for $\Delta H^0_a(SiO_3^{2-})_{mon}$, using the valence structure model for silicon-oxygen anions proposed by Lebedev [11], and the related computation with the theoretical formula of Yatsimirskii [2].

Silicon and oxygen have only one valence in the metasilicates. Calculation of the formation energy W_2 (using Yatsimirskii notations) of a monomeric SiO_3^{2-} radical form the simplest gaseous ions, $Si^+ + 3O^- = SiO_3^{2-}{}_{(gas)} + W_2$, is performed by the equation

$$W_2 = W_1 + nD_{E-O}, \qquad (4)$$

where W_1 is the electrostatic interaction energy of the ions, and D_{E-O} is the energy of a covalent bond in the radical.

The equation for calculating the standard heat of formation of SiO_3^{2-} is derived from the thermodynamic cycle and is as follows:

$$\Delta H^0_a(SiO_3^{2-}) = \Delta H^0(Si^+) + \Delta H^0(O^-) - W_2. \qquad (5)$$

The first two terms on the right in Eq. (5) are the standard heats of formation of the gaseous Si^+ cation and O^- anion taken from [9]: 277 kcal/g-ion and −13.12 kcal/g-ion, respectively.

The electrostatic interaction energy W_1 of a SiO_3^{2-} radical was computed with the equation

$$W_1 = kNe^2 \left[\frac{z_\kappa z_a n}{r}(1-\rho) - S_p \frac{z^2 n}{r} \right]. \qquad (4a)$$

Here $kNe^2 = 329.7$ kcal/Å; z_c and z_a are the charges on a cation ($Si^+ = 1$) and anion ($O^- = 1$); n is the number of anions coordinated to a central cation in the radical [$n = 3$ for $(SiO_3)^{2-}$]; r is the interatomic spacing (1.6 Å for Si—O); ρ is the repulsion term (equal to 0.22); and S_p is a shielding coefficient (equal to 0.58 for triangular ions).

W_1 equals 122 kcal/g-ion for a SiO_3^{2-} anion. The energy of a covalent bond (D_{Si-O}), is taken by our estimate as ~93 kcal. The total energy is $W_2(SiO_3^{2-}) = 122 + 93.3 = 401$ kcal/g-ion. Substituting the resulting value into (5), we obtain the value of the standard heat of formation as: $\Delta H^0_a(SiO_3^{2-}) = 277 - 39.3 - 401 = -163$ kcal/g-ion.

It is worth noting that, in spite of the use of approximate equations, the heat of formation values obtained for SiO_3^{2-} by the different methods are in good agreement. The average value calculated with formulas (2) and (5) is taken as the heat of formation of $(SiO_3)^{2-}$, $-(192 + 163)/2 = -178$, and is rounded off as $\Delta H^0_a(SiO_3^{2-}) = -180$ kcal/g-ion.

Figure 2 depicts graphically the character of several thermodynamic properties of four types of silicon-oxygen anions having $Si:O$ ratios of $1:4$, $1:3.5$, $1:3$, and $1:2.5$. As shown in Fig. 2, the heats of formation of the free anions in one silicon-oxygen tetrahedron ($\Delta H_a^0 / nSi$) change uniformly from a high positive value for SiO_4^{4-} to a negative value for the neutral, gaseous SiO_2 molecule. The curve describing the dependence of heats of formation in solution per Si tetrahedron ($\Delta H_{aq}^0 / nSi$) or per valence unit ($\Delta H_{aq}^0 (nSi)/z$), has the minimum energetic effect for the $Si_2O_7^{6-}$ anion, where at the same time a maximum value is observed for the heat of hydration per valence unit $[-\Delta H_h(L_a)/z]$.

The behavior described suggests that the $Si_2O_7^{6-}$ and SiO_3^{2-} anions are the most stable in solution compared to the other silicon-oxygen anions.

CONCLUSIONS

1. The thermochemical radii of triangular monomeric BeF_3^- and SiO_3^{2-} anions were determined as 2.06 Å and 2.16 Å, respectively.

2. The heats of formation of the alkali fluoberyllates Me_2BeF_4, $MeBeF_3$, and $MeBe_2F_5$ ($Me \rightarrow Li - Cs$) were found by methods of comparative calculation.

3. The standard heats of formation of free polymeric radicals were determined as follows, using theoretical equations and thermodynamic cycles: $\Delta H_a^0 (BeF_3^-)_{pol} = -286$ kcal/g-ion, $\Delta H_a^0 (Be_2F_5^-)_{pol} = -479$ kcal/g-ion, and $\Delta H_a^0 (SiO_3^{2-})_{pol} = +120$ kcal/g-ion. The values determined for monomeric radicals of analogous composition are $\Delta H_a^0 (BeF_3^-)_{mon} = -365$ kcal/g-ion and $\Delta H_a^0 (SiO_3^{2-})_{mon} = -180$ kcal/g-ion. The standard heats of formation in solution and heats of hydration were computed for metasilicate and fluoberyllate ions as $\Delta H_{aq}^0 = -331$ kcal/g-ion, $\Delta H_h = -67$ kcal/g-ion (monomeric) for BeF_3^- and $\Delta H_{aq}^0 = -248$ kcal/g-ion and $\Delta H_h = -258$ kcal/g-ion (monomeric) for SiO_3^{2-}.

4. It was shown on the basis of BeF_3^- and SiO_3^{2-} that the energetics of formation of these anions in the standard state in solution is practically independent of their degree of polymerization, since the heats of formation of the free radicals vary as a function of structuro-chemical parameters and, for the case of the SiO_3^{2-} anion, differ from one another quite sharply.

LITERATURE CITED

1. R. G. Grebenshchikov, Zh. Neorg. Khim., Vol. 9, No. 5, p. 1038 (1964).
2. K. B. Yatsimirskii, Thermochemistry of Complex Compounds [in Russian], Izd. Akad. Nauk SSSR, Moscow (1951).
3. R. G. Grebenshchikov and N. A. Toropov, Izv. Akad. Nauk SSSR, Neorg. Mat. (in press 1967).
4. M. Kh. Karapet'yants, Methods of Comparative Computation of Physicochemical Properties [in Russian], Izd. Nauka, Moscow (1965).
5. R. G. Grebenshchikov and N. A. Toropov, Dokl. Akad. Nauk SSSR, Vol. 158, No. 3, p. 710 (1964).
6. V. P. Kolesov, M. M. Popov, and S. M. Skuratov, Zh. Neorgan. Khim., Vol. 4, No. 6, p. 1233 (1959).
7. H. A. Skinner, Ann. Rev. Phys. Chem., Vol. 15, p. 449 (1964).
8. W. M. Latimer, Oxidation Potentials, 2nd ed., Prentice-Hall, New Jersey (1952).
9. F. D. Rossini et al., Selected Values of Chemical Thermodynamic Properties, Circ. 500, NBS (1952).
10. M. Kh. Karapet'yants, Chemical Thermodynamics [in Russian], Goskhimizdat, Moscow (1953).
11. V. I. Lebedev, The Fundamentals of Energetic Analysis of Geochemical Processes [in Russian], Izd. Leningr. Gos. Univ. im. A. A. Zhdanova (1957).
12. S. Aléonard and J. LeFur, Bull. Soc. Franc. Crist., Vol. 89, No. 4, p. 425 (1966).

METHOD FOR COMPUTING THE ELECTROSTATIC ENERGY OF THE COMPLEX AND DEFECT LATTICES OF IONIC COMPOUNDS

A. G. Boganov, I. I. Cheremisin,
and V. S. Rudenko

A number of refractory oxides (ZrO_2, HfO_2, Nd_2O_3, Pr_2O_3, Tb_2O_3, etc.) may be considered under certain assumptions as having predominantly ionic bonding. This circumstance considerably simplifies the approach to computing the lattice energies of such compounds, since it allows the lattice energy to be represented as a sum of the electrostatic and repulsive energy components. The Coulomb interaction energy can be determined only by a computational method.

A number of articles have lately appeared in the Soviet, and especially in the foreign literature, which are also devoted to computation of the energies of the repulsive forces [1-3]. Covalency corrections can be applied to compounds of ionic bonding, and these have already been evaluated [4].

In the particular case of the present study it was of interest to assess, within confines of an ionic model, the comparative stability of the low- and high-temperature modifications of the rare-earth sesquioxides (e.g., the C- and A-forms of Pr_2O_3). It was also of interest to evaluate the lattice energies of the intermediate, nonstoichiometric phases of the compound PrO_x ($1.5 \leq x \leq 2$), which have defect structures.

It should be stated that the thermodynamic stability of the low-temperature forms of several of the rare-earth oxides at sub-transition temperatures depends (as determined experimentally [5, 6]) on minute disruptions in stoichiometry.

The existing methods of computing the Madelung constant for ionic lattices (Ewald method, Born base potential method, Madelung method, etc.) are either complicated and tedious when applied to a random, complex crystal lattice, or have been perfected only for the simple NaCl- and CsCl-type lattices (methods of Evjen, Heiendal, and Franck).

In this connection, we developed a direct method [5] for computing the lattice sums of distorted and complex cubic lattices, and a special superpositional method for computing the Madelung constant of defect lattices.

The electrostatic interaction energy of the ions in a crystal lattice can be represented as a series:

$$U^{\cdot} = \frac{1}{2} \sum_i \sum_j {}' \frac{z_i z_j}{r_{ij}} e^2.$$

(1)

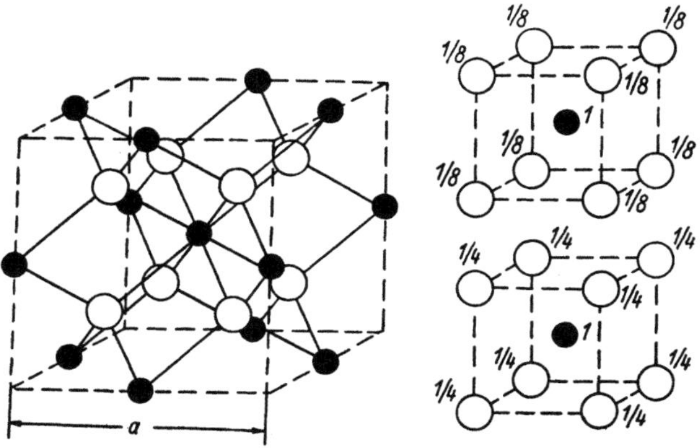

Fig. 1. Unit cell of CaF_2. ● Cu^{2+} ions; ○ F^- ions.

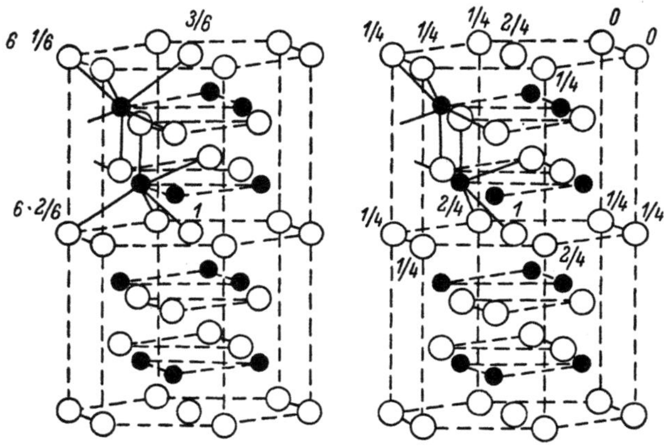

Fig. 2. Computational cell of $A-Pr_2O_3$. ● Pr^{3+} ions; ○ O^{2+} ions.

As shown by Evjen [7], the convergence of series (1) in the direct method computations can be sharply improved if summation over the coordination spheres is replaced by summation over the concentric surfaces congruent to the unit cell at intervals which are lattice constant multiples, with the ion whose energy is being calculated at the center.

Rapid convergence of series (1) occurs in the direct method calculations only on condition of lattice electroneutrality.

The fractions of ions located on a face, edge, and corner of a cubic lattice cell are respectively assumed (according to Evjen) as $\frac{1}{2}$, $\frac{1}{4}$, and $\frac{1}{8}$, since these ions belong simultaneously to two, four, or eight adjacent cells. Here the potential induced by the unit cell multifield decreases with distance proportional to r^{-7} for the NaCl-, r^{-5} for the CsCl-, and r^{-4} for the ZnS-type lattice (r is the distance from the center of the cell to the given point). This can be shown by resolving the potential induced by the unit cell multifield into a series of spherical Legendre functions.

The study shows that the direct method in the form proposed by Evjen cannot be used in calculating distorted or complex cubic lattices, since certain difficulties arise here relative to the choice of the basic computational cell and relative to the weights given adjacent ions.

Let us assume, for example, that we have a cubic fluorite lattice (Fig. 1). If the weight of the bordering F^- ions in a CaF_2 cell having a Ca^{2+} ion at its center and edges $a/2$ is taken as $\frac{1}{8}$ (in accordance to Evjen), then the charge on the cell will be $(+1)$. An absence of electroneutrality results for the Ca^{2+}-centered cells with edges of $3a/2$, $5a/2$, $7a/2$, ..., and for the F^--centered cells with edges of $a/2$, $3a/2$, $5a/2$, ...(as well as for the hexagonal cells of type $A-Pr_2O_3$, $\alpha-Al_2O_3$, etc.).

It is easy to see, however, that neutralization of the above computational cells can be achieved if the weight of the bordering ions is determined on the basis of a coordinational principle, i.e., by considering the weight of a bordering ion as equalling the fraction of the computational cell ions in the first coordination sphere of a given ion.

For CaF_2, we actually have that the weight of the bordering F^- ions will not be $\frac{1}{8}$, but $\frac{1}{4}$, since, of the four Ca^{2+} ions in the first coordination sphere, only one belongs to the cell. Under these conditions the charge on the cell becomes equal to zero.

In the other case, which is characteristic of lattices whose ionic coordination numbers do not correspond to compound stoichiometry ($A-Pr_2O_3$), still another difficulty arises, which is also related to the choice of bordering ion weights. As shown in Fig. 2, the Pr^{3+} ions have a sevenfold coordination with the O_I^{2-} and O_{II}^{2-} ions, while the O_I^{2-} ions have a sixfold, and the O_{II}^{2-} ions a fourfold coordination with the Pr^{3+} ions.

Computing weights in correspondence with the above rule does not guarantee the electroneutrality of the computational cells (Fig. 2a).

For example, the $A-Pr_2O_3$ unit cell has a $+\frac{6}{7}$ charge. It turns out, nonetheless, that even in this case the disruption of computational cell neutrality can be neglected if the lattice is made stoichiometric while arriving at the coordinational computation of the ionic weights.

We assume, for example, that of the three O_I^{2-} ions, only two belong to the Pr^{3+} sphere, and we consistently arrange to select these two throughout the whole lattice. Then all of the Pr^{3+} ions will automatically have a sixfold coordination with the oxygen ions $(4O_{II}^{2-} + 2O_I^{2-})$, while the O_I^{2-} and O_{II}^{2-} ions will have a fourfold coordination with Pr^{3+} ions. The computational cells become electroneutral (Fig. 2b), and rapid convergence of the computational series is ensured. The second rule for calculating the bordering ion weights for the direct method calculation of ionic lattice energies therefore consists of making the coordinational environment of each ion uniformly stoichiometric throughout the lattice.

Using a modification of the direct method, the authors computed the lattice energies of the cubic and hexagonal modifications of Pr_2O_3 (C- and A-forms) and also those of $\alpha-Al_2O_3$ and CaF_2 (to check the method).

1. For CaF_2, agreement with the calculations of Emersleben (Born base potential method) was obtained within the limits of 0.003% in computing the lattice energy on the basis of a quadruple cell (the respective Madelung constant values are 5.81842 and 5.81828).

2. For $\alpha-Al_2O_3$, the calculation gave an energy value of $4.337 \cdot 10^3$ kcal/mole against $4.33 \cdot 10^3$ kcal/mole from the calculation of Shmaeling (Ewald method), even though the computation was performed throughout using only a double unit cell.

3. For $A-Pr_2O_3$ (triple unit cell):

$$U^\bullet = -62.39203\, \frac{e^2}{c} = -10.38482\ \text{Å}^{-1}e^2.$$

4. For $C-Pr_2O_3$ (triple unit cell):

$$U^{\bullet} = -56.53320\,\frac{e^2}{a} = -10.16784 \text{ Å}^{-1}e^2;$$

here, $a = A/2$, where A is the $C-Pr_2O_3$ lattice parameter.

It is clear from the above examples that the most essential merit of the direct method is its simplicity, combined with an accuracy adequate for most problems. We primarily use the method for summation of the generalized Madelung series,

$$S = \sum_{x,\,y,\,z} (-1)^{x+y+z}\,\frac{x^m + y^m + z^m}{(x^2 + y^2 + z^2)^{1/2(m+1)}}, \tag{2}$$

which generally cannot be computed by analytical methods, or for that matter by the Ewald method.

The second part of the present work consisted of the calculation of the Coulomb energies of nonstoichiometric compounds of variable composition, e.g., of compounds in the PrO_x, TbO_x, etc., systems in the $1.5 \le x \le 2$ composition range.

To accomplish this task, a new method was developed for calculating the electrostatic energy of defect ionic lattices. A circumstance in the development and use of this method is associated with the fact that on formation of defects in a lattice (especially where they are small in number), the lattice parameter increases many times (e.g., 10 times and more).

Thus, even in the simplest lattice energy calculation by the direct method, the number of ions in a basic defect lattice cell is so large that all advantages of the method are lost due to the considerable expanse of computation.

The concept of the proposed superpositional method brings into consideration some original, defect-free lattice and a lattice of imposed defects. When these two complementary lattices are combined, the changes resulting at the sites should produce a defect ionic lattice, whose energy can be computed. In this case (as shown below), the energy of the defect lattice can be represented as a linear combination of the energy of the original lattice and the energy of the lattice of imposed defects.

Correspondingly, the energies to these two complementary lattices either are presumed as known, or can be computed with the direct method already illustrated.

Several schemes for defect formation in an ionic lattice can evidently be proposed, as well as a scheme for vacancy compensation. We shall examine the two individual cases which appear most interesting to us, namely,

(1) Vacancies form in the sublattice M of an ionic compound M_nN_m and are compensated by a change in the charge of part of the ions of the same sublattice. The overall charge of the sublattice does not change. Such a model can be proposed for defect formation in the FeO lattice, for example.

(2) Vacancies form in the sublattice M of an ionic compound M_nN_m and are compensated by a change in charge of part of the ions in the sublattice N. The charge of both sublattices correspondingly changes to preserve overall electroneutrality. In particular, this can represent the scheme for defect formation in the PrO_2 lattice on transition from PrO_2 to Pr_2O_3.

Let us examine the first case. Assume we have a compound M_nN_m. All changes in the lattice — vacancies and their compensations — occur in the sublattice M. If the total number of vacancies is k and the occurrence of each vacancy is compensated by the change in charge of a corresponding number of ions (a per vacancy), then the charge of such ions should change from m to a value $\mu = m/a$ and become equal to $(m + \mu)$.

By successive transformations an expression can be obtained for the energy of a lattice with vacancies.

$$U = \frac{1}{2}\left[\sum_i U_{i0} + \sum_{l,\,\alpha}^{k,\,a}(U_{Ml\alpha})_0 + \sum_l^k (U_{\otimes l})_0\right] +$$
$$+ \frac{1}{2}\left\{\sum_s^k \sum_l^{k}{}' U_{\otimes l \otimes s} - 2\,\frac{\mu}{m}\sum_s^k \sum_{l,\,\alpha}^{k,\,a} U_{\otimes s Ml\alpha} + \frac{\mu^2}{m^2}\sum_{l,\,\alpha}^{k,\,a}\sum_{s,\,\beta}^{k,\,a}{}' U_{Ml\alpha Ms\beta}\right\}. \tag{3}$$

Here U is the energy of the defect lattice, U_{i0} is the energy of the i-th ion (undisturbed by vacancy formation) in the original defect-free lattice; $(U_{Ml\alpha})_0$ is the energy of an ion α (compensated by a vacancy l) in the original defect-free lattice; $(U_{\otimes l})_0$ is the energy of an ion residing at the site of an l-th vacancy in the original lattice; and $U_{\otimes s \otimes l}$ is the Coulomb interaction energy between ions situated on the sites of s-th and l-th vacancies in the original lattice, etc.

The expression in brackets represents twice the energy of the original lattice, while that in braces is twice the interaction energy of the $(-m)$ charges at the sites of vacancies of $(+\mu)$ charge situated at compensating ion sites.

Thus the energy of the defect lattice is actually the sum of the energies of the original "ideal" lattice and of a lattice of imposed defects of charge $(-m)$ or $(+\mu)$ at the sites of vacancies and compensating ions. Generally speaking, the Madelung constant for the defect lattice is always greater in absolute value than the Madelung constant for the original defect-free lattice, if the overall charge of the sublattice is to be preserved.

It is impossible to confirm by the above scheme the assertions of Ariya et al. [8, 9], that the Madelung constant of FeO will not change on increase in the number of vacancies in the iron sublattice according to the latter scheme. Actually, the constant increases.

The conclusion of Bertaut [10] appears justified, that vacancies and compensating ions will be packed in the lattice with a minimum of self-potential energy (maximum in absolute value).

To exemplify the case under consideration, we shall perform a calculation of the Madelung constant for the CaF_2 and ZnS lattices, using known values for NaCl and CsCl.

a) The CaF_2 lattice may be considered as a defect lattice relative to the CsCl lattice, i.e.: the original lattice is the CsCl (parameter a), the lattice of imposed defects is the NaCl (parameter $2a$), the resulting lattice is CaF_2 (parameter $2a$).

For each two molecules in the CsCl lattice there is one molecule in the NaCl lattice and one molecule in the CaF_2 lattice. Using Eq. (3) for the energy of the defect lattice, we obtain

$$M_{CaF_2} \cdot 2 \cdot 1 \cdot \frac{e^2}{2a} = 2M_{CsCl} \cdot 1 \cdot 1 \cdot \frac{e^2}{a} + M_{NaCl} \cdot 1 \cdot 1 \cdot \frac{e^2}{2a};$$
$$M_{CaF_2} = \frac{4M_{CsCl} + M_{NaCl}}{2} = \frac{4 \cdot 2.035356 + 3.495115}{2} = 5.818270$$

(5.81828 ± 0.00003 — according to Emersleben [11]).

b) Analogously for ZnS: the original lattice is CaF_2 (parameter A), the lattice of imposed defects is NaCl (parameter A), and the resulting defect lattice is ZnS (parameter A). Then

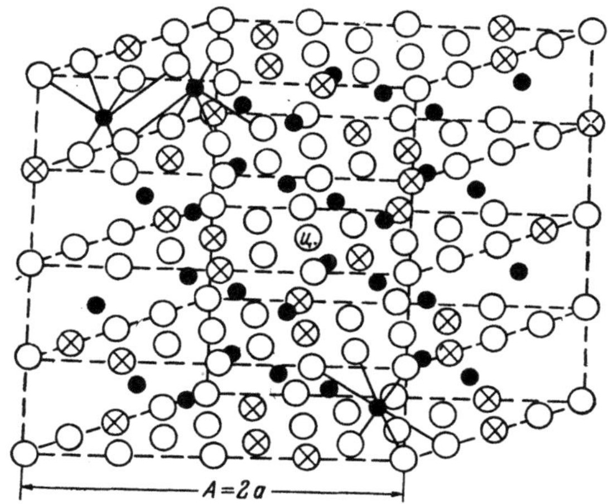

Fig. 3. Unit cell of C-Pr_2O_3. ● Pr^{3+} ions; ○ O^{2-} ions; ⊘ vacant sites in the oxygen sublattice.

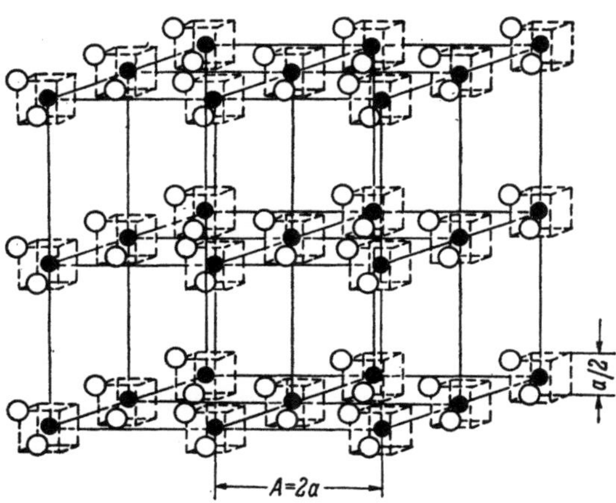

Fig. 4. Complementary lattice of imposed defects (*). ● Point charges of (+2e); ○ point charges of (−e).

$$M_{ZnS} = \frac{2M_{CaF_2} + M_{NaCl}}{4} = \frac{2 \cdot 5.818270 + 3.495115}{4} = 3.782914$$

(3.782920 ± 0.000015 — according to Emersleben [11]).

Let us now examine a second case.

Let the original compound have the formula M_nN_m and have vacancies in its sublattice M, each of which is compensated by b ions in the sublattice N. The charge of these ions changes from n to $(n - \nu)$, where $\nu = |m|/b$.

Carrying out the corresponding computations as in the first case, we obtain the following expression for the energy of the defect lattice:

$$U = \frac{1}{2}\left[\sum_i U_{i0} + \sum_{l,\beta}^{k,b}(U_{Nl\beta})_0 + \sum_l^k(U_{\otimes l})_0\right] - \sum_l^k(U_{\otimes l})_0 - \frac{\nu}{n}\sum_{l,\beta}^{k,b}(U_{Nl\beta})_0 +$$
$$+ \frac{1}{2}\left\{\sum_l^k\sum_s^k{}'U_{\otimes l\otimes s} + 2\frac{\nu}{n}\sum_l^k\sum_{s,\beta}^{k,b}U_{Ns\beta\otimes l} + \frac{\nu^2}{n^2}\sum_{l,\beta}^{k,b}\sum_{s,\delta}^{k,b}{}'U_{Nl\beta Ns\delta}\right\}. \quad (4)$$

We use the expression obtained for calculating the defect lattice of the compound PrO_x in the composition range $1.5 \leq x \leq 2$, i.e., for defect formation in the PrO_2 lattice. In this case, vacancies occur in the oxygen sublattice $M = O^{2-}$ (m = 2), and are compensated by a decrease (by two for each vacancy) in charge of the Pr^{4+} ions, in the $N = Pr^{4+}$ sublattice (n = 4). Since b = 2 and $\nu = 1$, $n - \nu = 4 - 1 = 3$, and formula (4) has the form

$$U = U_0 - \sum_l^k(U_{\otimes l})_0 - \frac{1}{4}\sum_{l,\beta}^{k,2}(U_{Pr\,l,\beta})_0 + U_{dil} \quad (5)$$

or

$$U = U_0 - k(U_{PrO_2\,mol})_0 + U_{dil}, \quad (6)$$

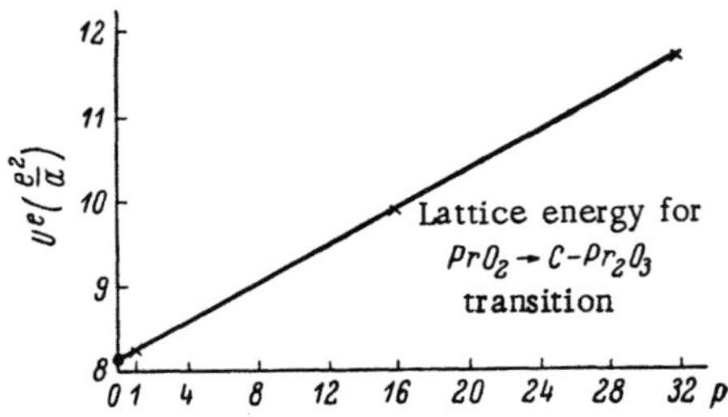

Fig. 5. Coulomb energy for lattices of imposed defects, obtained by combination of p(*) lattices per "molecule" $[(+2e)(-e)_2]$.

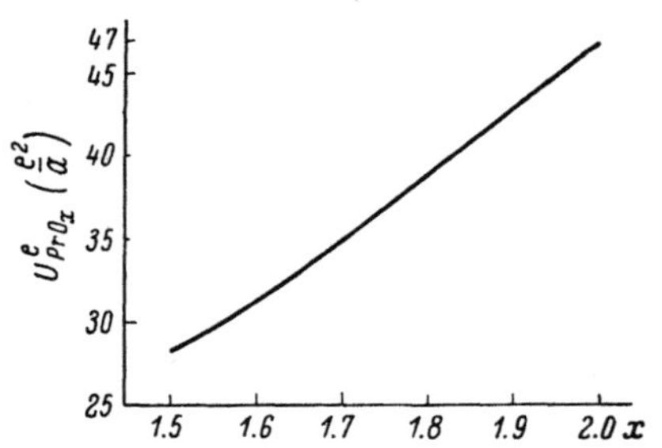

Fig. 6. Electrostatic component of the lattice energy of PrO_x (per molecule of PrO_x).

where k is the number of vacancies in the oxygen sublattice. The latter defect formation scheme is based on the concept that removal of one-fourth of the oxygen ions from the lattice is accompanied by a onefold charge decrease for all the Pr^{4+} ions on the transition from PrO_2 to $PrO_{1.5}$ (Fig. 3).

The lattice of imposed defects consists of (+2) charges at the sites of vacancy formation in the oxygen sublattice, and (−1) charges at the sites of the compensating ions. The energy of the imposed defect lattice, on a per-molecule basis, can be found by the direct method using the form assigned to this lattice.

It is nevertheless possible to use a somewhat different method for the case of the $PrO_2 \rightarrow PrO_{1.5}$ transition. Let us consider the (*) lattice shown in Fig. 4. It is easy to convince oneself that 16 such lattices superimposed form the lattice of imposed defects for the transition from the PrO_2 to $C-Pr_2O_3$ lattice.

Thirty-two (*) lattices form the CaF_2 lattice with the lattice constant $a = A/2$ (A is the lattice constant of $C-Pr_2O_3$ and a is that of PrO_2). By successive defect formation in the CaF_2 lattice, resulting from imposition of the 1, 2, . . . , 31 (*) lattices, energies of the imposed defect lattices can be found which are composed of the respective 31, 30, . . . , 2, 1 (*) lattices. The energy of the (*) lattice itself can be found readily by the direct method.

In the latter way of consideration, expression (4) is written in the form

$$U = U_0 - k\,(2U_{\text{CaF}_2,\,\text{mol}})_0 + U_{\text{dil}}^{\bullet}. \tag{7}$$

Here U is the energy of the defect lattice; U_0 is the energy of the original CaF_2 lattice (m = 2, n = 1, b = 2, ν = 1, n − ν = 0); and k is the number of molecules in the imposed (*) lattice. The calculations established that the behavior of the energy of the imposed defect lattice, on a per-molecule basis, $[(+2e)(-e)_2]$ can be represented as a straight line, with energy and the number of combined (*) lattices as coordinates. The straight line shown in Fig. 5 passes through the value $U_* = -8.252e^2/a$ for p = 1, and $11.636e^2/a$ for p = 32 (CaF_2 lattice). Further extrapolation to zero (single defect) gives the value for the energy of a single defect situated at a (*) lattice site.

The results of the calculations on the PrO_x lattice energies are presented on a per-molecule basis in Fig. 6.

In conclusion, it follows again to stress that computation of the Coulomb energy for lattices having predominantly ionic bonding is only one aspect of the task of finding the total energy of such compounds. It should be accompanied by consideration of covalency corrections, of repulsive force energies, and entropy calculations for the most probable distribution of defects in defect lattices.

It is nevertheless a fact that exact computation of the electrostatic term, which comprises the chief contribution to the lattice energy, will be of considerable interest in those cases which are exemplified by consideration of two energetically similar states of matter, as they occur in polymorphism.

LITERATURE CITED

1. E. C. Baughan, Trans. Faraday Soc., Vol. 55, No. 5, p. 736 (1959).
2. E. C. Baughan, Trans. Faraday Soc., Vol. 55, No. 12, p. 2025 (1959).
3. M. L. Huggins and Y. Sakamoto, J. Phys. Soc. Japan, Vol. 12, No. 3, p. 241 (1957).
4. S. Asano and Y. Tomishima, J. Phys. Soc. Japan, Vol. 11, p. 352 (1956).
5. A. G. Boganov and V. S. Rudenko, Dokl. Akad. Nauk SSSR, Vol. 161, No. 3, p. 590 (1965).
6. V. S. Rudenko and A. G. Boganov, this volume, p. 85.
7. H. M. Evjen, Phys. Rev., Vol. 39, p. 675 (1932).
8. S. M. Ariya, E. Vol'f, T. Grossman, and M. P. Morozova, Zh. Obshch. Khim., Vol. 26, p. 2097 (1956).
9. S. M. Ariya and M. S. Soboleva, Fiz. Tverd. Tela, Vol. 3, No. 10, p. 3157 (1961).
10. E. F. Bertaut, Acta Cryst., Vol. 6, No. 6, p. 557 (1953).
11. O. Emersleben, Phys. Z., Vol. 24, No. 4, pp. 73-97 (1923).

SEVERAL LAWS IN THE FORMATION OF OXIDES OF TRANSITION METALS OF SUBGROUPS IVa AND Va

V. A. Tskhai, P. V. Gel'd, G. P. Shveikin, and V. A. Perelyaev

The intermediate oxides of transition metals of subgroups IVa and Va occur chiefly as monoxides (rocksalt structure), sesquioxides (corundum structure), and dioxides (distorted rutile structure).

All these structures have the metal atom octahedrally surrounded by oxygen atoms. This characteristic is evidently associated with the fact that the d-orbitals of transition metal atoms of subgroups IVa and Va, being populated by one electron, can play a maximum role in bond formation [1].

In actuality, this situation (Fig. 1) involves both Me—O σ bonds (participation of the metal atom d_γ-orbital) and Me—O π bonds (participation of the metal atom d_ε-orbital). The extent of the role of the π bonds increases with an increase in the oxygen content of the compound, i.e., with an increase in metal atom valence relative to that of oxygen. In this process the metal d_ε-orbitals not participating in π-bond formation go into the formation of Me—Me bonds.

The next peculiarity of the intermediate IVa and Va subgroup transition-metal oxides consists of the fact that, in contrast to the majority of the other transition-metal oxides, they display strong Me—Me exchange interaction as a result of the considerable d-orbital overlap [2-4].

Earlier we examined in detail the relationship between the valence-electron and defect-structure properties of the cubic oxides TiO_x ($0.89 < x < 1.20$) and VO_x ($0.85 < x < 1.25$) within limits of their regions of homogeneity. We likewise examined such properties in the series of cubic monoxides FeO, VO, TiO, and NbO [5, 6]. It was shown on the basis of an analysis of the defect properties of the latter monoxides that there is a substantial amount of anionic shielding of the Me—Me bond, which has a decisive influence on many of their properties [7].

The essence of the Me—Me bond shielding consists of the following. The d-orbital overlap of the adjacent metal atoms stipulating formation of the Me—Me bond presupposes an increase in the electron cloud density in those spatial regions where the filled anionic orbitals are nearest each other. This results in the forced exchange of electrons of parallel spin between their states in the Me—Me bond and in the filled anionic orbital states, i.e., it causes such electrons to avoid one another. The latter, generally speaking, should tend to increase the distance between the oxygen atoms. For the case of the NaCl-type structure, however, the increase in distance between the oxygen atoms is associated with the deformation of the Me—O and Me—Me bonds, i.e., with the considerable increase in the energy of the system. In this respect, the exchange interaction in a complete structure leads to an increase of the Me—Me bond

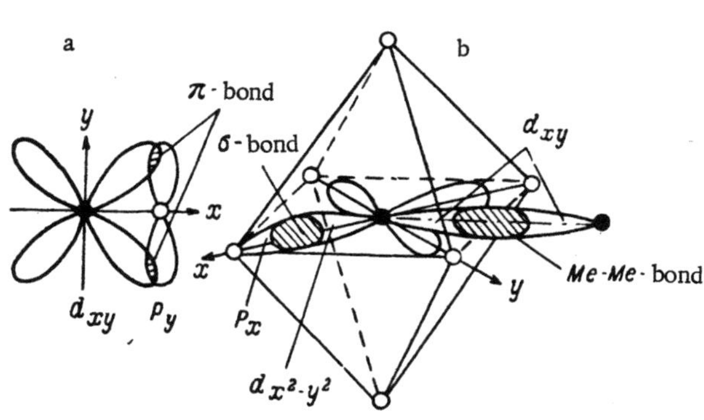

Fig. 1. a) Me—O π bonds; b) Me—O σ bonds and Me—Me bonds. ● Me atom; ○ oxygen atom.

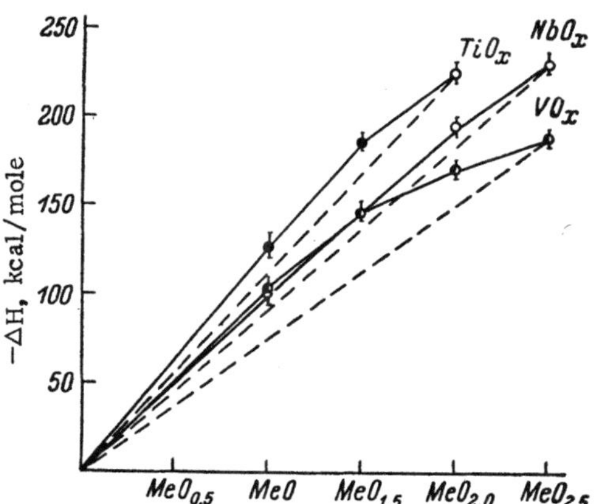

Fig. 2. Composition dependence of the enthalpy of formation (ΔH) for the oxides of Ti, V, and Nb.

electron gas pressure, and raises the energy levels of these electrons. Besides the exchange interaction, the same effect will result in Coulomb repulsion between the Me—Me bond electrons and the anions.

Thus both the exchange and Coulomb interactions raise the energy levels of the electrons in the Me—Me bonds of a complete oxide structure. This effect is called Me—Me bond shielding. The Me—Me bond shielding of Me—O bonds in oxides of noticeably ionic character is evidently stronger where there is greater d_ε-orbital overlap or a greater number of Me—Me bond electrons. The strong Me—Me bond shielding in the NaCl-type structures leads to vacancy formation both in formation of the oxygen and the metallic sublattices, and causes partial or complete transition of the shielded Me—Me bonds to the unshielded state. The authors used these ideas to explain (1962) the reason for the increase in vacancy concentration with the sequence FeO → VO, TiO → NbO [7].

Later Denker [8] mentioned similar considerations in judging the vacancy characteristics of TiO, TiN, and TiC (NaCl-type structure). He also examined molecular models for TiX_6 complexes (X = C, N, O). In TiN and TiO, which have, respectively, nine and ten outer electrons, one or two electrons fall into the disintegrating π orbitals of Me—O interaction. Since the state of these electrons is energetically unfavorable, vacancies form and part of the electrons go to their local levels. We note that the electrons in the disintegrating π-bond orbitals of TiX also are simultaneously the Me—Me bond electrons. The essential difference between the opinions of Denker and ourselves is the fact that the formal isolation of the TiX complex from a NaCl-type lattice automatically excludes the calculation of d_ε-orbital overlap, which is a very important parameter (besides the number of Me—Me bond electrons) in considering vacancy characteristics. The d_ε-orbital overlap of neighboring metal atoms essentially alters the electron density distribution along the Me—Me bond relative to that which would be observed with no overlap. In particular, the growth of d_ε-orbital overlap increases the negative electronic field at an anionic site. However, a simple increase of d_ε-orbital expanse throughout the MeX_6 complex leads to the inverse effect — attenuation of the negative field at anionic sites.

The Me—Me bond shielding essentially can define the thermodynamic conditions for formation of various intermediate oxide phases. For one or another intermediate oxide to be stable, it is necessary for its heat of formation (on a gram-atom of oxygen basis) to be larger than, or at least equal to, the heat of formation of a subsequent oxide of higher oxygen content. In the

Table 1. Structural Characteristics and Overlap Integrals $S(d_\varepsilon, d_\varepsilon)$
for Several Transition Metal Oxides

Oxide comp.	Structural type	Me—Me distance, Å	$S(d_\varepsilon, d_\varepsilon)$	No. Me—Me bond trans.	Structure properties
VO	NaCl	2.868	0.008	3	Defect character 15.5%
TiO	»	2.956	0.018	2	Defect character 16%
NbO	»	2.976	0.057	3	Defect character 25%
ZrO	»	3.14	—	2	Cubic metastable [10]
[ZrO]	»	3.18	0.077	2	—
[TaO]	»	2.98	0.108	3	—
[HfO]	»	3.18	0.128	2	—
V_2O_3	Corundum	2.70 / 2.88	0.009 / 0.005	2	O—O distance on the same face of two given octahedra is greater than for α-Fe_2O_3
Ti_2O_3	»	2.59 / 2.99	0.031 / 0.012	1	
$[Nb_2O_3]$	»	2.83	0.051	2	—
$[Zr_2O_3]$	»	2.88	0.094	1	—
$[Ta_2O_3]$	»	2.83	0.112	2	—
$[Hf_2O_3]$	»	2.88	0.139	1	—
TiO_2	Rutile	2.959	—	0	Undeformed.
VO_2	Distorted rutile	2.65 / 3.12	0.008 / 0.002	1	O—O distance on the same edge of two given octahedra is greater than for TiO_2
NbO_2	Same	2.80 / 3.20	0.051 / 0.021	1	
$[TaO_2]$	»	2.80	0.100	1	—

Note. The hypothetical oxides are shown in brackets. The $S(d_\varepsilon, d_\varepsilon)$ integrals for Ti_2O_3 and V_2O_3 were computed for the two nearest distances in directions parallel and perpendicular to the C axis. For VO_2 and NbO_2 they were computed for the two nearest distances in the [C] axial direction.

reverse case, formation of only the higher oxide would occur due to the large decrease of free energy in the system (if here the small entropy change is neglected). The above is illustrated graphically in Fig. 2, which shows the change in the heats of formation of oxides in the Ti—O, V—O, and Nb—O systems. As seen in the figure, the broken lines joining the schematic points corresponding to oxides of different composition have a convex shape, indicating a decrease in heats of formation (per gram-atom of oxygen) consequent to an increase in oxygen concentration.

The Me—Me bond shielding in the intermediate IVa and Va subgroup transition-metal oxides, where considerable metal atom d_ε-orbital overlap can occur, can substantially affect the thermodynamic conditions for formation of such oxides. Since their formation depends on the number of Me—Me bond electrons, and particularly on the extent of d_ε-orbital overlap, it is of interest to make a comparative evaluation for the different oxides.

Calculation of the overlap integrals was performed by a method earlier described [6]. The radial portions of the d-orbital cationic wave functions were described in the manner of Slater. The nuclear-charge shielding parameters for the outer cationic electrons of the monoxides were obtained by Hartree graphs for dipositive ions [9]. The cationic shielding constants in the sesquioxides and dioxides (according to Slater's rule) were respectively decreased to 0.35 and 0.70, relative to the monoxide.

We note that such an evaluation of the nuclear charge shielding parameters does not ensure precise numerical results. This is chiefly the result of the undetermined degree of ionic character in the Me—O bond of oxides. However, the overlap integrals calculated for a series of oxides of one type (e.g., MeO or Me_2O_3) will at least give the correct relative character of the compounds.

For corundum-structure oxides (Me_2O_3), the calculation of the overlap integrals was performed both for the nearest Me—Me distances (parallel to the C axis), as well as for distances perpendicular to the axis. In oxides of MeO_2 composition, they were computed for two distances along a chain of metal atoms. The nearest Me—Me distances for unstable oxides can be evaluated approximately using the following concepts.

As was shown in [8], computing of the TiO lattice constant with the assumption of ionic bonding (i.e., as the sum of the Ti^{2+} and O^{2-} diameters) gives approximately the same result as assuming covalent Ti—O bonding (i.e., computing it as the sum of the atomic radii of Ti and O). Both results are close to that observed. In connection with this, the difference between Me—Me distances in isostructural pure metals will remain nearly the same for all oxides of the same structure. The Me—Me distances expected according to the above law were evaluated for the intermediate oxides of Zr, Hf, Ta, and for Nb_2O_3. The values calculated for the overlap integrals $S(d\varepsilon, d\varepsilon)$ are presented in Table 1.

As the tabular data show, there is a direct connection between the lattice defect character and the values of overlap integrals and numbers of electrons. NbO is characterized by an ordered 25% site vacancy in both sublattices. Here all the Me—Me bonds in the NbO sublattice are unshielded, i.e., they form near oxygen vacancies. The definite ordering of vacant sites in the oxygen and metallic sublattices enables NbO to assume a definite structure. In it, each Nb atom is closely surrounded by four oxygen atoms in one plane, and by eight metal atoms in the second coordination sphere. The coordinational environment of oxygen atoms is analogous. Evidently the NbO structure is stable under the following conditions: (a) if the equilibrium distance of the adjacent metal atoms in the structure corresponds to that established by Me—O bonding; or, (b) if the Me—Me bond is considerably less effective than the Me—O bond, and does not result in significant displacement of the metal atom.

Comparison of $d\varepsilon$-orbital overlap data and the number of Me—Me bond electrons in the monoxides shows that the concentration of vacant sites for ZrO, HfO, and TaO should be larger than, or for ordered structures the same as, for NbO. This applies particularly to HfO and TaO. These oxides will evidently not form, due to their structural instability. We note that Schönberg [10] indicates the possibility of preparing a metastable cubic ZrO. Here the Me—Me distance evaluated using his data is somewhat smaller than that which we proposed for use in calculation of the overlap integral [whose computation should yield still larger $S(d\varepsilon, d\varepsilon)$ values and indicate a greater structural defect character].

The Me—Me bond shielding in the corundum structure results in an increase in the mutual spacing between the oxygen atoms on an octahedral face, through which the Me—Me interaction is realized via a triangle.

The substantially large $d\varepsilon$-orbital overlap on formation of the Nb, Zr, and especially Hf and Ta sesquioxides (corundum structure) renders these less stable thermodynamically than the higher oxides, due to the large Me—Me bond shielding.

If the exchange interaction between the Me—Me bond electrons and the anions actually leads to a separation of the anions in the corundum (or rutile) structure and to a sharp decrease in exchange, then the long-range Coulomb repulsive forces between Me—Me bond electrons and anions will continually increase with increase in $d\varepsilon$-orbital overlap. This leads to an increase in the energy of the oxide, with the result that its formation is thermodynamically less favorable than a formation of a final oxide.

An analogous picture is also observed for oxides of rutile structure, where the relatively larger d_ε-orbital overlap in NbO_2 (as opposed to VO_2) leads to a substantial increase in interanionic spacing. The absence of a stable tantalum dioxide also can be explained by its larger Me—Me bond shielding relative to VO_2 and NbO.

We also note that, on the whole, the d_ε-orbital overlap increases in the sequence of the isostructural oxides: $V \rightarrow Ti \rightarrow Nb \rightarrow Zr \rightarrow Ta \rightarrow Hf$. This, in turn, generally leads to an increase in Me—Me bond shielding, with the number of intermediate oxides decreasing accordingly. If the plot of heat of formation as a function of relative oxygen content is characterized for the V—O system by a greater curvature, such curvature will in turn subsequently decrease in the Ti—O and Nb—O systems (Fig. 2).

It is to be presumed that for the Zr—O, Hf—O, and Ta—O systems, which have a substantially larger d_ε-orbital overlap due to the larger Me—Me bond shielding, the dependency curve will assume a concave form, i.e., the heat of formation of the intermediate oxides (on a gram-atom of oxygen basis) is less than that for the final oxide. These circumstances can explain the absence of intermediate oxides in the latter systems.

LITERATURE CITED

1.   S. L. Altman, C. A. Coulson, and W. Hume-Rothery, Proc. Roy. Soc., Vol. A240, p. 145 (1957).
2.   J. B. Goodenough, Phys. Rev., Vol. 117, p. 1442 (1960).
3.   E. Morin, Bell Syst. Tech. J., Vol. 37, p. 1047 (1958).
4.   B. Marinder and A. Magneli, Acta Chem. Scand., Vol. 11, p. 1635 (1957).
5.   P. V. Gel'd and V. A. Tskhai, Zh. Strukt. Khim., Vol. 4, p. 275 (1963).
6.   V. A. Tskhai and P. V. Gel'd, Zh. Strukt. Khim., Vol. 5, p. 576 (1964).
7.   P. V. Gel'd and V. A. Tskhai, Several Questions of the Crystallochemistry of the Cubic Oxides and Carbides of Ti, V, and Nb. Proceedings of the Inter-Institute Colloquium on Solid Phases of Variable Composition, No. 32 (2) [in Russian], Leningrad (1962).
8.   S. P. Denker, Nuclear Metallurgy, Vol. 10, p. 51 (1964).
9.   D. R. Hartree, Calculation of Atomic Structures, John Wiley and Sons, New York (1957).
10.   N. Schönberg, Acta Chem. Scand., Vol. 8, p. 627 (1954).
11.   R. E. Newnhan and Y. M. Haan, Z. Krist., Vol. 117, p. 232 (1962).
12.   A. Magneli and G. Andersson, Acta Chem. Scand., Vol. 9, p. 1378 (1955).

SOME RESULTS OF THEORETICAL CALCULATIONS USING MATHEMATICAL STATISTICS AND ELEMENTS OF PROBABILITY THEORY AS APPLIED TO PHASE EQUILIBRIA

I. A. Bondar', N. A. Toropov, and L. N. Koroleva

The problem of comparing experimental data with theoretical calculations and finding the specific laws of different systems during phase equilibria studies is of considerable interest. The works of Pines and co-workers [1-4] give a few principles on the calculation of the simplest of equilibrium diagrams for binary metallic systems. These calculations require experimental data on phase mixing energies, and are applicable only to the simplest of metallic systems. References [7-9] contain the basic elements of the theory of probabilities and computational approximations.

Experimental data on heats of mixing are lacking for the silicates. Moreover, the matter is further complicated by the fact that the silicates contain two groups of ions which strongly differ in electronegativity, as opposed to the more similar electronegativities of metal atoms. The use of analytical expressions is therefore here very cumbersome and involves further considerations. We used the equations developed by Epstein and Howland [5], and particularly those in [6] to calculate the liquidus temperatures of binary oxide eutectic systems.

However, methods involving probability theory and mathematical statistics now supplement equations involving thermodynamic constants, for treatment of experimental material. Their use is exemplified in the book of L. M. Batuner and M. E. Pozin [10], the article of L. F. Syritso [11], and in the report of Yu. V. Granovskii, L. N. Komissarova, and V.I. Spitsyn at the 20th Congress on Theoretical and Applied Chemistry [12], etc.

The object of the present study is to establish a dependence between composition, melting and liquidus temperatures, and density for a number of oxide systems, and to determine the character of such a dependence.

Values corresponding to two quantities of x and y were obtained from several independent observations and experiments:

$$(x_1, y_1), (x_2, y_2), \ldots, (x_n, y_n).$$

These pairs of values are considered to be randomly selected from the aggregate of all possible values of x and y.

Table 1. Dependence of Density on Composition in the $(La, Sm)_2SiO_5$ System

Composition, m, mol.%	Density d, g/cm^3
$La_2O_3 \cdot SiO_2$	5.72
$70\% \, La_2O_3 \cdot SiO_2 + 30\% \, Sm_2O_3 \cdot SiO_2$	5.88
$50\% \, La_2O_3 \cdot SiO_2 + 50\% \, Sm_2O_3 \cdot SiO_2$	6.10
$30\% \, La_2O_3 \cdot SiO_2 + 70\% \, Sm_2O_3 \cdot SiO_2$	6.20
$Sm_2O_3 \cdot SiO_2$	6.36

Table 2. Dependence of the Compound Melting Temperature on Composition
in the $Ln_2O_3 - SiO_2$ Systems

Composition, m, mol.%		Melting temperature, T, °C			
Ln_2O_3	SiO_2	$Yb_2O_3 - SiO_2$	$La_2O_3 - SiO_2$	$Sm_2O_3 - SiO_2$	$Y_2O_3 - SiO_2$
100	0	2260	2315	2350	2410
50	50	1950	1930	1940	1980
40	60	1920	1975	1920	1950
33.3	66.7	1850	1750	1775	1775
0	100	1723	1723	1723	1723

Table 3. Dependence of Compound Melting Temperature on Composition in the $Y_2O_3 - Al_2O_3$ System

Comp., m, mol.%		Melting temp., T, °C
Y_2O_3	Al_2O_3	
100	0	2410
66.7	33.3	2020
50	50	1875
37.5	62.5	1930
0	100	2040

For the analysis, we examined the composition dependence of density for the $(La, Sm)_2SiO_5$ solid solution, of melting temperature for a number of rare earth silicates and aluminates, and of liquidus temperature for the $Sm_2O_3 - SiO_2$ system.

Tables 1-3 present the original data. A graphical analysis of the interrelation of the values in Tables 1-3 permits one group to be selected from among them whose two quantities have a nearly linear interdependence. This group contains the density values for the solid solution regions in the $(La, Sm)_2SiO_5$ system, and in the first approximation contains the melting temperatures of compounds in the $Yb_2O_3 - SiO_2$ system. Computation of a chosen correlation coefficient (r_n) with

$$r_n = \frac{\sum_{i=1}^{n} (x_i - \bar{x})(y_i - \bar{y})}{\sqrt{\sum_{i=1}^{n} (x_i - \bar{x})^2} \sqrt{\sum_{i=1}^{n} (y_i - \bar{y})^2}} \tag{1}$$

gives a linear interdependence between values of X and Y at r values numerically close to K ± 1. It is known that $|r| \leq 1$.

A value of $r_n > 0$ indicates an increase in one of the values on increase of the other, while a value of $r_n < 0$ indicates increase of one with decrease of the other. If r = 0, it can be affirmed that there is no linear relationship between x and y, even though relationships of another nature may exist. The nearer r is to unity, the more purely linear is the relationship between the investigated values.

The correlation coefficient calculated for the dependence of density (Y = d) on composition (X = m) is nearly unity (r = 0.99) and appears positive. These data suggest a linear dependence of density (d) on composition (m) in the solid solution region, where a simultaneous increase of d and m is observed. The dependence of compound melting temperature (T) on composition (m) in the $Yb_2O_3 - SiO_2$ system is in the same group. The correlation coefficient in this case is also nearly unity (r = −0.99); its negative sign indicates a reverse path of T relative to m: T decreases with an increase of m.

Later on, a regression function must be found and a linear regression constructed for the above two dependences. We use the equation

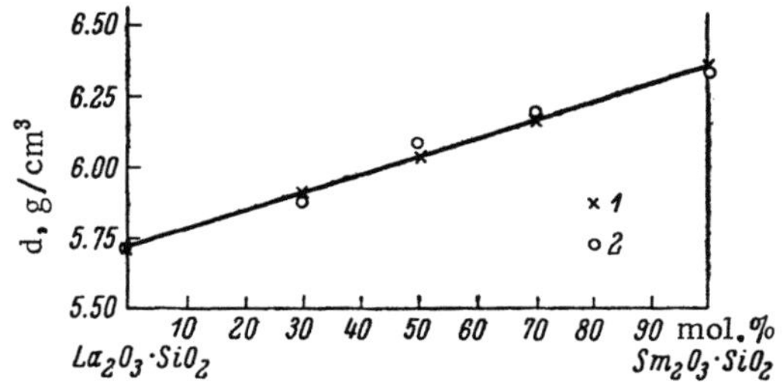

Fig. 1. Approximation of the dependence of density on composition in the region of $(La, Sm)_2SiO_5$ solid solutions using the linear formula (Y = AX + B). 1) Calculated data; 2) experimental data.

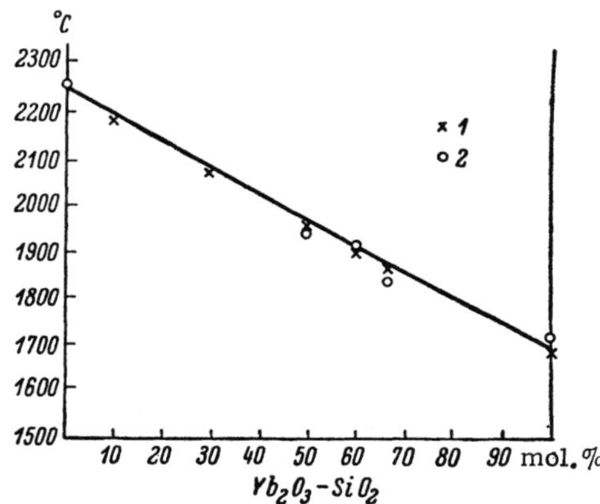

Fig. 2. Approximation, using a linear formula (Y = AX + B), of the dependence of compound melting temperature on composition in the $Yb_2O_3—SiO_2$ system. 1) Calculated data; 2) experimental data.

$$y - \bar{y} = r_n \frac{S_2}{S_1}(x - x), \qquad (2)$$

to obtain the chosen line of regression from Y to X, where

$$x \approx M(X); \quad x = \frac{\sum_{i=1}^{n} x_i}{n};$$

$$\bar{y} \approx M(Y); \quad \bar{y} = \frac{\sum_{i=1}^{n} y_i}{n};$$

M is the mathematical expectancy;
$S_1^2 \approx \sigma^2(X)$; S_1 is the dispersion of X;
$S_2^2 \approx \sigma^2(Y)$; S_2 is the dispersion of Y;

$$S_1 = \sqrt{\frac{\sum_{i=1}^{n}(x_i - \bar{x})^2}{n-1}};$$

$$S_2 = \sqrt{\frac{\sum_{i=1}^{n}(y_i - \bar{y})^2}{n-1}}.$$

We substitute the dependences of d and T on composition m into Eq. (2). Then the equation takes on the following form:

$$d = r\frac{S_2}{S_1}(m - \bar{m}) + d, \qquad (3)$$

$$T = r\frac{S_2}{S_1}(m - \bar{m}) + \bar{T}. \qquad (4)$$

After substitution of values from Tables 1 and 2, we obtain equations for the lines of regression.

$$d = 0.0066m + 5.72, \qquad (5)$$

$$T = -5.44m + 2242. \qquad (6)$$

The regression line for the dependence of density on composition is shown in Fig. 1.

The mean square error for d as a function of m in the region of $(La, Sm)_2SiO_5$ solid solutions is close to the experimental accuracy ($\varepsilon_{calc} = \pm 0.032$; $\varepsilon_{exp} = \pm 0.03$). The same thing is observed for the dependence of compound melting temperature (T) on composition (m) in the $Yb_2O_3—SiO_2$ system ($\varepsilon_{calc} = \pm 23.5°$, $\varepsilon_{exp} = \pm 30°$). This dependence is shown in Fig. 2. The amount of experimental error is within the limits of the accuracy of the observed values.

Further examination of experimental material, its graphical representation, and calculation of selected correlation coefficients for compounds in the $La_2O_3—SiO_2$, $Sm_2O_3—SiO_2$, $Y_2O_3—SiO_2$, and $Y_2O_3—Al_2O_3$ systems shows that there is not a linear, but a more complex (nearly parabolic) dependence between compound melting temperature and composition.

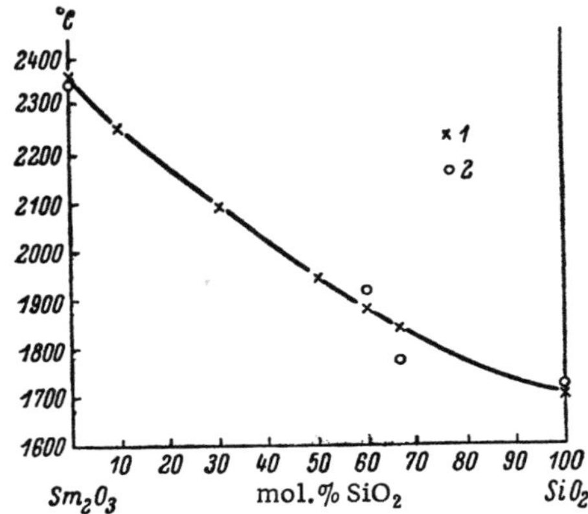

Fig. 3. Approximation, using the parabolic formula $Y = AX^2 + BX + C$, of the composition dependence of compound melting temperature in the Sm_2O_3—SiO_2 system. 1) Calculated data; 2) experimental data.

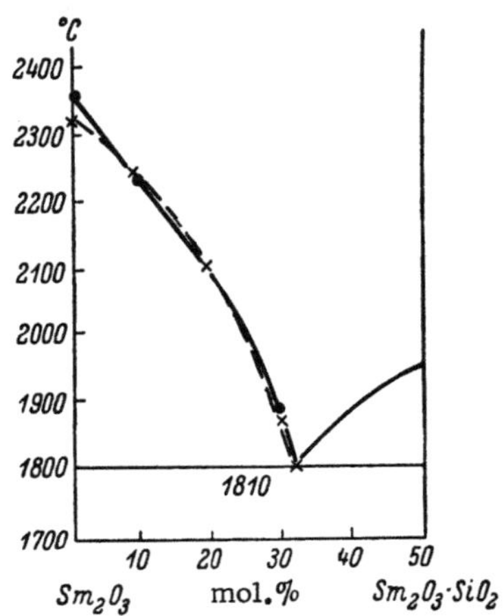

Fig. 4. Approximation, using the parabolic formula $Y = AX^2 + BX + C$, of the composition dependence of liquidus temperature in the Sm_2O_3—$Sm_2O_3 \cdot SiO_2$ system. 1) Calculated data; 2) experimental data.

The general equation for such a case is expressed as follows:

$$Y = AX^2 + BX + C. \tag{7}$$

For our case,

$$T = Am^2 + Bm + C. \tag{8}$$

We use the method of moments and the condition of a minimum sum for the squares of the deviations of the natural values of T from those calculated, namely,

$$\sum_{i=1}^{n} [T_i - (Am^2 + Bm + C)] = \min. \tag{9}$$

Condition (9) permits assembly of a system of equations for determination of the unknown coefficients A, B, and C:

$$\left. \begin{aligned} A \frac{\sum_{i=1}^{n} m_i^4}{n} + B \frac{\sum_{i=1}^{n} m_i^3}{n} + C \frac{\sum_{i=1}^{n} m_i^2}{n} &= \frac{m^2 T}{n}, \\ A \frac{\sum_{i=1}^{n} m_i^3}{n} + B \frac{\sum_{i=1}^{n} m_i^2}{n} + C \frac{\sum_{i=1}^{n} m_i}{n} &= \frac{mT}{n}, \\ A \frac{\sum_{i=1}^{n} m_i^2}{n} + B \frac{\sum_{i=1}^{n} m_i}{n} + C &= \frac{T}{n}. \end{aligned} \right\} \tag{10}$$

Later we substitute numerical values into formula (10); to determine the coefficients A, B, and C, we assemble and calculate the determinants Δ, during which

$$A = \frac{\Delta_1}{\Delta}, \quad B = \frac{\Delta_2}{\Delta}, \quad C = \frac{\Delta_3}{\Delta}.$$

Figure 3 depicts a curve representing the dependence of T on m, using the experimental points for compounds in the Sm_2O_3—SiO_2 system. In this instance, the formula is parabolic and of the following form:

$$T = 0.0357m^2 - 10m + 2357. \tag{11}$$

Calculation of the mean square error gives a value of $\pm 39°$, which is close to the experimental ($\varepsilon_{exp} = \pm 30°$).

The change of compound melting temperature with composition in the Yb_2O_3—SiO_2 system thus has a character other than that of the similar change in a number of other silicate systems (La_2O_3—SiO_2, Sm_2O_3—SiO_2, Y_2O_3—SiO_2). We also noted an unusual behavior in other characteristics of compounds of the Yb_2O_3—SiO_2 system (crystallo-optical, pyrochemical, etc.). This is associated with the lanthanide contraction in the La—Lu series, and with the presence of three structural subgroups among the rare earth silicates.

The dependence of liquidus temperature on composition also has a parabolic form in simple Ln_2O_3—SiO_2 eutectic systems.

We calculated the parabola for the Sm_2O_3—$Sm_2O_3 \cdot SiO_2$ system (Fig. 4) as being

$$T = -0.337m^2 - 4.938m + 2319. \tag{12}$$

The mean square error is $\pm 23.2°$.

The calculations presented show satisfactory agreement with experiment.

CONCLUSIONS

1. Performing calculations using mathematical methods enables their comparison with the results of experimental studies.

2. The character of the change of numerous properties with composition was completely defined.

3. A linear or parabolic dependence of density, melting, and liquidus temperature on composition was established for the $(La, Sm)_2SiO_5$, La_2O_3—SiO_2, Sm_2O_3—SiO_2, and Yb_2O_3—SiO_2 systems.

LITERATURE CITED

1. B. Ya. Pines, Zh. Éksp. Teor. Fiz., Vol. 13, Nos. 11-12, p. 411 (1943).
2. V. I. Danilov and D. S. Kamenetskaya, Zh. Fiz. Khim., Vol. 22, No. 1, p. 69 (1948).
3. B. Ya. Pines, Zh. Neorg. Khim., Vol. 3, No. 3, p. 611 (1958).
4. Ya. E. Geguzin and B. Ya. Pines, Dokl. Akad. Nauk SSSR, Vol. 75, No. 3, p. 387 (1950).
5. L. F. Epstein and W. H. Howland, J. Am. Ceram. Soc., Vol. 36, No. 10, p. 334 (1953).
6. I. A. Bondar', Izv. Akad. Nauk SSSR, Ser. Khim., No. 11, p. 1921 (1964).
7. L. Z. Rumshinskii, Elements of Probability Theory, 2nd ed. [in Russian], Fizmatgiz, Moscow (1963).
8. Ya. I. Lukomskii, Theory of Correlation and Its Application to Industrial Analysis [in Russian], Gosstatizdat, Moscow (1958).
9. M. Bezikovich and A. Fridman, Approximation Calculations, 2nd ed. [in Russian], Gostekhizdat, Moscow (1930).
10. L. M. Batuner and M. E. Pozin, Mathematical Methods in Chemical Technology [in Russian], Goskhimizdat, Moscow (1953).
11. L. F. Syritso, Zap. Vsesoyuzn. Mineral. Obshch., Ser. II, Vol. 92, No. 4, p. 434 (1963).
12. Yu. V. Granovskii, L. N. Komissarova, and V. I. Spitsyn, Theses of the Report to the 20th Congress on Theoretical and Applied Chemistry [in Russian], Izd. Akad. Nauk SSSR, Moscow (1965).

THERMODYNAMIC ANALYSIS OF
SOLID-PHASE REACTIONS

B. F. Yudin

The thermodynamic analysis of solid-phase processes (i.e., processes of solid-phase interaction) has several characteristics basically related to specific aspects of such processes. The published literature does not contain an adequate, exhaustive account on the principles of the thermodynamic analysis of such processes. The necessity therefore arose to formulate the fundamental character of the thermodynamic analysis of solid-phase processes, using general thermodynamic relationships as the basis in considering specific aspects.

Determination of the parameters characterizing the equilibrium state of a system under study is the principal task of thermodynamic analysis. It is therefore especially necessary to define the criterion of the equilibrium state.

Since a large number of the studied processes occur under isobaric conditions, a basic criterion for equilibrium (according to the second law of thermodynamics) is the change in a system's isobaric-isothermal potential, designated ΔZ in international thermodynamic terminology.

Let us assume that a system undergoes the reaction

$$\nu_A A + \nu_B B = \nu_C C + \nu_D D. \tag{1}$$

Then the equation defining the isotherm of such a reaction is

$$\Delta Z = RT \left[\ln \frac{a_C^{\nu_C} \cdot a_D^{\nu_D}}{a_A^{\nu_A} \cdot a_B^{\nu_B}} - \ln K_a \right], \tag{2}$$

where a_i is the activity of the i-th component, K_a is the equilibrium constant of reaction (1), T is temperature in °K, and R is the universal gas constant. The activity values in Eq. (2) are for the original state of the interacting components. The isotherm equation shows how ΔZ changes when the system deviates from the equilibrium state. The value of K_a is found through the equality

$$\left(\frac{a_C^{\nu_C} \cdot a_D^{\nu_D}}{a_A^{\nu_A} \cdot a_B^{\nu_B}} \right)_{\Delta Z=0} = K_a. \tag{3}$$

If $\Delta Z > 0$, reaction (1) can go only in the reverse direction; if $\Delta Z < 0$, it can go only forward.

Let us now assume that the original state of components A and B is standard (i.e., a_A and a_B are 1) and that the final state is also standard (i.e., a_C and a_D are 1). During this we consider that the amounts of A, B, C, and D in the original mixture are of such a magnitude that the transformation of ν_A and ν_B moles of A and B into ν_C and ν_D moles of C and D has no effect on the whole of the reaction components. Then the first term in Eq. (2) reduces to 0 and the entire equation (2) becomes

$$\Delta Z^0 = -RT \ln K_a. \tag{4}$$

Here ΔZ^0 is the change in standard isobaric–isothermal potential corresponding to the complete transformation of ν_A and ν_B moles of A and B into ν_C and ν_D moles of C and D. The A, B, C, and D components are always in the standard state ($a = 1$) during the entire process.*

The value of ΔZ^0 can be determined from the value of ΔH^0 and ΔS^0, using the relation:

$$\Delta Z^0 = \Delta H^0 - T\Delta S^0. \tag{5}$$

Values of ΔH^0 and ΔS^0 can usually be calculated from tabulated values of the thermodynamic properties of specific materials (cf., for example, [1, 2]).

It is clear from the above that it is impossible to ascribe an identity to ΔZ^0 and ΔZ. ΔZ^0 corresponds to a reaction having component activities equal to one which completely runs its course. Since ΔZ is the isobaric–isothermal potential of the actual process, it is almost impossible to judge the possibility for, and extent of a reaction by the value and sign of ΔZ^0. Data on the equilibrium compositions of the systems under study are a more objective basis for such judgments.

REACTIONS OCCURRING WITH FORMATION OF ONE GASEOUS PRODUCT

We shall separately consider reactions involving formation of only one gaseous product, with no formation of multicomponent phases. Reactions of this type are exemplified by reactions of dissociation, disproportionation, volatilization, sublimation, and many others. For the case in question, equilibrium can be accomplished only with one, strictly defined partial pressure of the gaseous reaction product. The general formature for the equilibrium constant of such a process is expressed by the equality

$$K_e = P_i^{\nu_i} \quad \text{or} \quad P_i = \sqrt[\nu_i]{K_e}. \tag{6}$$

Thus the equilibrium pressure is determined only by the equilibrium constant, whose magnitude depends only on temperature.

With uniform atmospheric pressure, the reaction in a closed system is considered to occur only in the first instant, since an equilibrium partial pressure of the component is assumed to establish itself at each step of the reaction. The process will not occur during this stage, since the chemical reaction isotherm ΔZ is zero, according to Eq. (2).

In an open system, the equilibrium is displaced by reaction product removal when the partial pressure of the latter in the surrounding gaseous environment is less than its equilibrium pressure. When the product partial pressure in the surrounding gaseous medium is greater than the equilibrium pressure, the reaction will not occur either in closed or open systems, since by Eq. (2) the isotherms of chemical reaction, ΔZ, are greater than zero. With

---

*The activity of solid, undissolved phases is always standard; the activity of ideal gases equals the partial pressure.

the reverse relationship of partial and equilibrium pressures ($\Delta Z < 0$), the reaction occurs simultaneously.

Let us now assume that the system has a number of processes, all involving formation of the same gaseous product. Then only one of such processes can occur, namely, that which results in the largest pressure of the latter component.

Thus the simultaneous course of a solid phase process involving formation of one gaseous product is determined chiefly by the partial pressure of the product in the surrounding gaseous medium. This fact explains the behavior of solid phase systems under high vacuum. As it is known, the rate and extent of solid phase processes involving generation of gaseous products greatly increase in a high vacuum.

TOPOCHEMICAL REACTIONS

Reactions occurring completely in the solid phase, i.e., when both the initial and final states of the components are essentially solid phases, are known as topochemical reactions. It is the characteristic of such reactions that the activities of both the reactants and the products remain constant at one during the entire process. Therefore, topochemical systems should be considered as double nonequilibrium systems. At the same time, topochemical reactions belong to the unusual type of process in which $\Delta Z = \Delta Z^0$. Since ΔZ^0 does not depend on pressure, the possibility for the simultaneous occurrence of the processes for a given case is determined exclusively by temperature. The temperature at which $\Delta Z^0 = 0$ is known as the equilibrium temperature. This is the only temperature at which coexistence of the reacting phase constituents is possible. Since absolute isothermality is unattainable in the system, we should consider all topochemical reactions as nonequilibrium reactions.

COMPLEX PROCESSES

Processes in chemical thermodynamics and kinetics are known as complex when they involve the parallel occurrence of several reactions. This type of process is the most widespread in the practice of scientific research.

The term "complex processes" is not arbitrary. The thermodynamic analysis of such processes involves specific assumptions on their reaction mechanism, leading to a complication of the methods for determining the equilibrium state.

For the general case, calculation of the equilibrium state in complex systems is performed by solution of a system of equations constructed on the basis of the law of mass action. The correct choice of the reactions characterizing all changes in the equilibrium state determines the correctness of the thermodynamic analysis of a complex system. It is expedient to consider the fundamentals of thermodynamic analysis of complex systems as exemplified by the process of dissociative volatilization of Al_2O_3 [3]. It is known from literature data that the Al_2O_3 molecule does not exist in the gaseous state. On volatilization Al_2O_3 decomposes into Al, AlO, and Al_2O. Thus the volatilization process of Al_2O_3 can be represented by the following reactions:

$$Al_2O_3 = Al_2O + O_2,$$
$$Al_2O_3 = Al_2O + 2O,$$
$$Al_2O_3 = 2AlO + \tfrac{1}{2}O_2,$$
$$Al_2O_3 = 2AlO + O,$$
$$Al_2O_3 = 2Al + \tfrac{3}{2}O_2,$$
$$Al_2O_3 = 2Al + 3O.$$

As we see, six fundamental reactions can occur in the system in a parallel manner. The equilibrium mixture of gases consists of five components, in a manner governed by the oxidation of the equilibrium composition of the four equations required to relate all components of the mixture. The decrease in number of required equations by one is a result of the mole fraction of each component being determined by the mole fractions of the remaining components

$$N_i = 1 - \sum_k N_{k \neq i}.$$

We shall now determine the reaction equations necessary to perform a thermodynamic analysis of the dissociative volatilization of Al_2O_3. We assume that the Al_2O_3 volatilization process occurs in two stages. The first stage involves formation of molecular oxygen and Al_2O_3 reduction products. The second stage involves dissociation of molecular oxygen into atoms. Then the Al_2O_3 volatilization process can be written by the following reactions:

$$
\begin{aligned}
&1.\ Al_2O_3 = 2AlO + {}^1\!/_2 O_2, && K_{e_1} = P^2_{AlO} P^{1/2}_{O_2}; \\
&2.\ Al_2O_3 = Al_2O + O_2, && K_{e_2} = P_{Al_2O} P_{O_2}; \\
&3.\ Al_2O_3 = 2Al + {}^3\!/_2 O_2, && K_{e_3} = P^2_{Al} P^{3/2}_{O_2}.
\end{aligned}
$$

In the latter there are also the concurrent reactions:

$$
4.\ O_2 = 2O, \qquad\qquad K_{e_4} = P^2_O / P_{O_2}.
$$

To solve this system, we construct a material balance equation for the process. The partial pressure of oxygen in the system is determined by the pressures of AlO, Al_2O, Al, and O in the following manner:

$$P_{O_2} = {}^1\!/_4 P_{AlO} + P_{Al_2O} + {}^3\!/_4 P_{Al} - {}^1\!/_2 P_O. \tag{7}$$

Since

$$P_{AlO} = \sqrt{\frac{K_{e_1}}{P^{1/2}_{O_2}}}, \quad P_{Al_2O} = \frac{K_{e_2}}{P_{O_2}}, \quad P_{Al} = \sqrt{\frac{K_{e_3}}{P^{3/2}_{O_2}}}, \quad P_O = \sqrt{K_{e_4} \cdot P_{O_2}}, \tag{8}$$

by substituting such expressions in equality (7), we obtain

$$P_{O_2} = {}^1\!/_4 \sqrt{\frac{K_{e_1}}{P^{1/2}_{O_2}}} + \frac{K_{e_2}}{P_{O_2}} + {}^3\!/_4 \sqrt{\frac{K_{e_3}}{P^{3/2}_{O_2}}} - {}^1\!/_2 \sqrt{K_{e_4} \cdot P_{O_2}}.$$

Solving this equation with respect to the equilibrium oxygen pressure, we determine, using relation (8) (by values of the equilibrium constants of the above four reactions), the partial pressures of all the components of the equilibrium gaseous mixture.

Based on the example shown, several aspects of thermodynamic analysis of complex processes can be formulated.

1. First, starting with the chemical composition possible for interaction products, it is necessary to construct a definite model of interaction, i.e., to divide the process figuratively into several stages.

2. It is necessary to choose the reactions characterizing the total interaction in keeping with the assumed model. Here the number of reactions should be one less than the number of equilibrium components.

3. A material balance equation is constructed according to the chosen scheme of interaction. Solution of such an equation permits determination of the equilibrium composition of

components. Values of the reaction equilibrium constants (determining the interaction according to the assumed model) are used in constructing the material balance equations.

CONCERNING THE BEGINNING TEMPERATURE OF INTERACTION

By itself, the term "beginning temperature of interaction" better applies to the field of
kinetics and not to thermodynamics, since a chemical process does not occur under conditions
of equilibrium (except, of course, in cases where activity changes can occur with no disruption
of equilibrium conditions). However, it is generally necessary in research and industrial practice, to be concerned with nonequilibrium processes. Then a measure for the possibility of a
process is the extent of a system's deviation from the equilibrium state, which is characterized
by the difference in thermodynamic potential of the system under given conditions as opposed
to conditions of equilibrium. Insofar as this value is temperature-dependent, we can accordingly speak of the "beginning temperature of interaction," or of that temperature at which $\Delta Z = 0$.
As was shown above, the value of ΔZ is also defined by a system's pressure. Consequently,
the "beginning temperature of interaction" depends largely on the pressure of the surrounding
gaseous environment.

We shall assume that under equilibrium conditions the total pressure of the gaseous interaction products is P. The pressure of the atmosphere is P' (isobaric conditions). In the case
where P < P' in a closed system, partial pressures are established for the equilibrium components, whose sum is equal to P, and the interaction is completed according to how P < P'. If
P > P', the chemical interaction of the reactants does not guarantee saturation of the gaseous
phase. Then the process behaves irreversibly. Thus the "beginning temperature of interaction"
for the given case can be equated to that temperature at which P = P'.

These conclusions, justified for closed systems, can apply also (to a certain extent) to
open systems having mass exchange with the environment. At P < P', the removal process is
limited by the molecular diffusion of the reaction products. At P > P', the removal process of
gaseous products is determined by forced diffusion, which occurs at much faster rates than
molecular diffusion. It should be noted, of course, that the molecular diffusion process occurs
at considerable speed in open systems with sufficiently high reaction-product equilibrium pressure. In this case, the condition P = P' adequately fits the interaction ensuing.

A number of authors determine the "beginning temperature of interaction" by the equality
$\Delta Z^0 = 0$. It was shown above that this condition is applicable only to topochemical reactions. If
$\Delta Z^0 = 0$, it follows here that the equilibrium constant is one. This means that the stoichiometric
product of the activities of the reaction products is equal to the stoichiometric product of the
activities of the original materials. Such a condition in solid phase systems generally corresponds well to the interaction already developed.

CONCLUSIONS

We examined the basic concepts of chemical thermodynamics from the point of view of
their use for solid phase processes. As we see it, thermodynamic analysis of a system involving solid phase processes requires close assessment of the conditions under which such processes occur. The following are required to perform an adequate, comprehensive thermodynamic analysis:

1. Knowledge of the complete chemical composition of the system at the moment of
equilibrium.

2. Determination, respecting the probable chemical composition of the system at the moment of equilibrium, of a possible cycle of reactions which will produce the given chemical composition at the moment of equilibrium.

3. Determination, considering the specific aspects of these reactions, of the possibility for their occurrence based on the actual conditions existing in the system (pressure, temperature, and the chemical composition of the surrounding medium).

4. Determination, on the basis of the methods of thermodynamic analysis, of the equilibrium composition of the interaction products and the extent of one or another reaction. If possible, determination of the "beginning temperature of interaction" for the given system.

The present work does not consider problems concerning the thermodynamics of reactions involving solid solutions, since the thermodynamics of solutions is a special division of physical chemistry. However, it should be noted that formation of solid or liquid solutions during interaction can be essentially treated through the equilibrium composition of solid phase interaction products.

LITERATURE CITED

1. V. P. Glushko, Thermodynamic Properties of Individual Materials [in Russian], Izd. Akad. Nauk SSSR, Moscow (1962).
2. A. N. Krestovnikov, L. P. Vladimirov, B. S. Gulyanitskii, and A. Ya. Fisher, Handbook on Calculation of Equilibria of Metallurgical Reactions [in Russian], Metallurgizdat, Moscow (1963).
3. B. F. Yudin and A. K. Karklit, Zh. Prikl. Khim., No. 3, p. 547 (1966).

PHYSICAL INTERPRETATION OF
HIGH-TEMPERATURE CREEP IN OXIDES AND SILICATES

N. V. Solomin

Three different mechanisms exist for creep (inelastic deformation) in oxides and silicates at high temperatures.

1. **Creep of Glassy Oxides and Silicates.** Here the creep mechanism is analogous to the flow process in a viscous liquid. The several deviations from the rules characterizing flow of an ideal, viscous liquid are explained by the formation of a supramolecular structure. Creep in materials of a mixed variety, which contain numerous glassy and crystalline phases, generally depends more on the viscosity of the glassy phase.

2. **Creep in Polycrystalline Silicates and Oxides.** Here the character of inelastic deformation is primarily determined by the crystal lattice imperfection in the boundary regions separating single crystal grains. It depends on the differences between the chemical state of the boundary region material and the average chemical state of the material.

3. **Creep in Oxide and Silicate Single Crystals.** In this case, the reasons for inelastic deformation appear to be: a) crystal lattice microdefects forming during single crystal growth and, b) microdefects of a variable nature originating from thermal fluctuations in the material structure. One can also speak of the dislocation aspects of inelastic deformation in oxide and silicate single crystals.

The coefficient of viscosity (i.e., coefficient of Newtonian internal friction), is easily determined for glassy oxides and silicates. A coefficient of effective viscosity can be determined in the same units for the other two creep mechanisms. This is the viscosity at every given stress and at each determined moment of time from the beginning of an inelastic deformation process.

Thus, for example, we reevaluated the data of other researchers on the inelastic deformation rates of polycrystalline Al_2O_3 and obtained an effective viscosity of the order of 10^{17} poise at 1300°C under stresses of 130–490 kg_f/cm^2.

We also measured and calculated the effective viscosity of a number of other refractory materials under a stress of 2 kg_f/cm^2.

Further, knowing a value for the coefficient of viscosity of a substance, we can compute the inelastic deformation rate of real objects under actual loading conditions, which is very important to current technology.

With this goal in mind, we (and researchers abroad) derived special equations for the various inelastic deformation types, i.e., tension, compression, bending, torsion, and shear.

We recently found that the most diverse types of deformation can utilize one modulus to describe the transition from elastic to inelastic deformation, according to the equation

$$d_\tau = Sd, \tag{1}$$

where d_τ is the inelastic deformation of a body after a specified interval of time, d is the elastic deformation of the same body, and S is the modulus for transition from elastic to inelastic deformation.

By comparing equations for the elastic and inelastic deformation of materials, we determined the magnitude of S as

$$S = \frac{\tau E}{3\eta}, \tag{2}$$

or, more exactly,

$$S = \frac{\tau G}{\eta} = \frac{\tau E}{2(1+\mu)\eta}, \tag{3}$$

where τ is the period of inelastic deformation, E is the first-order modulus of elasticity (Young's modulus), η is the coefficient of viscosity for an amorphous substance or the coefficient of effective viscosity of substances having structure, G is the shear modulus, and μ is Poisson's coefficient.

PART II

THERMO- AND ELECTROPHYSICAL PROPERTIES OF OXIDE SYSTEMS

THERMAL CONDUCTIVITY OF
REFRACTORY OXIDE SOLID SOLUTIONS

I. I. Vishnevskii and V. N. Skripak

1. Study of the thermal conductivity of metal oxides at intermediate and high temperatures has been neglected from both theoretical and experimental standpoints. The available data are chiefly of an applied nature and nearly always neglect such important questions as the dependence of the thermal conductivity on the chemical and phase composition, and on the structure and perfection of the crystal lattice.

The study of thermal characteristics of defect crystal structures is even more necessary because oxide-base solid solutions are widely used in technology as semiconducting thermocouples, various ferrites, refractories, etc.

The present work considers the results of investigations of the temperature and concentration dependences of the thermal conductivity during the formation and decay of several refractory oxide solid solutions. The chief aim of these studies was to explain the effect of a broader distribution of point defects (including substituted chemical impurities, cationic and anionic vacancies) on the thermal resistivity of a crystal lattice above 300°K.

Oxides having different crystal structures were chosen as objects of the studies. It was possible, in some of their solid solutions, to produce vacancy concentrations some ten to one hundred times exceeding the equilibrium values for thermal activation.

The thermal conductivity was measured under steady-state conditions using the cylinder method and the apparatus described in [1].

2. The thermal conductivity of a continuous crystalline dielectric or semiconductor of low electrical conductivity ($\sigma \leq 10^3 \ \Omega^{-1} \cdot m^{-1}$) can be expressed as

$$\varkappa = \varkappa_L + \varkappa_f, \tag{1}$$

where $\varkappa_L$ is the thermal conductivity due to heat transfer by atomic lattice vibrations (phonons) and $\varkappa_f$ is the photonic thermal conductivity stipulated by the ease with which the body transmits infrared radiation [2]:

$$\varkappa_f = \frac{16 n^2 \sigma_0 T^3}{3\alpha}$$

(n is the index of refraction, σ_0 is the Stefan—Boltzmann constant, and α is the absorption coefficient).

It can be assumed [3-5] that at temperatures $\leq 1800°K$, $\varkappa_f$ for refractory oxides does not exceed $10^{-2}\varkappa$, so that $\varkappa_L$ is the principal term in (1).

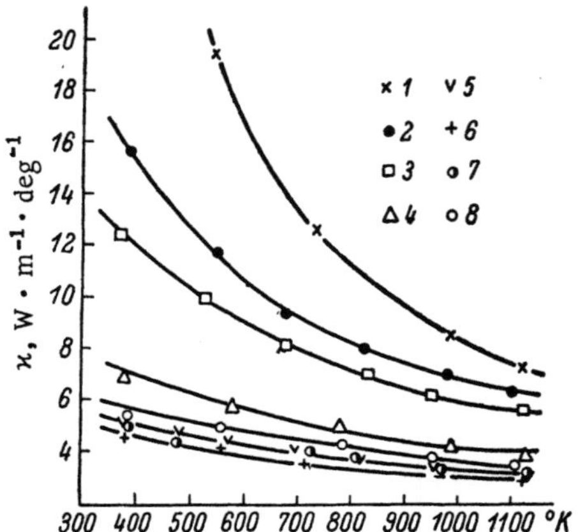

Fig. 1. Temperature dependence of thermal conductivity of Al_2O_3—Cr_2O_3. Cr_2O_3 content (mol.%): 1) 0; 2) 5; 3) 10; 4) 30; 5) 50; 6) 70; 7) 90; 8) 100.

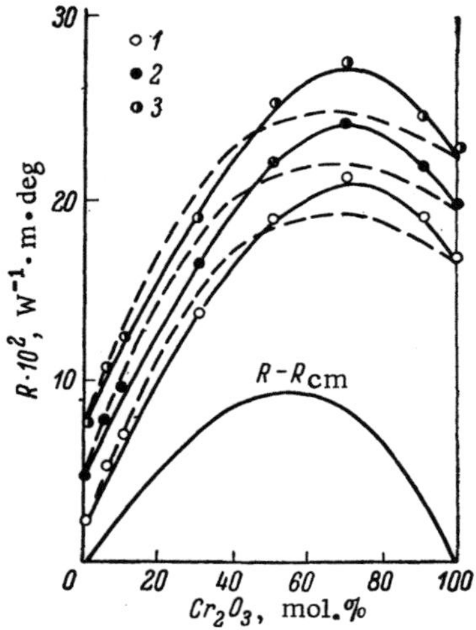

Fig. 2. Concentration dependence of the thermal resistivity of Al_2O_3—Cr_2O_3 at: 1) 300°K; 2) 500°K; 3) 700°K.

The effect of specimen porosity can be calculated by converting to a zero porosity using known formulas [6].

The thermal resistivity of an insulating crystal is written (in the approximation of Matthieson's rule) as

$$R_L = R_{an} + R_d, \qquad (2)$$

where R_{an} is the anharmonic thermal resistivity of phonon—phonon interaction, and R_d is the thermal resistivity due to phonon scattering at defects in the crystal lattice. Point defects from chemical impurities of substitutional or interstitial nature and vacancies of both signs are of greatest importance.

Most experimental facts substantiate that at $T \gtrsim (0.3-0.5)\theta_D$ (θ_D is the Debye temperature), R_{an} about equals T (Eiken's rule), and that R_d is independent of temperature. Therefore, the thermal resistivity of a solution of concentration c can be written in the form [7]

$$R(c, T) = R_d(c) + k(c) F(T). \qquad (3)$$

Here F (T) is a linear function at high temperatures and k(c) is the angular coefficient of this line. It generally turns out well to choose a temperature region in which k(c) is nearly constant. Then the difference $\Delta R = R(c, T) - R(0, T)$ (which is a function only of concentration) is used to determine the contribution of point defects to the lattice thermal resistivity.

If the point defects do not cause a pronounced change in the lattice energy, the thermal resistivity of a solution can be described by the formula [12]

$$\frac{\Delta R}{R(0)} = S_1 c^2 (1 - c) + S_2 c (1 - c)^2, \qquad (4)$$

which can be thought of as a Nordheim formula expressing the dependence of the physical properties of a dispersed alloy on the concentration of its components. Here S_1 and S_2 characterize the scattering contributed to the solution by each of the components.

It was shown in [7] that a large number of vacancies changes the phonon spectrum of the crystal, decreasing the size of the average wave vector. As a result of this, the concentration dependence of thermal resistivity becomes stronger than for the case of substitutional solutions, and can be approximately described by an alternating-sign fourth-order polynomial:

$$\frac{\Delta R}{R(0)} = \varphi(c) = a_1 c - a_2 c^2 + a_3 c^3 - a_4 c^4. \qquad (5)$$

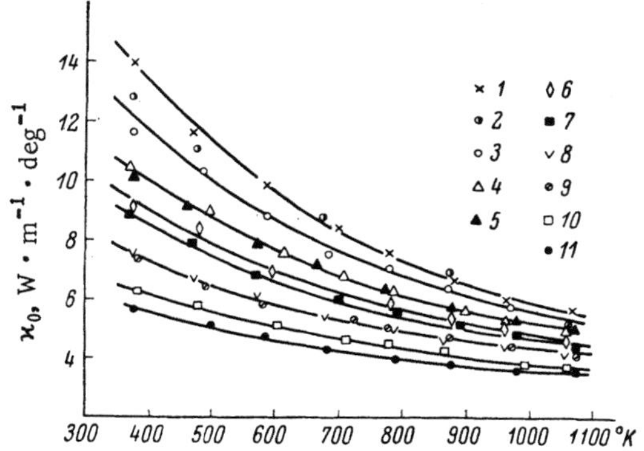

Fig. 3. Temperature dependence of thermal conductivity in $MgAl_2O_4$–Al_2O_3 solid solutions. Concentration of Al_2O_3 (mol.%): 1) 0; 2) 0 (according to data of [5]); 3) 5; 4,5) 10; 6) 20; 7) 30; 8,9) 50; 10) 60; 11) 70.

The stronger concentration dependence qualitatively agrees with the computation of [8], according to which the cross section of phonon scattering by vacancies is 4-5 times greater than that by substitutional ions.

Under the conditions $\varphi''(0) < 0$, curve (5) determines the maximum of thermal resistivity of defect structure around a component (relative to the metal ions).

3. Phonon scattering by impurity centers in substitutional solid solutions of isomorphous crystals (rhombohedral and spinel structure) was studied, using the Al_2O_3–Cr_2O_3 system and the chromium spinellide sesquioxides as examples.

Figure 1 shows the temperature dependence of the thermal conductivity of Al_2O_3–Cr_2O_3 specimens at different phase compositions. Here and below, the value of $\varkappa$ was standardized to a zero porosity using the formula $\varkappa_0 = \varkappa (1 - p)^{-1}$.

Figure 2 shows the R values at temperatures of 300, 500, and 700°K as a function of Cr_2O_3 content, as well as their approximation with the third order computed by the method of least squares (solid lines). The dotted lines show the approximation of experimental data by Nordheim equations of the type

$$R = R_{Cr_2O_3}c + R_{Al_2O_3}(1 - c) + 0.449c(1 - c)^2 + 0.267c^2(1 - c). \tag{6}$$

It is a fact that the approximation by Nordheim equations (6) is worse than by the least squares method. However, even here the maximum discrepancy does not exceed 10%.

We note that the ratio $S_{Cr_2O_3} : S_{Al_2O_3} \sim 1.7$ to some extent characterizes the greater phonon scattering by a Cr^{3+} as opposed to an Al^{3+} ion in the hexagonal, corundum-type structure.

Measurement of the thermal conductivity of a chromium spinellide subjected to thermochemical treatment in atmospheres of differing oxygen partial-pressure shows that its thermal conductivity increases with increase in the degree of oxidation [9]. Since oxidation involves separation of sesquioxides from chromium spinellide (chiefly Fe_2O_3 [10, 11]), it would appear that the divalent Fe ions of fourfold coordination are the additional scattering centers. Computation of the number of scattering centers in formation of a substitutional solution in a tetrahedral region of the spinel lattice leads to the expression [9]

$$R(c) = R(0)(1 + 4c + 37c^2 - 20c^3), \tag{7}$$

where c is the concentration of Fe^{2+} ions.

The concentration dependence of the thermal resistivity of spinel solutions of the type studied is therefore also expressed by a third-degree polynomial, which is characteristic of isomorphous substitutional structures [12].

With relatively small impurity concentrations, the $R = f(c)$ dependence becomes linear.

4. A study of phonon scattering by cation vacancies in the spinel lattice of the $MgAl_2O_4$–γ-Al_2O_3 system was performed in [13]. The $\varkappa = f(T)$ curves for the solutions are shown in Fig. 3 for the concentration interval 0-65 mol.% γ-Al_2O_3 (limit of vacancy concentration over 10%). The temperature dependence of the thermal resistivity attests to the fact that scattering

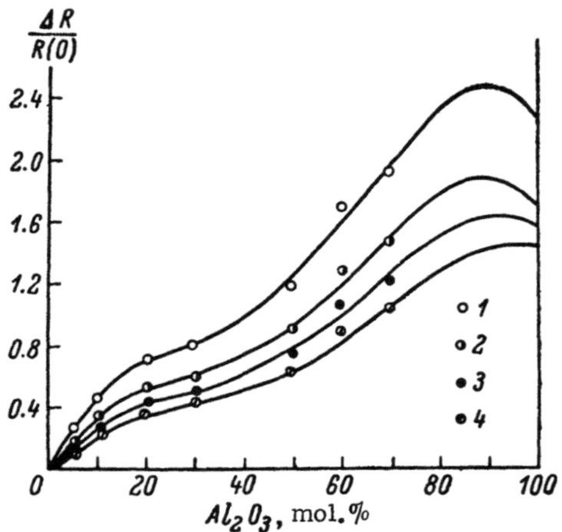

Fig. 4. Concentration dependence of the relative thermal resistivity of $MgAl_2O_4$– Al_2O_3 solutions at different temperatures (°K): 1) 300; 2) 400; 3) 500; 4) 600.

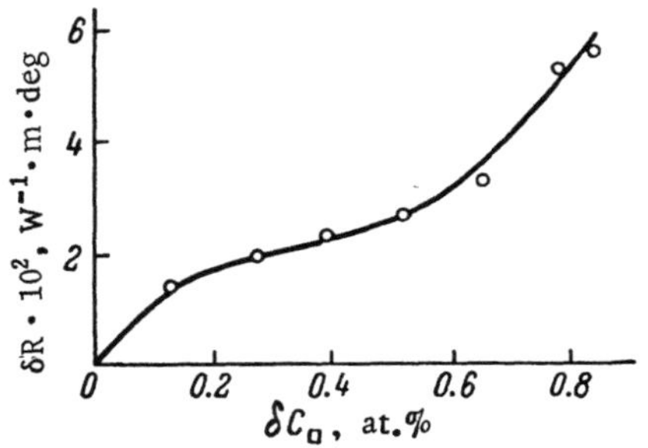

Fig. 5. Correlation between the additional resistivity of MgO–$MgCr_2O_4$ solutions in comparison to that of MgO–$(Mg^{2+},Fe^{2+})\cdot Fe_2^{3+}O_4$ solutions, and the excess vacancy content at the same concentration of substituted Mg^{2+} ions (T = 300°K).

by impurities is actually temperature independent and that $\Delta R = R(c, T) - R(0, T)$ is a measure of the contribution to thermal resistivity from point defects of Al^{3+}-type and from vacancies.

The $\Delta R/R(0) = f(c)$ curves for 300, 400, 500, and 600°K (Fig. 4) are characterized at $\varphi''(0) < 0$ by a sharp bend at about 30%. Since the thermal resistivity should be maximum in the interval $0 < c < 1$, such conditions indicate that the $\varphi(c)$ function should be described by at least a fourth-order polynomial.

The experimental $\Delta R/R(0)$ values were approximated by fourth-order curves, whose coefficients were computed by the least squares method. The succession of coefficient signs is the same as for (5). The thermal resistivity maximum at $c \sim 0.9$ and the thermal conductivity for γ-Al_2O_3 of $\sim 5W \cdot m^{-1} \cdot deg^{-1}$ at T = 300°K were determined by extrapolation into a region of imaginary solid solutions. This value was compared to that calculated using the Dugdale– McDonald formula [14]

$$x = \frac{ru}{3\chi\gamma^2 T}.$$ (8)

Here r is the nearest interatomic distance, u is the speed of sound, χ is the coefficient of compressibility, and γ is the Grüneisen constant. Calculation of the approximate ratio between the phase and group speeds of phonons gave satisfactory agreement with experimental data for the case of $MgAl_2O_4$ and $MgFe_2O_4$ [7], as well as with values found by extrapolation for γ-Al_2O_3. We note that the thermal resistivity of γ-Al_2O_3 is a linear function of temperature, as indeed it should be to be in keeping with Eiken's law.

5. An analogous method was used to study the influence of cationic vacancies on thermal conductivity in the NaCl-type solid solutions of magnesium ferrites and magnesium chromites in magnesium oxide.

A vacancy concentration of 4.4 at.% was attained in the $MgO-(Mg,Fe^{2+})Fe_2^{3+}O_4$ system and one of 2.2 at.% in the $MgO-MgCr_2O_4$ system. In both cases the concentration dependence of the thermal resistivity is characterized at $\varphi''(0) < 0$ by a sharp bend in the low-concentration region, and is well approximated by fourth-order curves.

At the same concentration ξ of substituted Mg^{2+} ions, the effective scattering diameter calculated by the formula in [12] is

$$\sigma_{eff} = \frac{\Delta R c_o u a^3}{3\xi}.$$ (9)

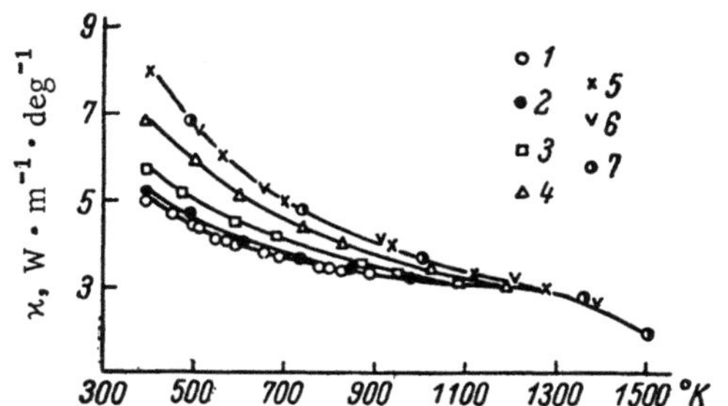

Fig. 6. Temperature hysteresis (T = 300°K) of the thermal conductivity of a MgO—(Mg, Fe^{2+})$Fe_2^{3+}O_4$ solid solution at 16.8 mol.% magnesio-ferrite content. Cooling from: 1) 600, 700, 800, 900; 2) 970; 3) 1100; 4) 1200; 5) 1280; 6) 1390; 7) 1500°K.

(c_v is the heat capacity and a is the lattice parameter.) This also means that the thermal resistivities of MgO—$MgCr_2O_4$ solutions are three times higher than the corresponding values for the MgO—(Mg, Fe^{2+})$Fe_2^{3+}O_4$ system, the result of a higher rate of vacancy formation on increase of concentration in magnesio-chromite solutions ($c_\square = 0.33\xi$ for MgO—$MgCr_2O_4$; $c_\square = 0.2\xi$ for MgO—$MgFe_2O_4$ [15]). The correlation between the additional thermal resistivity and the excess amount of vacancies at constant ξ is shown in Fig. 5.

6. Measurement of thermal conductivity in cubic ZrO_2—CaO compositions containing 6–9 at.% anionic vacancies showed, that while dependent on concentration of stabilizing additive and amounts or kind of other impurities, $\varkappa$ is almost independent of temperature. Its absolute values are very low (1.5–2 W · m^{-1} · deg^{-1}) and differ very little [16, 17].

The high lattice symmetry and adequately dense ionic packing should guarantee for cubic ZrO_2 a higher thermal conductivity and a dependence of $\varkappa$ on the reciprocal of temperature similar to that observed in ThO_2 and UO_2 [17]. The latter oxides have a room-temperature thermal conductivity eight times that of ZrO_2, even though thorium and uranium are twice as dense as zirconium.

It is apparent that the concentration of CaO necessary for stabilization of the cubic modification assures preferential phonon scattering at substitutional ions and ionic vacancies up to the highest temperatures (~2300°K). The extremely low thermal conductivity values correspond to a phonon free path length of 8–10 Å, i.e., one to two lattice parameters. Consequently, scattering occurs in almost every unit cell — signified by the absence of a temperature dependence on $\varkappa$ [18].

7. Since oxide solid solutions are often in the nonequilibrium state at working temperatures (especially when they have limited solubility), it was of interest to study the thermal conductivity during their decay process and also to study the conditions for heat transfer in a two-phase system.

The temperature dependences of the thermal conductivities obtained on cooling from regions of solid solution decay display a hysteresis which is dependent on the decrease in impurity defect content (Fig. 6).

The thermal resistivity of a partially decayed solid solution can be written in the form of [19]:

$$R_{cm} = [\alpha_1 + (\alpha_2 - \alpha_1)V]T + R(c)(1-V) + bV. \tag{10}$$

Here V is the volumetric concentration of the second phase, R(c) is the temperature-independent contribution to solid-solution thermal resistivity, and α_1 and α_2 are the temperature coefficients of thermal resistivity of the original components.

This expression was used to describe the thermal resistivity of partially decayed solutions of magnesium ferrites and chromites in magnesium oxide, and also of Al_2O_3 in $MgAl_2O_4$. In all cases, the cooling curves from different temperatures form a set of parallel lines, whose slope is determined by the amount of undissolved spinel.

The temperature dependence of thermal resistivity computed by (10) for solid solutions in $MgAl_2O_4$ + Al_2O_3 specimens agrees well with the experimental lines for single-phase solutions of corresponding concentration.

Phase compositions of $MgO-MgFe_2O_4$ specimens were determined using (10), giving satisfactory agreement with values found by magnetic measurements. The threshold temperatures for decay are indicated by breaks in the heating curves, and are approximately the same as the temperatures exhibiting a sharp increase in magnetizability, stipulated by separation of free magnesioferrite.

The results of the experiments and the computer calculations established with authenticity the relationship between the nature of a material's cooling curve and the appearance of a temperature hysteresis in its thermal conductivity [20]. The internal temperature of any material whose thermal conductivity has a positive hysteresis, will increase on cooling from an equilibrium state, and it will do so more the lower its thermal resistivity.

CONCLUSIONS

Substitutional solid solutions are characterized by a cubic dependence of thermal resistivity on temperature.

In substitutional solid solutions, the high concentration of structural vacancies leads to a stronger concentration dependence. This can be approximately described by a fourth-order polynomial, which determines the maximum of thermal resistivity around a defect-structure component.

In concentrated solutions, the thermal conductivity decreases to limiting lower values which correspond to the free path length of a phonon (several lattice constants), and becomes practically independent of temperature and amount (or kind) of other impurities.

The temperature dependence of the thermal conductivity obtained on cooling from solid-solution decay regions is characterized by a hysteresis. The thermal resistivity of such a two-phase system is additively made up of the resistivities of the separate phases.

LITERATURE CITED

1. I. I. Vishnevskii and V. N. Skripak, Ogneupory, No. 5, p. 227 (1964).
2. L. Gensel, Z. Phys., Vol. 135, p. 177 (1953).
3. J. C. Jamieson and A. W. Lowson, J. Appl. Phys., Vol. 29, p. 1313 (1958).
4. D. W. Lee and W. D. Kingery, J. Am. Ceram. Soc., Vol. 43, p. 594 (1960).
5. W. D. Kingery, Property Measurements at High Temperatures, John Wiley and Sons, New York (1959).
6. A. F. Chudnovskii, Thermophysical Characteristics of Dispersed Materials [in Russian], Fizmatgiz, Moscow (1962).
7. I. I. Vishnevskii and B. Ya. Sukharevskii, Fiz. Tverd. Tela, Vol. 6, p. 2168 (1964).
8. P. G. Klemens, Proc. Roy. Soc., Vol. A208, p. 108 (1951); Vol. A68, p. 1113 (1955).
9. I. I. Vishnevskii, A. S. Frenkel', and V. N. Skripak, Fiz. Tverd. Tela, Vol. 5, p. 2691 (1963).
10. B. Ya. Sukharevskii, Rentgeno'grafiya, Mineral. Syr'ya, No. 2, p. 11 (1962).
11. V. V. Goncharov, E. A. Prokof'eva, et al., Dokl. Akad. Nauk SSSR, Vol. 124, p. 638 (1959); Vol. 140, p. 648 (1961).
12. A. V. Ioffe and A. F. Ioffe, Izv. Akad. Nauk SSSR, Ser. Fiz., Vol. 20, p. 65 (1956); Fiz. Tverd. Tela, Vol. 2, p. 781 (1960).
13. I. I. Vishnevskii and V. N. Skripak, Fiz. Tverd. Tela, Vol. 7, p. 2925 (1965).

14. J. S. Dugdale and D. K. C. McDonald, Phys. Rev., Vol. 98, p. 1751 (1955).
15. I. I. Vishnevskii and B. Ya. Sukharevskii, Dokl. Akad. Nauk SSSR, Vol. 160, p. 642 (1965).
16. D. M. Shakhtin and I. I. Vishnevskii, Zavodsk. Lab., No. 8, p. 927 (1957).
17. M. Adams, J. Am. Ceram. Soc., Vol. 37, p. 74 (1954).
18. A. F. Ioffe, Fiz. Tverd. Tela, Vol. 1, p. 160 (1959).
19. I. I. Vishnevskii and V. N. Skripak, Dokl. Akad. Nauk SSSR, Vol. 163, p. 418 (1965).
20. A. S. Frenkel', I. I. Vishnevskii, and V. N. Skripak, Inzh.-Fiz. Zhur., Vol. 7, No. 12 (1964).

DEPENDENCE OF THE COEFFICIENT OF THERMAL EXPANSION OF STABILIZED ZIRCONIUM DIOXIDE ON THE RATIO OF CUBIC AND MONOCLINIC PHASES

A. G. Karaulov, A. A. Grebenyuk, V. Ya. Belik, and I. N. Rudyak

The intense development of technological processes involves the problem of preparing refractories with special engineering properties. A property of materials often measured is the coefficient of thermal expansion. It is particularly important to know changes of the thermal expansion coefficient of polymorphous materials as a function of temperature.

Although ZrO_2 has a high melting point (2700°C) and good chemical stability, it cannot be used as a construction material in the pure state because of its polymorphism. The opposite exception is seen in formation of solid solutions between ZrO_2 and several oxides. However, ZrO_2 solid solutions have a large thermal expansion coefficient, reflecting the lack of thermal stability in objects made completely of stabilized ZrO_2 [1, 2].

The technical literature of recent years contains complete recommendations for manufacturing ZrO_2 pieces having satisfactory or increased thermal stability. These recommendations are based on the principle of securing two modifications (cubic or tetragonal, and monoclinic) in the pieces [1, 3-6]. The increase in thermal stability of such pieces is here explained by the decrease in the thermal expansion coefficient of a specific "microcrack" structure [7-9].

The presence in ZrO_2 pieces of two phases of typically unstated ratio indeed explains the large divergence among the thermal expansion coefficients (7.2 to $12.7 \cdot 10^{-6}$/deg) published in the literature for this material by different authors [1, 10-14].

The coefficients of thermal expansion of ZrO_2 are generally given for the temperature range 20-1000°C, and at best up to 1500-1800°C [11, 13, 14].

Since the practical use of ZrO_2 pieces should be at temperatures above 2000°, it was of paramount interest to study the ZrO_2 thermal expansion coefficients at temperatures up to 2100-2300°C, and to note possible changes with specimen phase composition [18], which indeed occurred in the present work.

For study of the thermal expansion coefficient as a function of ZrO_2 phase composition, specimens were prepared from a granular mass of ZrO_2, stabilized at 1750°C by various amounts of CaO, and having monoclinic ZrO_2 additive.

About 50-60% of the charge during specimen preparation was coarse-grained (2-0.5 mm) and 50-40% was dust (less than 0.088 mm). To attain the necessary amount of the monoclinic phase in the specimens, monoclinic ZrO_2 additive was added to the charge in a finely divided

Table 1. Properties of ZrO_2 Specimens Annealed at 1750°C

Charge No.	Amt. of CaO in stabilized grain, %	Amt. of < 0.002 monoclin. addit., %	Phase, according to x-ray structure analysis data,%		Porosity (open),%	Weight by vol., g/cm³	Compr. strength limit, kg/cm²	Thermal stability, thermal change 20-160-20°
			cubic	mono-clinic				
12	6.35		100	—	19.4	4.48	486	2
P-26	6.2	20	80—85	15—20	22.9	4.36	428	12—14
P-47	6.2	30	70—75	30—25	20.5	4.48	453	15—16
11	3.05	—	60	40	22.4	4.37	264	14—16
11-M	3.05	10	55—60	45—40	21.3	4.40	317	30
11-2M	3.05	20	40—45	60—55	18.6	4.57	348	30

Table 2. Dependence of the Thermal Expansion Coefficient on the Monoclinic Phase Content in Specimens for Different Temperature Intervals

Charge No.	Amt. mono-clinic phase, by x-ray struc. anal. data,%	Temp. in-terval of measure-ments, °C′	Coefficient of thermal expansion, 10^{-6}/°C		second determination on the same specimen	
			primary determination			
			heating	cooling	heating	cooling
12	0	25—1000	14	10.8	—	—
		1000—2200	15	15.2	—	—
P-26	15—20	25—1000	7.8	6.4	8.0	8.0
		1000—2200	13.1	13.1	13.4	13.4
P-47	25—30	25—900	7.5	5.0	7.5	6.0
		1100—2200	16.3	14.3	13.6	14.3
11	40	25—900	7.4	0.6	—	—
		1100—2200	nonlinear size change	15.4	13.4	13.4
11-M	45	25—900	3.1	0.4	—	—
		1100—2300	nonlinear size change	12.3	13.8	12.3
11-2M	55—60	300—850	3.5	0.0	8.0	2.5
		1000—2300	nonlinear size change	11.6	13.3	11.6

state (< 2 μ). The specimens were pressed at 1000 kg/cm² and annealed 6 h at 1750°C. The properties and phase composition of the specimens are given in Table 1.

As seen from the table, by using ZrO_2 stabilized with different amounts of CaO (3 and 6 wt.%) and monoclinic ZrO_2 additive, it was possible to prepare specimens containing two phases: cubic (from 100 to 40%), and monoclinic (from 0 to 60%).

The investigation of the ZrO_2 thermal expansion coefficient was performed on two sets of apparatus. It was determined up to 1200°C (in air) on a dilatometer using specimens 8 mm in diameter and 20 mm in length; * over the temperature interval 1000-2300°C, it was determined in a high-temperature vacuum apparatus with argon atmosphere at a pressure of 1 to 1.05 atm.†

The high-temperature apparatus for determination of thermal expansion coefficients of oxide specimens at temperatures to 2000-2300°C was described earlier [15, 16].

---

*The determinations were performed by S. Ya. Tsypin.

†Determinations performed by D. M. Shakhtin, S.V. Levintovich, T.L. Pivovar, and G.G. Eliseeva.

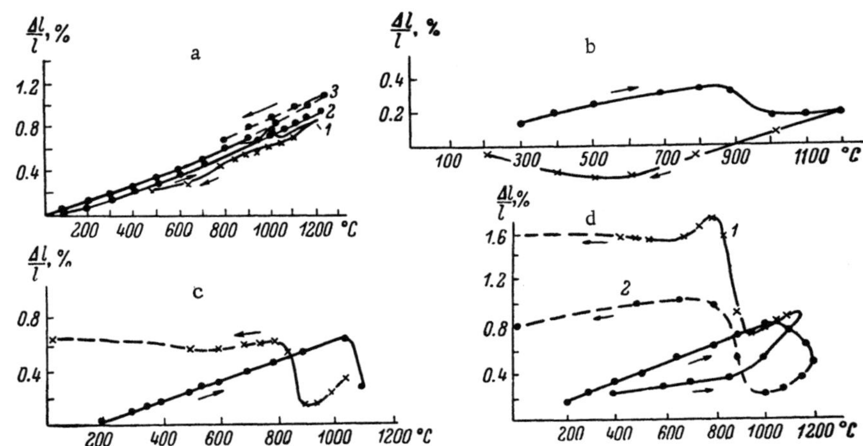

Fig. 1. Expansion and contraction curves for ZrO_2 specimens in the temperature range 20-1200°C. Specimens contain: a) 15-20% monoclinic phase (charge R-26) on first (1), second (2), and third (3) heating; b) 25-30% monoclinic phase (charge R-47); c) 40% monoclinic phase (charge 11); d) 55-60% monoclinic phase (charge 11-2M) on first (1) and second (2) heating.

The studies conducted established that both the absolute value of the thermal expansion coefficient, and its temperature dependence, vary essentially as a function of the amount of monoclinic phase in the specimens. With increase of the monoclinic phase from 0 to 45-60%, a uniform decrease is observed in the thermal expansion coefficient (from $14 \cdot 10^{-6}$/deg to $2.5 \cdot 10^{-6}$/deg) when heating specimens in the temperature range 25-900°C (Table 2, Fig. 1).

This behavior also holds for cooling. However, the absolute value of the thermal expansion coefficient is less, and at 40% or more monoclinic phase content the specimen thermal expansion coefficient is generally nearly zero.

ZrO_2 specimens containing more than 30% of the monoclinic phase display a hysteresis between the curves for thermal expansion and contraction (Fig. 1b, c, d).

The hysteresis loop grows with increase in the amount of the monoclinic phase. This is related to the fact that from 1000-1100°C the transition from the monoclinic to the tetragonal phase occurs with a volume decrease (7%). This results in a diminished thermal expansion coefficient and a break in the curves.

However, cooling the specimen from 1200°C to room temperature effects a reverse transition of the tetragonal to monoclinic phase, with an increase in volume. This causes a cracking and an increase in dimensions of specimens. Therefore, the thermal expansion coefficient on cooling is somewhat less than that on heating. Further specimen expansion is observed at large monoclinic phase contents (40-60%), and the thermal expansion coefficient then becomes zero (Fig. 1c, d). With repeated heating of 60% monoclinic phase specimens at 25-1200°C, the relative size of specimen expansion decreases, although the thermal expansion coefficient hysteresis is preserved. The second and third thermal expansion determinations on the same ZrO_2 specimen containing 15-20% monoclinic phase (Fig.1a), show little change.

The thermal expansion coefficient did not change and was just as large ($15 \cdot 10^{-6}$/deg) for specimens not containing the monoclinic phase at 1000-2300°C (Table 2). With specimens containing 15-20% of the monoclinic phase, the thermal expansion coefficient, though having linear

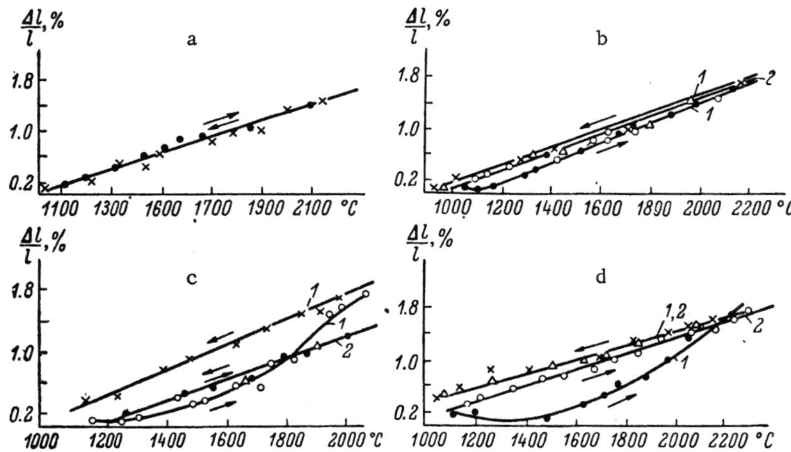

Fig. 2. Expansion and contraction curves of ZrO_2 specimens in the
temperature range 1100–2300°C. Specimens contain: a) 15–20%
monoclinic phase; b) 25–30% monoclinic phase on initial (1) and secondary (2) heating; c) 40% monoclinic phase on initial (1) and secondary (2) heating; d) 55–60% monoclinic phase on initial (1) and
secondary (2) heating, and on initial and secondary cooling (1, 2).

character in the temperature range 1000–2300°C, has a somewhat steeper slope there than at 25–
1000°, with $\alpha = 13.1 \cdot 10^{-6}/\deg$ (Table 2, Fig. 2a).

For specimens containing 25–30% or more monoclinic phase, the thermal expansion coefficient for initial heating to 2300°C now displays a nonlinear temperature dependence (Fig. 2b,
c, d), evidently associated with irregularity of the polymorphic transformation process in ZrO_2
at 850–1200°C, as well as with diminished sample expansion, conforming to a decrease in its
dimensions during transformation from the monoclinic to tetragonal phase at up to 2000°.

A linear dependence is observed between change of sample dimensions and temperature
during specimen cooling in the temperature range 2300–1000°C. Here the thermal expansion coefficient of all repeatedly heated specimens was approximately the same: 13.3–$13.8 \cdot 10^{-6}/\deg$
(Table 2). A decrease of the thermal expansion coefficient from 13.4–$14.3 \cdot 10^{-6}/\deg$ to $11.6 \cdot$
$10^{-6}/\deg$ was observed on cooling with the increase in monoclinic phase content of original specimens from 40 to 60%.

The disappearance of expansion anomalies on heating and contraction anomalies on cooling for specimens containing an initially increased amount of the monoclinic phase is associated
with a redistribution of the stabilizing CaO additive. It is also associated with the additional
stabilization of ZrO_2, which occurs on heating specimens to 2200–2300°C during the thermal expansion measurements. This is confirmed by x-ray structure analysis data. Specimens initially containing up to 40% monoclinic phase were composed of only cubic ZrO_2 solid solution after
annealing to 2300°C. Specimens initially containing up to 45–60% monoclinic phase had only
10% monoclinic phase and 90% cubic ZrO_2 solid solution after annealing to 2300°C. Since specimens containing up to 20% of monoclinic phase have a thermal expansion coefficient which is a
linear function of temperature, a disappearance of the thermal expansion anomaly is under-
standable in specimens having additional stabilization and a decrease of monoclinic phase content from 60 to 10%.

CONCLUSIONS

1. It was established from work performed that there is an interrelation between the temperature dependence of the thermal expansion coefficient and the ratio of cubic and monoclinic phases in ZrO_2 specimens. This occurs independently of whether the monoclinic phase results from incompletely stabilized specimens or from introduction of monoclinic phase additives.

2. It was established that the maximum value of the thermal expansion coefficient over the range 25-900°C occurs in specimens composed only of stabilized ZrO_2 (14-15 · 10^{-6}/deg). This decreases to 2.5-3.1 · 10^{-6}/deg with increase of monoclinic phase content of the specimen to 40-60%.

3. It was shown that the presence of more than 20% monoclinic phase in specimens of 19-22% porosity causes abrupt change in volume in the temperature range 850-1200°C. The magnitude of these changes increases with increased amount of the monoclinic phase, and results in a hysteresis, with expansion of specimens on cooling from 1200° to room temperature.

4. At 30-60% monoclinic phase content, an expansion anomaly is observed in ZrO_2 specimens at temperatures up to 2000°C. However, the amount of monoclinic phase in such specimens decreases from 40-60% to 10% (as a result of CaO redistribution and additional ZrO_2 stabilization) on specimen anneal at 2000-2300°C. This causes disappearance of the expansion anomaly.

LITERATURE CITED

1. C. E. Curtis, J. Am. Ceram. Soc., No. 6, p. 180 (1947).
2. R. Pampuch, Prace Inst. Hutniczych, No. 13, p. 229 (1961).
3. G. K. Basalova, A. V. Stovbur, and O. M. Margulis, Byull. Izobret., No. 10, p. 79 (1958).
4. O. I. Whittemore, J. Canad. Ceram. Soc., Vol. 28, pp. 43-48 (1959).
5. H. C. Wagner, U. S. Patent No. 2,937,102, May 17, 1960.
6. O. I. Whittemore and D. W. Marshall, J. Am. Ceram. Soc., Vol. 35, No. 4, p. 85 (1952).
7. O. M. Margulis, in: Scientific Transactions of the Ukrainian Scientific Research Institute for Refractories, No. 4 (1960), p. 50.
8. P. S. Mamykin, Ogneupory, No. 9, p. 405 (1947).
9. G. V. Kukolev and I. I. Nemets, Ogneupory, No. 2, p. 85 (1963); No. 5, p. 214 (1964); No. 8, p. 23 (1965).
10. I. H. Koenig, Materials in Design Engr., Vol. 47, No. 5, p. 121 (1958).
11. P. P. Budnikov and A. I. Cherepanov, Zh. V. Khim. Obshch. im. D. I. Mendeleeva, Vol. 8, No. 2, p. 141 (1963).
12. W. D. Kingery, Property Measurements at High Temperatures, John Wiley and Sons, New York (1959).
13. T. H. Nisen and M. H. Leipold, J. Am. Ceram. Soc., Vol. 47, No. 3, p. 155 (1964).
14. O. I. Whittemore and N. N. Ault, J. Am. Ceram. Soc., Vol. 39, No. 12, p. 443 (1956).
15. A. N. Lyulichev et al., Zavodsk. Lab., No. 10, p. 1282 (1958).
16. D. M. Shakhtin et al., Ogneupory, No. 7, p. 37 (1965).
17. L. P. Kachalova and A. I. Avgustinik, Zh. Prikl. Khim., No. 7, p. 1451 (1959).

PHYSICOCHEMICAL PROPERTIES OF
REFRACTORY OXIDE COMPOUNDS
OF THE RARE-EARTH ELEMENTS

K. I. Portnoi and N. I. Timofeeva

The oxides of the rare-earth elements are widely used in different areas of contemporary technology. A number of new rare-earth oxide compounds have lately been synthesized, among which the ones having greatest practical significance are those with high melting points. Some of the latter have already found industrial application. For example, the titanates, tungstates, and molybdates of cerium are industrially used for coloring porcelain [1]; europium titanate [2], gadolinium aluminate [3], and a europium molybdate $(Eu_{5.3}MoO_{11})$ [4] are used in nuclear technology; several rare-earth iron garnets are used in electronics [5]. Many rare-earth oxide compounds have prospective uses in the ceramic and radiotechnical industry.

However, until now, information on the properties of most refractory rare-earth-oxide compounds has been either inadequate or of a contradictory nature; preparation methods have often not been developed for these compounds. This acts as a deterrent to the broad use of such compounds in technology.

The present article is devoted to the problem of synthesis, and the study of several properties, of the rare-earth chromites and monoaluminates. The data obtained are compared with the properties of the rare-earth sesquioxides.

Publications have appeared lately which are devoted to the study of the phase diagrams of oxide systems. The results of a study of the $La_2O_3-Cr_2O_3$ system revealed the presence of a chemical compound which melts congruently at 2430°C [6]. Investigation of the phase diagrams of the $Al_2O_3-La_2O_3$ [7], $Al_2O_3-Nd_2O_3$ [8], $Al_2O_3-Cr_2O_3$ [9], and $Al_2O_3-Gd_2O_3$ [10] systems showed that all such systems contain high-melting compounds — the monoaluminates.

We investigated two methods for preparing the rare-earth chromites and monoaluminates: (1) direct annealing of the original oxides, and, (2) decomposition of the binary nitrates [11, 12]. The second route proved most advantageous because compound formation occurs at a lower temperature, and also because the resulting products have high homogeneity and fine dispersion.

Sesquioxides of the rare earths and metallic chromium or aluminum (or their nitrates) were used as starting materials in preparation of the chromites and monoaluminates. The more stable oxides Pr_6O_{11} and Tb_4O_7 were used in preparing the praseodymium and terbium compounds. The purity of the starting materials corresponded to the government standard grades I and II.

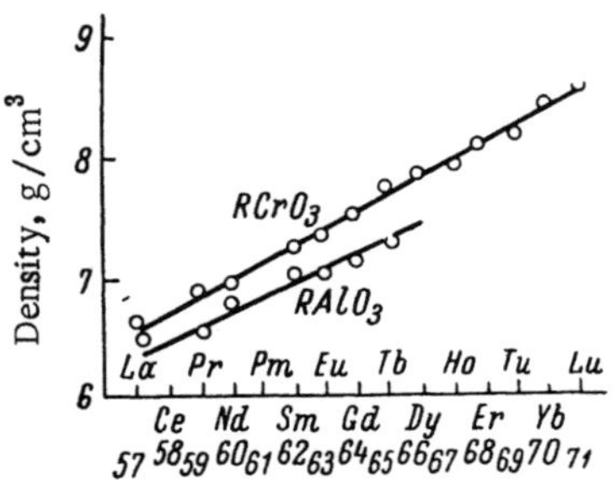

Fig. 1. Density change of rare-earth chromites and monoaluminates as a function of atomic number.

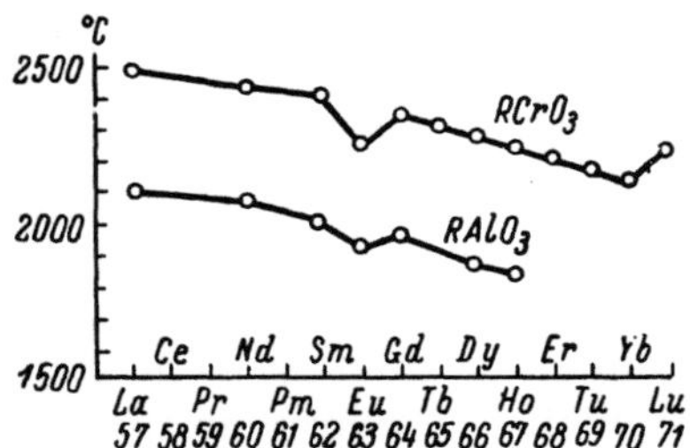

Fig. 2. Melting temperature of rare-earth chromites and monoaluminates.

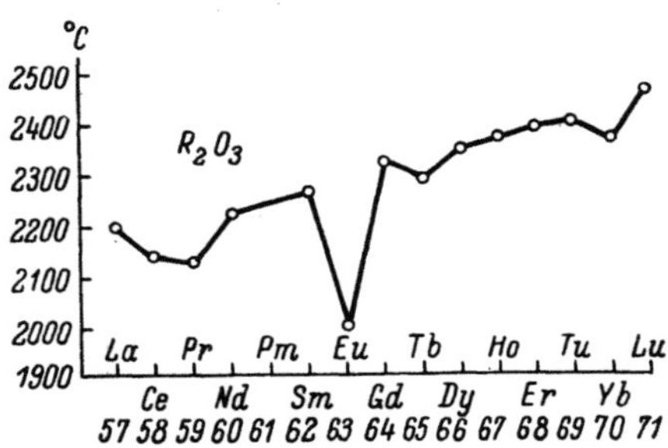

Fig. 3. Melting temperature of rare-earth sesquioxides.

The compound synthesis consists of dissolving equimolar ratios of the starting constituents in nitric acid, volatilizing the resulting solution until salt crystallization, and heating the salts in air for 6-7 h at a temperature of 1000-1100°C.

It was established that the nitrate salts vigorously decompose at 450-500°C. The decomposition is completed at 800-900°C. Formation of the complex oxides was monitored by x-ray structure- and chemical phase-analysis methods.

The phase analysis was based on the differing solubility of the original rare earths and the resulting compounds in dilute acids. It was determined that the chromites are very stable in stirred solutions of hydrochloric acid (1:1) [13], and that the monoaluminates are stable in acetic acid. The rare-earth oxides at the same time readily dissolve in both hydrochloric and acetic acids. The completeness of the oxide interaction can be judged by the rare-earth content of the solution.

The chromites of all the rare-earth elements (from lanthanum to lutecium), and the monoaluminates of lanthanum, neodymium, praseodymium, samarium, and europium were prepared by calcining the nitrates for 7 h at 1000-1100°C. Use of a higher temperature and pelletization of powders were required for synthesis of the monoaluminates of the rare earths having atomic numbers of 64 and higher.

The monoaluminates of gadolinium, terbium, and dysprosium were prepared at 1400°C (annealing time: 2-3 h). The monoaluminates of the other rare earths exist in a narrow temperature range, and can evidently be obtained with higher annealing temperatures. The monoaluminates of ytterbium and lutecium do not exist.

It was not possible, using the given technique, to obtain the chromite or monoaluminate of cerium. Synthesis of these compounds requires heating in a reducing atmosphere, as shown in [14] for the case of cerium monoaluminate.

The compounds obtained had nearly a stoichiometric composition and a distorted perovskite structure.

Figure 1 shows the density change of those chromites and monoaluminates synthesized, as a function of rare-earth atomic number. The density of the powders (<1 μ in size) was determined by pycnometer using toluene as medium. The experimental data proved close to the calculated.

The character of the melting temperature change of the chromites and monoaluminates is shown as a function of rare-earth atomic number in Fig. 2. The chromites have a higher melting

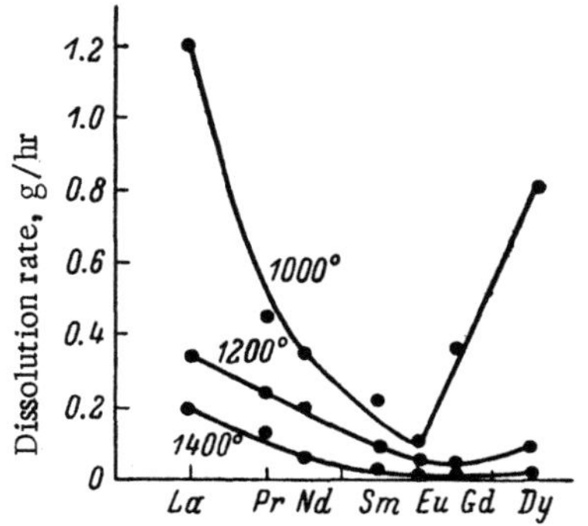

Fig. 4. Rate of dissolution of monoaluminates in HCl (sp. gr. = 1.19).

temperature than the monoaluminates. The following
law is observed for the change of melting temperatures
of the compounds: the melting temperature of rare-
earth chromites and monoaluminates decreases with
rare-earth atomic number.

The reverse relationship is observed in the rare-
earth sesquioxides: the melting temperature of the
oxides increases with rare-earth atomic number (Fig. 3).
Deviations from this general rule are observed for several
elements (europium and ytterbium).

The melting temperatures of all compounds were
determined by one method, utilizing a tungsten resistance furnace and contact-free specimen heating. The
melting temperatures of Ce_2O_3, Pr_2O_3, and Tb_2O_3 were
measured in a hydrogen atmosphere, while those of the other compounds were done in argon. The
change of melting temperature was followed using an optical pyrometer with an accuracy of $\pm 50°$.

Study of the crystal structure of the fused chromites and monoaluminates showed that they
melt without decomposition. The structure of the fused compounds was identical to the structure of the compounds in their original state.

The microhardness of the fused chromites and monoaluminates is 1600-1700 kg/mm^2;
the microhardness of the cerium subgroup oxides is 350-400 kg/mm^2, while that of the oxides
of the ytterbium subgroup is 600-700 kg/mm^2.

The chemical stability of the rare-earth monoaluminates largely depends on their annealing temperature. Hydrochloric acid appears to be the best solvent for the monoaluminates.
Figure 4 shows the rate of dissolution of several monoaluminates in hydrochloric acid as a
function of their mode of preparation. Monoaluminates annealed at 1700-1800°C are very stable
to acids and alkalis, and a potassium bisulfate melt was used to effect their total dissolution
for analysis.

The chromites and monoaluminates have higher chemical stability toward water and acids
than do the rare-earth oxides. Investigations showed that the complex oxides are practically
insoluble in concentrated acids at room temperature, and are only partially soluble on heating.

Chromites of all the rare earths easily dissolve on heating either in perchloric acid or a
1 : 1 mixture of perchloric and hydrochloric acids. They also dissolve easily in a 2 : 1 mixture
of concentrated sulfuric and phosphoric acids containing potassium permanganate or dichrom-
ate additive.

All compounds synthesized are very stable in boiling water and in water at high temperature under pressure. The studies showed that the crystal structure of the chromites and monoaluminates does not change during interaction with boiling water (100-h duration) or with water
at 350°C (24 h). Dissolved rare earth was not detected by either chemical or spectroscopic
methods. In this respect the rare-earth chromites and monoaluminates differ advantageously
from the rare earth sesquioxides, which easily hydrolize and transform to the hydroxides [15].

CONCLUSIONS

The chromites and monoaluminates of the majority of the rare-earth elements were synthesized and several of their properties were studied (density, melting temperature, chemical
stability, etc.).

It was shown that all compounds synthesized have high melting points and considerable stability (compared to the sesquioxides), toward boiling water and many chemical reagents.

LITERATURE CITED

1. Rare-Earth Metals [Russian translation], IL, Moscow (1957), p. 408.
2. B. I. Kogan, in: Questions on the Theory and Application of Rare-Earth Metals [in Russian], Izd. Nauka, Moscow (1964), p. 15.
3. H. Lehmann and K. Mueller, Ber. Dtsch. Keram. Ges., Vol. 40, p. 109 (1963).
4. K. Reaver, Nucl. Chem. Abstr., No. 18, p. 4177 (1963).
5. R. A. Mintorn, New Scientist, Vol. 14, p. 285 (1962).
6. S. G. Tresvyatskii and V. N. Pavlikov, in: Questions on the Theory and Application of Rare-Earth Metals [in Russian], Izd. Nauka, Moscow (1964), p. 159.
7. I. A. Bondar' and N. V. Vinogradova, Izv. Akad. Nauk SSSR, Ser. Khim., No. 5, p. 785 (1964).
8. N. A. Toropova and T. P. Kiseleva, Zh. Neorg. Khim., Vol. 6, p. 2353 (1961).
9. A. I. Leonov, in: Silicates and Oxides in High-Temperature Chemistry [in Russian], Izd. Akad. Nauk SSSR, Moscow (1963), p. 95.
10. S. G. Tresvyatskii, V. I. Kushakovskii, and V. S. Belevantsev, Atomnaya Énergiya, Vol. 9, p. 219 (1960).
11. K. I. Portnoi and N. I. Timofeeva, Izv. Akad. Nauk SSSR, Neorg. Mat., Vol. 1, No. 9, p. 1593 (1965).
12. K. I. Portnoi and N. I. Timofeeva, Izv. Akad. Nauk SSSR, Neorg. Mat., Vol. 1, No. 9, p. 1598 (1965).
13. K. I. Portnoi, N. I. Timofeeva, and V. I. Fadeeva, Zh. Neorg. Khim., Vol. 9, p. 2041 (1965).
14. A. I. Leonov and É. K. Keler, Izv. Akad. Nauk SSSR, Ser. Khim., No. 11, p. 1905 (1962).
15. K. I. Portnoi, V. I. Fadeeva, and N. I. Timofeeva, Atomnaya Énergiya, Vol. 10, p. 559 (1963).

CHROMITES OF THE RARE-EARTH ELEMENTS
AND SOME OF THEIR PHYSICOCHEMICAL PROPERTIES

V. N. Pavlikov, A. V. Shevchenko, L. M. Lopato, and S. G. Tresvyatskii

It is known from present literature data on rare-earth sesquioxide—chromium oxide interaction that chromium oxide forms chemical compounds of the general formula $A^{3+}B^{3+}O_3$ [1-3]. The rare-earth chromites have the structural type of a broad class of compounds called perovskites. Keith and Roy [1] cite data on the sintering of sesquioxide mixtures and the results of crystallographic studies of the phases forming. X-ray structure studies on chromites were performed by a number of authors [2-4].

An attempt was made to interpret the phase transformations in the Ln_2O_3—Cr_2O_3 systems in [3], which generalized the subsolidus phase relations in the binary sesquioxide systems.

The present communication presents the results of a study on rare-earth chromite single crystals from six phase diagrams of Ln_2O_3—Cr_2O_3 type (Ln = La, Nd, Sm, Gd, Y, Sc).

The study of phase transformations in Ln_2O_3—Cr_2O_3 systems was conducted using the anneal-quench method from temperatures of 1600-2500°C. An exact description of the experimental technique was given in [5, 6]. Phase diagrams of the above systems were constructed on the basis of experimental data. The determining points are presented in Table 1. All systems studied (with the exception of the Sc_2O_3—Cr_2O_3 system) belong to one type, and are distinguished primarily by their compound melting temperatures and eutectic-point coordinates. The chromites of lanthanum, neodymium, samarium, ytterium, and gadolinium melt congruently at temperatures of 2300-2450°C. The phase diagram of the Gd_2O_3—Cr_2O_3 system (Fig. 1a), reflects the general behavior of the phase transformations in the above systems. It appears general for all systems studied that there is no apparent solid solubility. This fact confirms the opinion [7] that solid solutions are not characteristic of systems of this type, which involve formation of compounds with perovskite structure.

The stability of the perovskite structure is governed by the ratio of the ionic radii of the cations and anions from which it is formed (known as the "Goldschmidt tolerance factor" [8]). For chromites this tolerance factor is within the limits $0.86 > t > 0.73$. This sets the stability range of the perovskite structure.

The ionic radii differences in the Sc_2O_3—Cr_2O_3 system (Fig. 1b) are very small for formation of the perovskite structure — small enough for solid-solution formation. The existence of

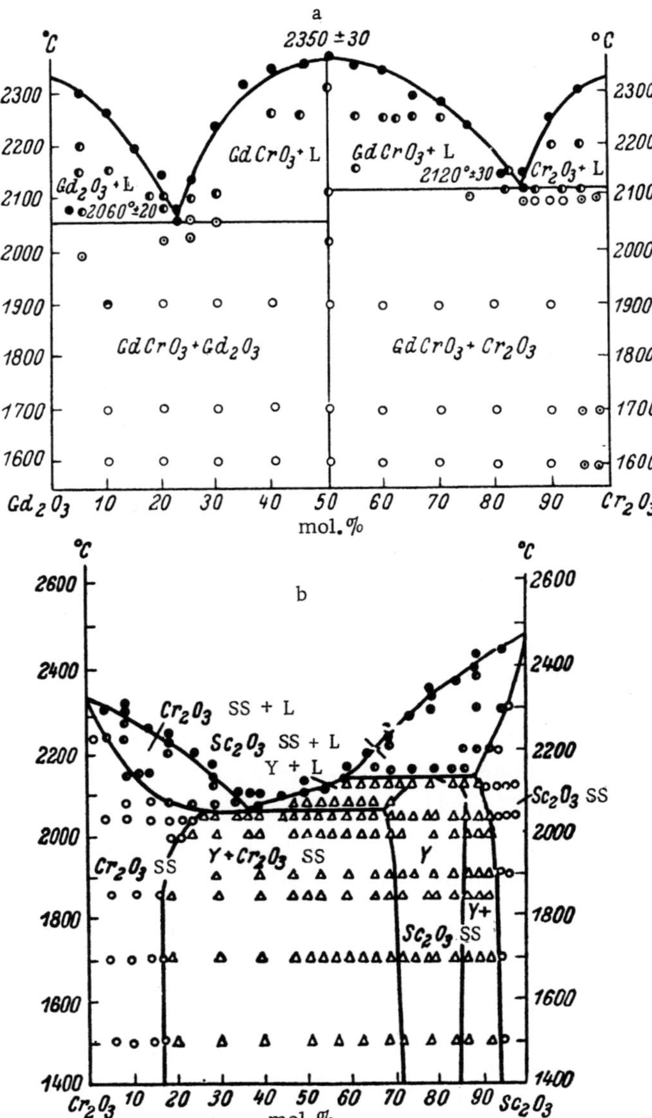

Fig. 1. Phase diagram of the $Gd_2O_3-Cr_2O_3$ (a) and $Sc_2O_3-Cr_2O_3$ (b) systems.

broad regions of solid solubility between the original oxides and an incongruently melting compound was established in this system. The single-phase region of γ-base solid solutions (Fig. 1b) decomposes at 1300°C over the range 73-85 mol.% Sc_2O_3. Thus, the probable compound composition is $1Cr_2O_3 : 3Sc_2O_3$ or $1Cr_2O_3 : 4Sc_2O_3$.

By preliminary evaluation the compound may belong to a structural class of medium symmetry (distorted cubic structure). To study some of their physicochemical properties, the rare earth chromites were grown as single crystals by spontaneous crystallization from PbO and PbF_2 solutions. The possibility for formation of compounds of garnet-type structure in the $Ln_2O_3-Cr_2O_3$ systems was also of interest.

Reference [9] hypothesized the existence of solid solutions of garnet structure in the $Y_2O_3-Cr_2O_3$ system. However, Cr^{3+} cations never occupy a tetragonal site in the garnet structure [10], as do Fe^{3+}, La^{3+}, and Al^{3+}. Our investigations established that the starting oxide

Table 1. Points Defining the Ln_2O_3—Cr_2O_3 (Ln = La, Nd, Sm, Gd, Y) and Sc_2O_3—Cr_2O_3 Systems

System	Phases	Process	Comp., mol.%		Temp., °C
			Ln_2O_3	Cr_2O_3	
La_2O_3—Cr_2O_3	$LaCrO_3 + Cr_2O_3$	Eutectic	22	78	2080 ± 30
	$LaCrO_3$	Compound	50	50	2430 ± 30
	$LaCrO_3 + La_2O_3$	Eutectic	80	20	2060 ± 30
Nd_2O_3—Cr_2O_3	$NdCrO_3 + Cr_2O_3$	Eutectic	22	78	2080 ± 30
	$NdCrO_3$	Compound	50	50	2330 ± 30
	$NdCrO_3 + Nd_2O_3$	Eutectic	76	24	2060 ± 30
Sm_2O_3—Cr_2O_3	$SmCrO_3 + Sm_2O_3$	Eutectic	80	20	1980 ± 30
	$SmCrO_3$	Compound	50	50	2300 ± 30
	$SmCrO_3 + Cr_2O_3$	Eutectic	16	84	2080 ± 30
Y_2O_3—Cr_2O_3	$YCrO_3 + Y_2O_3$	Eutectic	72	28	2020 ± 30
	$YCrO_3$	Compound	50	50	2290 ± 30
	$YCrO_3 + Cr_2O_3$	Eutectic	20	80	2070 ± 30
Gd_2O_3—Cr_2O_3	$GdCrO_3 + Gd_2O_3$	Eutectic	77	23	2060 ± 30
	$GdCrO_3$	Compound	50	50	2350 ± 30
	$GdCrO_3 + Cr_2O_3$	Eutectic	15	85	2120 ± 30
Sc_2O_3—Cr_2O_3	$Cr_2O_3 + \gamma +$ liquid	Eutectic	37	63	2060 ± 30
	$\gamma +$ liquid	Congruent melting	72—87	28—13	2130 ± 30
	$\gamma + Sc_2O_3$ SS + liquid	peritectic-	60	40	2130 ± 30

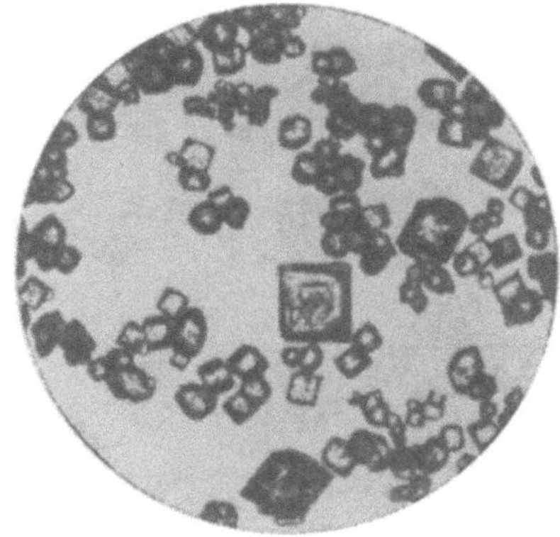

Fig. 2. Solid-solution single crystals between compounds in the Sc_2O_3—Cr_2O_3 system. Transmitted light, 30×.

ratios of 1 : 1 or $3Ln_2O_3$ and $5Cr_2O_3$ with experimental conditions used (dissolution temperature 1350°, cooling rate 10°C/h) lead to formation of only perovskite single crystals.

The size of given single crystals varies from 0.1 mm³ to 8 mm³. Contamination from lead inclusions during crystallization is 0.1 wt.%, from chemical and spectroscopic analysis data. An analogous experiment with a PbO melt (according to data of [11]) resulted in formation of stable lead chromites, which impeded growth of the rare-earth chromite single crystals.

Pure chromites of the cerium subgroup rare earths were obtained from a melt of $LnCl_3$ and K_2CrO_4 having a small excess of K_2CrO_4, by the reaction

$$LnCl_3 + K_2CrO_4 \rightarrow LnCrO_3 + 2KCl + \tfrac{1}{2}Cl_2 + \tfrac{1}{2}O_2.$$

Mixtures of the salts were melted, and then raised in temperature to 1200°C over a period of 5-6 h. The resulting crystalline powder was washed in hot water to remove the KCl and excess K_2CrO_4. The resulting crystal sizes were 100 to 150 μ. Figure 2 shows the solid-solution crystals obtained by this method, based on compounds in the Sc_2O_3—Cr_2O_3 system.

X-ray studies of the single-crystal chromites synthesized from oxide mixtures showed them to be completely identical, and independent of mode of preparation.

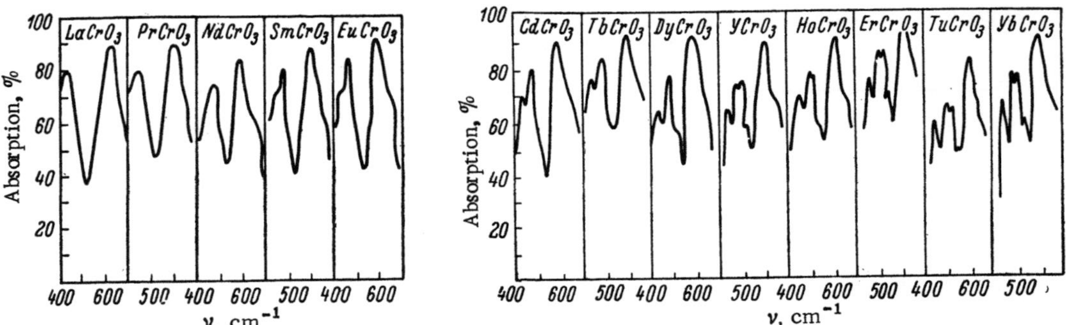

Fig. 3. Infrared spectra of the rare-earth chromites.

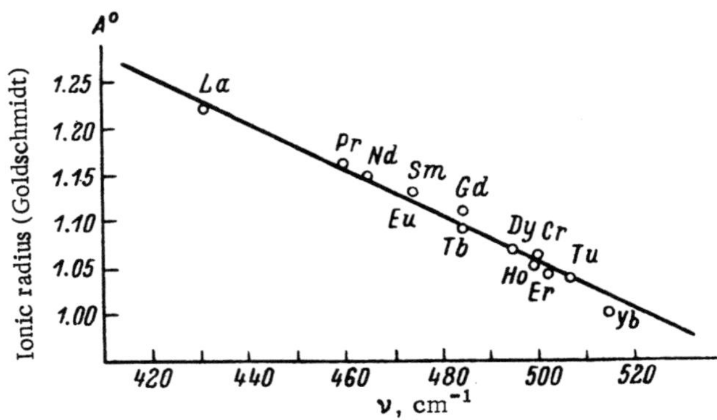

Fig. 4. Dependence of second absorption band wave-number of the rare-earth chromites on rare-earth ionic radius.

Table 2. Characteristic Frequencies of Rare-Earth Chromites.

Compound	Characteristic frequency, cm^{-1}	
	ν	other absorption fields
Cerium subgroup		
$LaCrO_3$	620	430
$PrCrO_3$	600	460
$NdCrO_3$	600	465, 445 (s)
$SmCrO_3$	595	475, 435 (s)
$EuCrO_3$	595	475, 440 (s)
Yttrium subgroup		
$GdCrO_3$	595	515 (s), 485, 445
$TbCrO_3$	590	520 (s), 485, 448
$DyCrO_3$	600	530 (s), 505, 485, 450
$\gamma\text{-}CrO_3$	605	535 (s), 510, 495, 455
$HoCrO_3$	600	535 (s), 510, 490, 450
$ErCrO_3$	600	540, 515, 490, 450
$TuCrO_3$	605	540, 520, 495, 445
$YbCrO_3$	600	550, 530, 500, 450

N o t e .   The letter s denotes a shoulder on the slope of the strongest band.

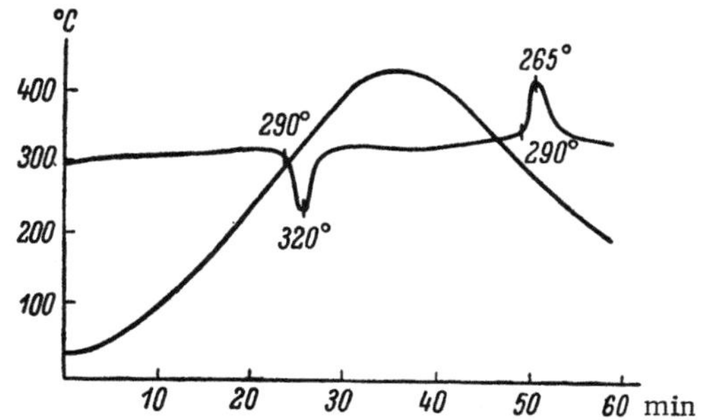

Fig. 5. DTA curves for lanthanum chromite.

Table 3. Coefficients of Linear Expansion
of Several Rare-Earth Chromites

Compound	Mean coeff. of linear expansion (room temp. to 900°C)
$LaCrO_3$	$85 \cdot 10^{-7}$
$PrCrO_3$	$85 \cdot 10^{-7}$
$NdCrO_3$	$83 \cdot 10^{-7}$
$SmCrO_3$	$86 \cdot 10^{-7}$
$Y-CrO_3$	$79 \cdot 10^{-7}$
$GdCrO_3$	$70 \cdot 10^{-7}$

The rare-earth chromite single crystals are transparent, green, tabular, square (occasionally rhomboid), anisotropic, biaxial negative, of linear extinction, of negative elongation, and of average birefringence. The indices of refraction of the single-crystal chromites are $2.37 > Ng, Np > 2.26$ for the series of compounds from La to Yb.

It was established that the rare-earth chromites undergo solid phase dissociation on exposure of $Ln_2O_3-Cr_2O_3$ compositions to high temperatures. The degree of high-temperature dissociation increases over the rare-earth series with decrease in rare-earth radius. Thus, a 30-min, 2100°C heat treatment in an argon atmosphere results in a 5-10 vol.% decomposition of lanthanum chromite, while the identical conditions cause a 70 vol.% decomposition for ytterbium chromite. If the duration of the $YbCrO_3$ heat treatment is increased to 3 h (same temperature), the specimen will consist of about 90% of a porous, cellular phase with an index of refraction of 1.940, corresponding to pure Y_2O_3.

The occurrence of rare-earth chromite dissociation is also attested by the fact that lanthanum and neodymium chromite specimens, heat-treated 2 h at 2100°C, decompose at room temperature in the presence of water vapor. This conforms to disintegration of the stoichiometric compound, resulting in formation of a layer of free lanthanum or neodymium oxide in the grain boundaries. On hydration, the layer volume increases because of hydroxide formation, causing sample disintegration. Absorption bands corresponding to $(OH)^-$ groups appear in the infrared spectra of specimens which have decomposed for 48 h.

We studied the infrared absorption spectra of yttrium and all rare-earth chromites (except the chromites of cerium and lutecium). The infrared absorption spectra were found identical and independent of the mode of sample preparation. The absorption frequencies characteristic of the 13 rare-earth chromites are shown in Table 2, and the infrared spectra in Fig. 3. As the figure shows, all rare-earth chromites have two strong absorption bands in the range 400-700 cm^{-1}. One of them remains constant over the series of chromites from $LaCrO_3$ to $YbCrO_3$. It lies in the region of 600 cm^{-1}. According to [12], it may represent the valence fluctuations of the Cr—O bond.

The second band in the cerium subgroup chromites appears to be a separate band which, for the ytterbium subgroup chromites, splits first into a doublet, and then into a triplet and a quartet. Lawson's data [13] reveal a band splitting which is often linked to a change in crystal symmetry. The latter also evidently occurs in the rare-earth chromites. As is known from the x-ray studies of [2], the rare-earth chromites have a perovskite-type lattice, with orthorhombic distortions (with such distortions increasing from lanthanum to lutecium). The structures of the chromites of the first members of the rare-earth series approximate the most ideal perovskite lattice. Absorption band splitting does not occur in these.

The second absorption band has still another characteristic: the magnitude of its first wave number varies with substitution of one rare earth by another. Figure 4 shows the dependence between the rare-earth radii and wave number of the second band of the chromite spectra. As the figure shows, there is nearly a linear dependence between rare-earth ionic radius and wave number. This absorption band is evidently associated with fluctuations of the Ln—O—Cr bond.

An investigation of phase transformations in the chromites of La, Nd, Pr, Sm, Gd, and Y was conducted in air over the temperature range 20 to 900°C. It was established by DTA, dilatometric, and high-temperature microscopic techniques that lanthanum chromite has a reversible polymorphic transformation at 290 ± 5°C. This transformation is accompanied by an endothermic heat effect (Fig. 5).

No transformations or associated heat and volume effects were noted for the chromites of Pr, Nd, Sm, Gd, and Y in the temperature range investigated. The coefficients of linear expansion computed from dilatometric measurement data are presented in Table 3.

High-temperature optical studies revealed that $LaCrO_3$ single crystals heated to 300°C undergo an intermittent change in interference coloration, which again repeats itself on cooling. A similar occurrence (noted in [14, 15]), was explained by the relative size change of the indicatrix of the optic axis of the newly forming modification. Single crystals of the other chromites maintain an unaltered interference coloration up to 900°C. It can be confirmed from these data that lanthanum chromite has a reversible, first-order polymorphic transformation, evidently associated with transition from orthorhombic to rhombohedral structure. This is in distinction to the orthorhombic lattices of the other chromites, which are stable in the temperature range studied. However, the experimental data from rare-earth aluminates, suggest the presence of a polymorphic transformation in the rare-earth chromite series (from $PrCrO_3$ to $LuCrO_3$) at temperatures above 900°C. This is also attested by the fact that compositions of 30 to 70 mol.% Cr_2O_3 in the Y_2O_3—Cr_2O_3, Sm_2O_3—Cr_2O_3, and Gd_2O_3—Cr_2O_3 systems have a porous structure and diminished mechanical strength.

LITERATURE CITED

1. M. L. Keith and R. Roy, Am. Mineral., Vol. 39, p. 1 (1954).
2. S. Geller, Acta Cryst., Vol. 10, p. 243 (1957).
3. R. S. Roth and S. J. Schneider, J. Res. Natl. Bur. Stds., Vol. 64A, p. 4 (1960).
4. S. Querel-Ambrunaz and M. Mareschal, Bull. Soc. Franc. Mineral. et Crystallogr., Vol. 86, p. 204 (1963).
5. S. G. Tresvyatskii and V. N. Pavlikov, Questions on the Theory and Application of the Rare-Earth Metals [in Russian], Izd. Nauka, Moscow (1964).
6. L. M. Lopato, S. G. Tresvyatskii, and Z. A. Yaremenko, Poroshkovaya Metallurgiya, No. 1, p. 29 (1964).
7. R. Roy. J. Am. Ceram. Soc., Vol. 37, No. 12, p. 580 (1954).
8. V. M. Goldschmidt, Skv. Norske Vidensk. Akad. Oslo, I Kl., No. 2 (1926).
9. S. J. Schneider, R. S. Roth, and J. Z. Waring, J. Res. Natl. Bur. Stds., Physics and Chemistry, Vol. 65A, No. 4, p. 345 (1961).
10. R. S. Roth and S. J. Schneider, J. Am. Ceram. Soc., Vol. 38, No. 11, p. 530 (1955).
11. J. R. Remeika, J. Am. Ceram. Soc., Vol. 78, p. 4595 (1956).
12. S. M. Ariya and M. V. Gomolzina, Fiz. Tverd. Tela, Vol. 4, No. 10, p. 2921 (1962).
13. K. E. Lawson, Infrared Absorption of Inorganic Substances, Reinhold, New York (1961).
14. A. I. Boikova and N. A. Toropov, Dokl. Akad. Nauk SSSR, Vol. 156, No. 6, p. 1428 (1964).
15. N. A. Toropov et al., Izv. Akad. Nauk SSSR, Ser. Khim., No. 7, p. 1158 (1964).

ELECTRICAL PROPERTIES OF THE RARE-EARTH OXIDES AND SEVERAL OF THEIR COMPOUNDS

A. V. Zyrin, V. A. Dubok, and S. G. Tresvyatskii

The oxides of scandium, yttrium, and the lanthanides, as well as many rare-earth oxide compounds, find numerous uses in nuclear technology as luminescent, thermoelectric, and semiconductor materials. This makes a study of their electrical properties important at this time.

The present article gives the results of electrical conductivity and thermoelectromotive force (thermo-emf) measurements as a function of temperature, performed on the pure oxides of scandium, yttrium, neodymium, and gadolinium at two oxygen partial pressures (P_{O_2} = 0.2 atm and P_{O_2} = 10^{-5} atm). Results on the chromites of the rare-earth series at P_{O_2} = 0.2 atm and at different concentrations of doped impurities are given.

The fraction of ionic conductivity in the pure oxides was also measured. All measurements were performed on polycrystalline samples prepared by usual ceramic methods.

The true porosity of the prepared samples did not exceed 3-4% for pure oxides and 10-12% for the rare-earth chromites. The total impurity content of the samples, by spectroscopic analysis data, was on the order of 0.1-0.3% for the pure oxides and less than 0.6% for the rare-earth chromites.

Electrodes for the specific resistivity measurements were applied by burning a platinum paste onto the end surface of the specimens.

TECHNIQUE OF MEASUREMENT

The specific resistivity of the samples was measured by the double probe method with a direct current and a 500-Hz alternating current. The results of the measurements agree within limits of experimental error. The error in resistivity measurements at all temperatures does not exceed ±15% for the high resistivity specimens, and ±7% for specimens with a resistivity below 10^2 Ω. The thermo-emf was measured relative to the platinum loops of two thermocouples. The error in measuring the coefficient of thermoelectromotive force thermo-emf did not exceed ±20%. The specific resistivity and thermo-emf measurements on the pure oxides were performed in air and in purified argon (P_{O_2} = 10^{-5} atm) at temperatures up to 1600°C. Measurements on the fraction of ionic conductivity in the pure oxides were performed by the concentrated emf method [1]. For measurement of the transference numbers, the measured emf was

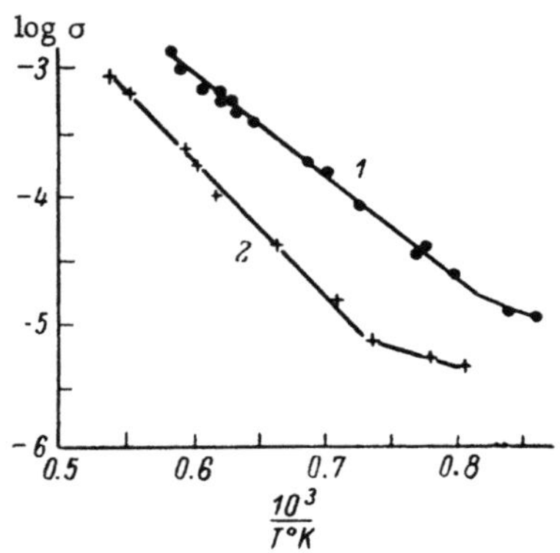

Fig. 1. Dependence of the logarithm of specific electrical conductivity of yttrium oxide on the reciprocal of temperature. 1) In air; 2) at $P_{O_2} = 10^{-5}$ atm.

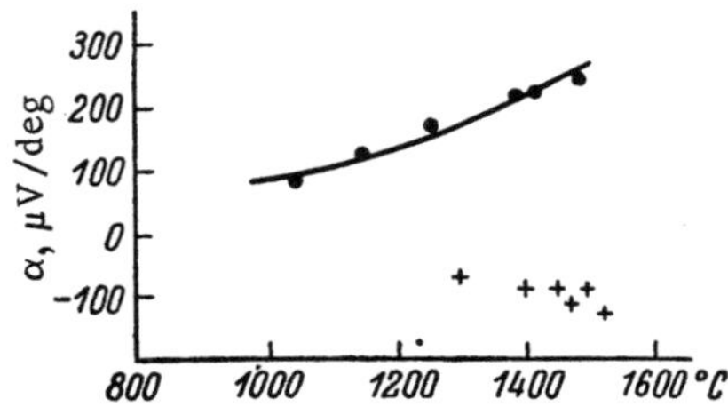

Fig. 2. Temperature dependence of the coefficient of thermo-emf of yttrium oxide. Legend as in Fig. 1.

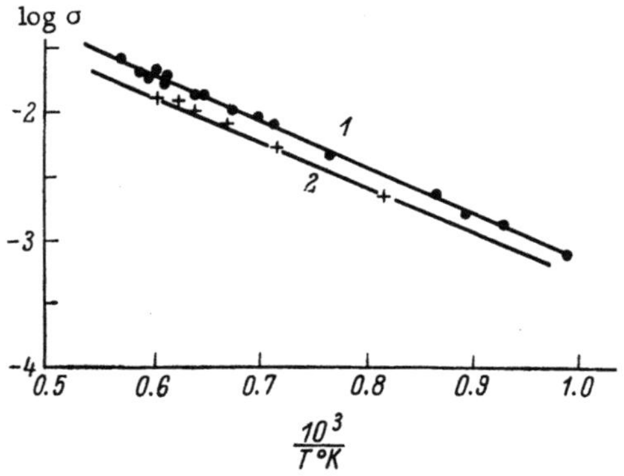

Fig. 3. Reciprocal temperature dependence of the logarithm of the specific electrical conductivity of neodymium oxide. Legend as in Fig. 1.

compared with that obtained under the same conditions using a solid electrolyte cell with cubic ZrO_2 + 15% CaO solid solution electrolyte, which is a pure ionic conductor [2].

For measurement of the Hall emf, an apparatus providing an alternating (50-Hz frequency) magnetic field was used. The sample alternating current had a frequency of 75 Hz. The Hall emf was measured at 25-Hz intervals using a selective range microvoltmeter.

RESULTS OF MEASUREMENTS

1. Scandium Oxide (Sc_2O_3)

The emf of a galvanic cell having solid Sc_2O_3 electrolyte at 1250°C is less than 2 mV (the limit of sensitivity of the measuring apparatus). The emf of a cell with solid ZrO_2 + 15% CaO electrolyte is 730 ± 20 mV under the same conditions. Thus, at 1250°C, less than 0.3% of the total current in Sc_2O_3 belongs to the ionic current fraction, and under these conditions Sc_2O_3 may be considered as a purely electronic conductor.

The results of the electrical conductivity measurements on Sc_2O_3 at temperatures above 1100°C are represented by a straight line on $\log \sigma$ versus $1/T$ coordinates, i.e., at these temperatures there is an exponential dependence of the electrical conductivity on temperature:

$$\sigma = \sigma_0 \exp\left[-\frac{E}{kT}\right], \tag{1}$$

where the activation energy E for electrical conductivity at $P_{O_2} = 0.2$ atm is 2.2 ± 0.2 eV. At 1600°C, the specific electrical conductivity in air is $1.6 \cdot 10^{-3} \ \Omega^{-1} \cdot cm^{-1}$.

The sign of the thermo-emf corresponds to hole conductivity (positive at the cold end). The coefficient of thermo-emf increases almost linearly with a rise in temperature. At 1450°C, the coefficient of thermo-emf (α) is 230 ± 30 μV/deg. A change in oxygen partial pressure to $P_{O_2} = 10^{-5}$ atm causes an eightfold decrease in the electrical conductivity at 1450°C, at which time there is no change in the sign of the thermo-emf, but a threefold increase in the magnitude of its coefficient (to 780 μV/deg).

The activation energy for electrical conductivity at $P_{O_2} = 10^{-5}$ atm is 2.6 ± 0.2 eV.

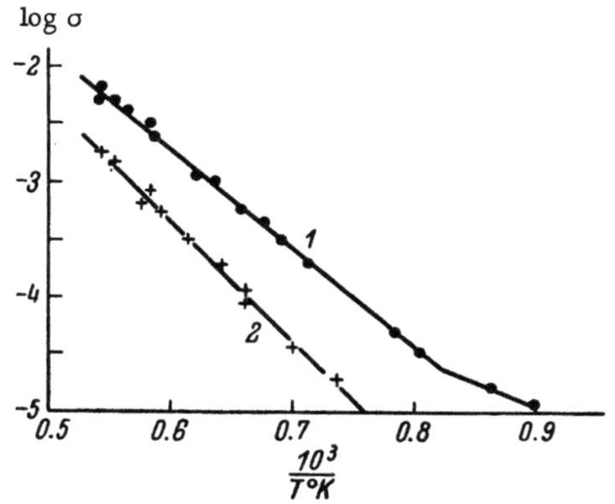

Fig. 4. Reciprocal temperature dependence of the logarithm of the specific electrical conductivity of gadolinium oxide. Legend as in Fig. 1.

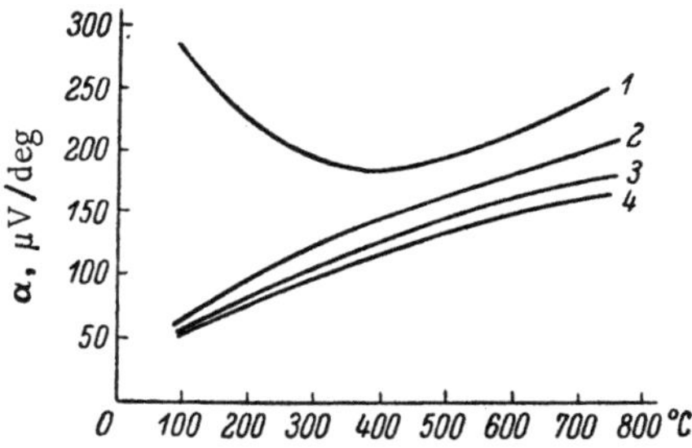

Fig. 5. Temperature dependence of the thermo-emf coefficient of a neodymium chromite containing different amounts of strontium dopant: 1) $NdCrO_3$; 2) $Nd_{0.99} \cdot Sr_{0.01}CrO_3$; 3) $Nd_{0.98}Sr_{0.02}CrO_3$; 4) $Nd_{0.96}Sr_{0.04}CrO_3$.

2. Yttrium Oxide (Y_2O_3)

The fraction of ionic conductivity in both Y_2O_3 and Sc_2O_3 is less than 0.3% at 1250°C. The high-temperature electrical conductivity at Y_2O_3 is also described by Eq. (1), with the transition to the high-temperature conductivity region beginning in air at about 1000°C, with an activation energy E of 1.7 ± 0.2 eV (Fig. 1). The sign of the thermo-emf at P_{O_2} = 2 atm is positive, corresponding to a p-type conductivity. The coefficient of thermo-emf is 270 ± 30 $\mu V/$deg at 1450°C.

In contrast to Sc_2O_3, a decrease in oxygen partial pressure to P_{O_2} = 10^{-5} atm leads to a decrease in the thermo-emf, which eventually passes through zero and changes to a negative sign corresponding to n-type conductivity. During this process the value of α at 1450°C is 100 ± 10 $\mu V/$deg (Fig. 2). The electrical conductivity of Y_2O_3 at the same temperature is 4.5 times lower under an atmosphere of P_{O_2} = 10^{-5} atm than in air. The activation energy (E) for electrical conductivity is 2.1 ± 0.2 eV.

3. Neodymium Oxide (Nd_2O_3)

The electrical properties of neodymium oxide resemble in nature the behavior of the properties of scandium oxide. The electrical conductivity of neodymium oxide at 1250°C is purely electronic (within limits of experimental error). The dependence of electrical conductivity on temperature is exponential: a transition to high-temperature conductivity occurs in air at about 700°C (Fig. 3). The sign of the thermo-emf denotes a hole conductivity. The thermo-emf coefficient increases with temperature.

A decrease of P_{O_2} to 10^{-5} atm at 1450°C results in a relatively small (approximately 1.4 times) change in electrical conductivity and to a significant increase in the thermo-emf coefficient (from 300 to 900 $\mu V/$deg). The sign of the thermo-emf does not change during this process.

The activation energy for electrical conductivity in neodymium oxide is 0.64 ± 0.06 eV, and does not change (within limits of experimental error) with decrease in P_{O_2}.

4. Gadolinium Oxide (Gd_2O_3)

The emf of a concentration galvanic cell having solid Gd_2O_3 electrolyte and gas-permeable platinum electrodes is 380 ± 20 mV at 1250°C, while that of a cell having ZrO_2 + 15% CaO electrolyte is 730 ± 10 mV under the same conditions. Under these conditions, more than half of the total current in Gd_2O_3 is therefore carried by ions.

The temperature dependence of the electrical conductivity in air above 1100°C approximates an exponential curve, with an activation energy E of 1.6 ± 0.1 eV (Fig. 4). The thermo-

emf coefficient of gadolinium oxide is positive and increases with rise in temperature analogou
ly to the thermo-emf of the other pure oxides investigated. On decrease of oxygen partial pressure to 10^{-5} atm, the electrical conductivity of Gd_2O_3 at 1450°C undergoes a fourfold decrease.
The thermo-emf sign remains positive, but the thermo-emf coefficient decreases from 600 to
200 $\mu V/deg$. The activation energy for electrical conductivity in Gd_2O_3 increases to 2.0 ± 0.2
eV at $P_{O_2} = 10^{-5}$ atm.

RARE-EARTH CHROMITES

To explain the effect of the form of the rare-earth ion on the electrical properties of
(rare earth) chromites doped with Sr or Ca impurity, a series of compounds were synthesized
having the general formula $Ln_{0.95}Ca(Sr)_{0.05}CrO_3$ (Ln = La, Pr, Nd, Sm, Eu, Gd, Er, Tm, and Yb).
These chromites, with exception of those of samarium and thulium, have approximately the
same room-temperature specific resistivity (10-50 Ω-cm. The samarium and thulium chromites have anomalously high specific resistivity values — up to 10^4 Ω-cm. As a rule, the stronti-
um-doped rare-earth chromites have higher specific resistivities than do the same chromites
doped with calcium.

A more detailed study on the effect of the amount of divalent-metal-atom impurity on the
electrical conductivity and thermo-emf coefficient was performed for neodymium chromite, of
which samples were prepared which had a composition corresponding to the formula
$Nd_{1-x}Sr_xCrO_3$ (where x equals 0, 0.01, 0.02, 0.04, and 0.08).

Two linear portions with a break near 380-420°C are observed in all compositions for the
temperature dependence of specific resistivity constructed using $\log p = f(1/T)$ as coordinates.

The room-temperature electrical conductivity is proportional to the dopant concentration.
This confirms the fact that the current carrier concentration on doping is proportional to the
concentration of dopant atoms introduced; the carrier mobility does not depend on dopant concentration.

The dependence of the thermo-emf coefficient α on temperature for specimens of different composition is shown in Fig. 5. In all doped specimens, α rises with temperature over the
temperature range investigated, with the sign of α denoting hole conductivity. For pure neodymium chromite, α first decreases with an increase in temperature, but then rises at about
the same rate as for doped specimens. The minimum of α is at 380-400°C, corresponding to
the break on the $\log p = f(1/T)$ dependence.

The sensitivity of mobility measurements (10^{-2} cm^2/sec) was calculated for the apparatus
used in studying the Hall effect, with the values obtained for specimen electrical conductivities
at room temperature. The Hall effect in neodymium chromite could not be observed in the latter apparatus, indicating the lower current carrier mobility of $NdCrO_3$.

DISCUSSION OF RESULTS

1. The resulting data attest to the fact that the electrical conductivity of all the pure rare-
earth oxides studied has an impurity character, and is associated with nonstoichiometric defects
in these oxides. With all pure oxides (except yttrium oxide), the entire investigated range of
oxygen partial pressures displays deviations of stoichiometry to the side of excess oxygen content in the oxide lattice. All these oxides have p-type conductivity, which increases with an increase in the oxygen partial pressure in the surrounding atmosphere.

Yttrium oxide, in distinction to Sc_2O_3, Nd_2O_3, and Gd_2O_3 in the same pressure range, displays an amphoteric behavior of a different nature in the range $P_{O_2} = 0.2-10^{-5}$ atm. There is a
a surplus of oxygen at large oxygen partial pressures in the Y_2O_3 lattice, and the properties of
Y_2O_3 are analogous to the properties of the other oxides studied.

At small P_{O_2}, an n-type conductivity occurs in Y_2O_3, i.e., under these conditions Y_2O_3 contains an excess of metal relative to the stoichiometric composition. The electrical conductivity should rise with a decrease of P_{O_2}.

The stoichiometric composition and an electrical conductivity minimum is evidently observed in Y_2O_3 at an intermediate oxygen partial pressure between 0.2 and 10^{-5} atm.

2. The observed dependence of the electrical conductivities and the activation energies for electrical conductivity on P_{O_2} indicates that the pure rare-earth oxides belong to the class of so-called intrinsic defect semiconductors [3, 4], which evidently includes most binary or more complex compounds which do not contain impurities. In such semiconductors, the concentration and degree of ionization of intrinsic defects responsible for the electrical conductivity of such semiconductors varies as a function of the temperature and the properties of the surrounding environment according to a law which is often nearly exponential. Such parameters as the energy of formation, energy of defect ionization, partial elasticity of components, etc., are also involved in the activation energy for electrical conductivity, and determine the width of the forbidden zone. It is therefore frequently impossible to determine the width of the forbidden zone for the oxides studied from the slope of the $\log \sigma = f(1/T)$ plot [5].

3. The occurrence of significant ionic conductivity in gadolinium oxide can be interpreted as suggestive of the fact that the holes responsible for the electrical conductivity of Gd_2O_3 (and probably of other rare-earth oxides) belong not to the 2p zone of oxygen, but to the 4f zone of the cations. The process of hole formation is in this case associated with the transition: $Gd^{3+}+\oplus \rightarrow Gd^{4+}$. Since the outer electronic shell of the Gd^{3+} ($4f^7$) ion is stable, disruption of this configuration requires a very large activation energy. This decreases the effective hole mobility and electronic conductivity, and results in more apparent ionic conductivity.

The large difference in the activation energies for hole conductivity observed for different rare-earth oxides of the same crystal lattice and similar cation size also indicates that hole conductivity occurs in the 4f cation zone of all the rare-earth oxides, in addition to Gd_2O_3.

4. An $\alpha(T)$ dependence similar to that obtained by experiment for the pure rare-earth oxides can be expected for intrinsic defect semiconductors [6]. However, an exact calculation of α is lacking for this case. The thermo-emf determination can therefore be used only for telling the sign of the current carrier and for a qualitative appraisal of the change in carrier concentration. The $\alpha(P_{O_2})$ dependence in Gd_2O_3 is probably explained by the ionic contribution to the thermo-emf.

5. As the results of the electrical conductivity measurements on various rare-earth chromites showed, there is no clear connection between the magnitude of the electrical conductivity and the form of the rare-earth ion in the chromite (with the exception of the samarium and thulium chromites, whose decreased electrical conductivity could not be explained). The transition from one rare-earth atom to another evidently does not effect significant changes in the chromium-atom sublattice which determines the electrical conductivity. The electrical conductivity decrease in the chromites on replacing a strontium impurity by a calcium one can be understood on comparing the sizes of the ionic radii of calcium and strontium: with the latter, the ionic radius differs considerably more from the ionic radius of the replaced rare-earth ion than is so for calcium.

The conductivity activation energies computed using the expression $\rho = \rho_0 \exp(E_a/kT)$ decrease with an increase in the amount of impurity. For the low-temperature portion of the curve, this is from 0.33 eV to 0.09 eV for pure $NdCrO_3$ (for a composition with x = 0.08). For the high-temperature portion, this is correspondingly from 0.16 to 0.06 eV.

The results of the electrical conductivity and thermo-emf coefficient determinations on neodymium chromite indicate that insertion of divalent ions in the chromite crystal lattice leads to the same amount of (hole) carrier current. The mobility of these carriers is very small (by Hall effect measurement, less than 10^{-2} cm^2/sec; by calculation from electrical conductivity values, $6 \cdot 10^{-4}$ cm^2/sec).

The increase of electrical conductivity with rise in temperature can be ascribed to the increase of carrier mobility, but not to an increase in carrier concentration.

The increase in the activation energy for conductivity with decrease in temperature is probably associated with the transition from low-temperature conductivity (with strong polaron scattering at phonons) to high-temperature conductivity (with classic superbarrier jumping from one site to another) [7].

The increase in the thermo-emf of pure neodymium chromite samples on lowering temperatures below 350°C is evidently associated with a decrease in carrier concentration. Annealing such specimens in air leads to the formation of lattice defects: the trivalent chromium tends to be oxidized under these conditions, since the chromium ions in the lattice have an increased valence, i.e., the lattice has holes, and to compensate the excess positive charge, the lattice absorbs oxygen in excess of stoichiometric composition.

Above 400°C, the electrical conductivity mechanism for pure neodymium chromite is analogous to that of the doped specimens. This indicates approximately the same rate of thermo-emf coefficient increase with increase of temperature.

LITERATURE CITED

1. C. Wagner, Z. Phys. Chem. (Abt. B), Vol. 21, p. 25 (1933).
2. A. D. Neuimin and S. F. Pal'guev, in: Silicates and Oxides in High-Temperature Chemistry [in Russian], Izd. Akad. Nauk SSSR, Moscow (1963), p. 253.
3. H. Schmalzried, J. Chem. Phys., Vol. 33, No. 3, p. 438 (1960).
4. F. A. Kröger and H. J. Vink, Solid State Physics, Vol. 3, New York (1956).
5. K. W. Böer, Z. Naturforsch., Vol. 10a, p. 898 (1955).
6. V. N. Chebotin, The Electronic Conductivity of Ionic Crystals in Equilibrium with the Gas Phase [in Russian], Abstract of Candidate's Dissertation, Sverdlovsk (1964).
7. P. C. Newman, Proc. Phys. Soc., Vol. 79, No. 6(512), p. 1299 (1962).
8. H. C. Lang and Yu. A. Firsov, Zh. Éksp. i Teor. Fiz., Vol. 43, p. 1843 (1962).

IONIC AND ELECTRONIC CONDUCTIVITY OF SAMPLES FROM THE $ZrO_2-PrO_{1.83}$ SYSTEM

Z. S. Volchenkova

Earlier we measured the total electrical conductivity of samples from the praseodymium oxide—zirconium oxide system as a function of composition and temperature [1]. The investigations showed that the change in magnitude and character of the electrical conductivity corresponds to a change in specimen structure.

We could compare the electrical conductivity measurements on a number of samples from this system by studying the electrical conductivity character by the emf method [2, 3]. In such a way we could separate the total electrical conductivity into its component parts: ionic and electronic conductivity. In the present work, we set out to study the dependence of the ionic and electronic components on temperature and sample composition.

The specimens were prepared by intimate mixing of ZrO_2 and $PrO_{1.83}$ oxide starting materials by a method described in [1]. The final sample anneal was performed in air at a temperature of 1550°C. We attached platinum electrodes to the ends of the fired specimens, which were in the form of pellets. Such pellets were used as the electrolyte in the electrochemical cell. The pellets further acted as a barrier to separate the cell volume into two regions of different oxygen partial pressures. We calculated the thermodynamic value of the electromotive force of the cell (Pt, oxygen$|ZrO_2-PrO_{1.83}|$air, Pt) from the equation

$$E_0 = \frac{RT}{4F} \ln \frac{P_1}{P_2}.$$

The ratio of the emf measured (E) to its theoretical value (E_0) at a given temperature, $E/E_0 \cdot 100\%$, corresponds to the magnitude of the ionic transport numbers $(\bar{t}_a + \bar{t}_c)$, since $E = E_0(\bar{t}_a + \bar{t}_c)$ for the case of purely ionic conductivity and $E = E_0[1 - (\bar{t}_e + \bar{t}_0)]$ for the case of mixed ionic and electronic conductivity. Here $\bar{t}_e$ is the average transport number for electrons and $\bar{t}_0$ is the average transport number for electronic holes [4].

The values we obtained for the ionic transport numbers (or percent ionic conductivity) for specimens of the studied system are quoted in Table 1 for different temperatures. The sample composition is based on the composition of the initial components (ZrO_2 and $PrO_{1.83}$). Weight loss experiments on specimens containing 0 to 50 mol.% $PrO_{1.83}$ forced the assumption that praseodymium is in the trivalent state in this composition region, i.e., we actually are dealing with the $ZrO_2-Pr_2O_3$ system. In this case, the composition $0.66\,ZrO_2 \cdot 0.33\,Pr_2O_3$ corresponds to an initial composition of $0.5\,ZrO_2 \cdot 0.5\,PrO_{1.83}$. Since we did not know the exact

Table 1. Values of $E/E_0 \cdot 100\%$ of a (Pt, oxygen$|ZrO_2 - PrO_{1.83}|$air, Pt) Cell at Various Temperatures

Starting comp., mol.%		Temperature, °C								
ZrO_2	$PrO_{1.83}$	600	650	700	750	800	850	900	950	1000
100.00	—	89.8	88.8	87.8	87.7	89.3	91.0	89.2	89.7	79.8 [2]
95.00	5.00	0.3	2.6	4.6	5.9	7.5	9.3	11.2	13.0	17.8
90.80	9.20	51.1	56.8	60.4	64.3	69.9	74.5	77.5	80.0	80.9
85.00	15.00	71.0	72.0	74.4	77.1	78.9	81.0	82.4	84.6	85.5
74.70	25.30	72.6	75.5	83.8	87.5	89.8	92.0	92.5	94.0	94.4
65.00	35.00	2.4	14.8	34.6	53.0	64.0	69.3	74.4	78.8	86.1
60.70	39.30	22.5	37.4	52.2	62.4	70.0	74.5	81.7	83.0	84.0
50.98	49.02	0.7	2.1	2.7	4.1	5.1	6.4	7.8	9.4	11.2
36.33	63.67	—	—	—	0.5	0.6	0.8	1.3	1.7	2.3
30.00	70.00	—	—	—	—	0.5	0.7	0.7	1.3	1.3

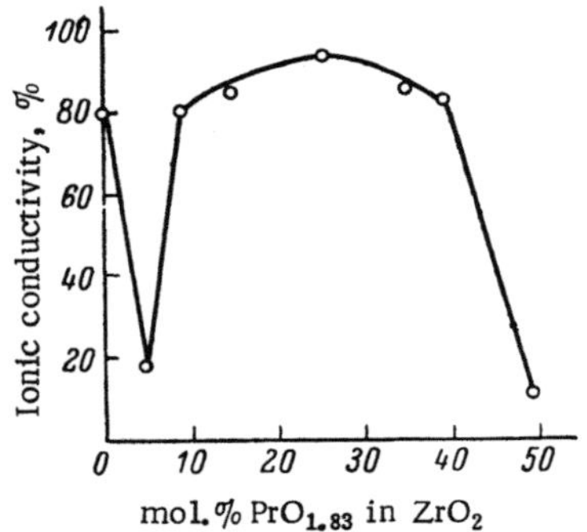

Fig. 1. Change in percent ionic conductivity with change of starting composition for samples from the $ZrO_2 - PrO_{1.83}$ system at 1000°C.

sample composition after sintering from the sample characteristics, we use in Table 1 the composition of the initial components, ZrO_2 and $PrO_{1.83}$.

Figure 1 shows a curve defining the percent ionic conductivity as a function of specimen composition at 1000°C. As seen from Fig. 1 (and also from Table 1), the ionic conductivity fraction has a minimum at 5 mol.% $PrO_{1.83}$ in ZrO_2. It then rises to nearly 100% ($0.25\,PrO_{1.83} \cdot 0.75\,ZrO_2$), responding to compound formation near the original composition of $0.50\,ZrO_2 \cdot 0.50\,PrO_{1.83}$. The ionic conductivity of samples of this compound is only about 10% of the total electrical conductivity at 1000°C. Even with increase of praseodymium oxide content in ZrO_2, the fraction of ionic conductivity (1-2%) is practically negligible: in this case the sample conductivity is predominantly electronic [1]. The ionic conductivity fraction, moreover, decreases with lower temperatures. For each given temperature, the dependence of the ionic conductivity fraction on composition remains analogous to Fig. 1. With a lowering of the isotherm temperature, $E/E_0 \cdot 100\%$ is shifted downward in a parallel fashion. A decrease in the ionic conductivity fraction occurs at lower temperatures with a reduced praseodymium oxide content in ZrO_2.

The presence of the extremes (minimum at 5 mol.% and maximum at 25 mol.% $PrO_{1.83}$) explains the interrelation between the ionic and electronic components of sample electrical conductivity. The isotherms of the ionic and electronic conductivity for the $ZrO_2 - PrO_{1.83}$ system are shown in Fig. 2, based on original composition. The electronic conductivity rises sharply at 5 mol.% praseodymium oxide in ZrO_2. The ionic conductivity remains almost changeless or has a minimum depending on the specimen temperature. As a result, the ionic conductivity fraction has the lowest value.

At the extreme, an increase in ionic conductivity occurs with increase of praseodymium oxide. The latter passes through a maximum at 25 mol.% $PrO_{1.83}$, exceeding the electronic component. The isotherm of the electronic component increases quite sharply only at the start; its growth then slows near the equimolar composition. Measurements of specimen electrical conductivity in air and in oxygen show that the electronic component has an n-type behavior in the

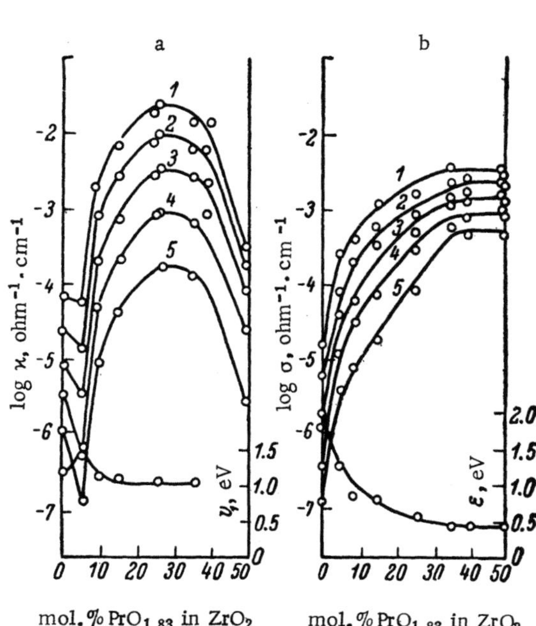

Fig. 2. Isotherms of ionic and electronic conductivity for samples from the ZrO$_2$—PrO$_{1.83}$ system. 1) 1000°; 2) 900°; 3) 800°; 4) 700°; 5) 600°C. a) Ionic conductivity; b) electronic conductivity. U) Activation energy for ionic conductivity; ε) temperature coefficient of electronic conductivity.

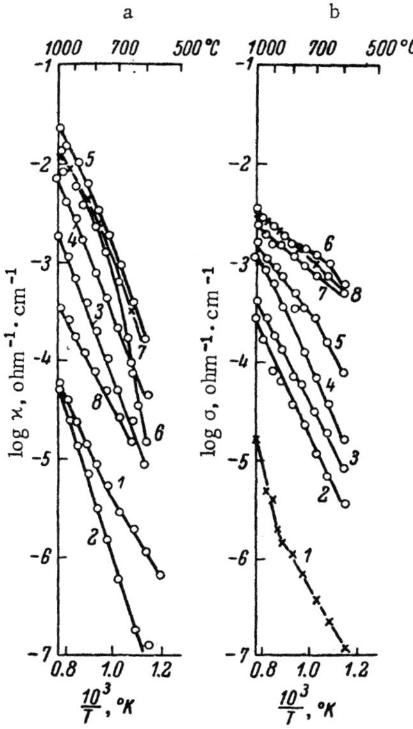

Fig. 3. Temperature dependence of ionic and electronic conductivity in samples of different composition from the ZrO$_2$—PrO$_{1.83}$ system. 1) ZrO$_2$; 2) 95 ZrO$_2$ · 5 PrO$_{1.83}$; 3) 90.8 ZrO$_2$ · 9.2 PrO$_{1.83}$; 4) 85 ZrO$_2$ · 15 PrO$_{1.83}$; 5) 74.7 ZrO$_2$ · 25.3 PrO$_{1.83}$; 6) 65 ZrO$_2$ · 35 PrO$_{1.83}$; 7) 60.7 ZrO$_2$ · 39.3 PrO$_{1.83}$; 8) 51.0 ZrO$_2$ · 49.0 PrO$_{1.83}$. a) Ionic conductivity; b) electronic conductivity.

Table 2. Change of Activation Energy for Oxygen-Ion Conductivity for Samples from the ZrO$_2$—PrO$_{1.83}$ System, as a Function of Composition

Original comp., mol.%		U_1, eV	Trans. temp. for $U_2 \rightarrow U_1$, °C	U_2, eV
ZrO$_2$	PrO$_{1.83}$			
100.00	—	1.19	750	0.72
95.00	5.00	1.45	—	—
90.80	9.20	1.1	—	—
85.00	15.00	1.02	800	1.2
74.69	25.31	1.03	850	1.30
75.00	25.00	1.09	700	1.74
65.00	35.00	1.03	725	2.42
60.70	39.30	0.95	750	1.40
50.98	49.02	0.95	—	—
36.33	63.67	1.0	—	—

composition range 0-25 mol.% PrO$_{1.83}$. At higher content (down to pure praseodymium oxide at 800-1000°C) it has p-type.

Figure 3 shows the temperature dependence of the ionic and electronic conductivities of samples of varying composition in the ZrO$_2$—PrO$_{1.83}$ system. They are represented as straight lines with breaks at temperatures of 700-800°C. We computed (from each linear portion of these plots) the temperature coefficients of an equation describing the temperature dependence of ionic and electronic conductivity:

$$\varkappa = \varkappa_0 \exp\left(-\frac{U}{kT}\right), \qquad (1)$$

Table 3. Change of Temperature Coefficient of Electronic Conductivity for Samples from the ZrO_2—$PrO_{1.83}$ System as a Function of Composition

Original comp., mol.%		ε_1, eV	Trans. temp. for $\varepsilon_2 \to \varepsilon_1$, °C	ε_2, eV
ZrO_2	$PrO_{1.83}$			
100.00	—	2.03	850	0.9
95.00	5.00	1.25	850	0.93
90.80	9.20	0.87	850	0.82
85.00	15.00	0.82	850	1.1
74.69	25.31	0.58	700	1.0
65.00	35.00	0.46	800	0.26
60.76	39.29	0.41	—	—
50.98	49.02	0.4	—	—
36.33	63.67	0.4	800	0.49
20.00	70.00	0.29	700	0.42

$$\sigma = \sigma_0 \exp\left(-\frac{\varepsilon}{kT}\right). \qquad (2)$$

Here $\varkappa$ and σ are, respectively, the ionic and electronic conductivities, U is the temperature coefficient (or activation energy) of ionic conductivity, and ε is the temperature coefficient of electronic conductivity.

The data of these calculations and the temperatures of the breaks in the curves are given in Tables 2 and 3. The reason for the breaks appears to be a change in character of some aspect of the specimen electrical conductivity. The breaks are also indicated by the character of the total specimen electrical-conductivity temperature dependence [1].

The activation energy for ionic conductivity, being dependent on the mobility of oxygen ion vacancies [1], has a minimum at 25 mol.% $PrO_{1.83}$ in ZrO_2. This minimum corresponds to the maximum of ionic conductivity. We have an analogous correspondence in the CeO_2-base systems with BeO, MgO, CaO, BaO [5], as well as La_2O_3, Nd_2O_3, and Y_2O_3 [6] additives. This agrees quite well with theory [7]. The ZrO_2-base systems earlier described had a noticeable increase in the activation energy with rising oxide impurity content in the high concentration regions (regions of a small amount, ~5 mol.%, of additive). This effect is distorted by the occurrence of a phase transition [8-11].

In ionic-crystal solid solutions, defects theoretically affect the activation energy for oxygen ion mobility in such a way that each carrier is surrounded by defects enabling it to jump to a nearby stable position, with no simultaneously adjacent defects which encumber the jump [7]. The change in the temperature coefficient of electronic conductivity also corresponds to a change in path of the electrical conductivity isotherm. The temperature coefficients for n- and p-type conductivity differ considerably.

CONCLUSIONS

1. The nature of the electrical conductivity of specimens fired in air at 1550°C was determined by the emf method for samples from the ZrO_2—$PrO_{1.83}$ system over a broad composition range (0–70 mol.%) and temperature range (600–1000°C).

2. The temperature dependence of the ionic and electronic conductivities was obtained for a number of samples from this system. The temperature coefficients (activation energies) of ionic and electronic conductivity were computed.

3. The isotherms were constructed for the ionic and electronic components of the electrical conductivity in the ZrO_2—$PrO_{1.83}$ system. The data obtained were considered in light of the theory on ionic-crystal solid solutions.

LITERATURE CITED

1.   Z. S. Volchenkova, in: Electrochemistry of Molten and Solid Electrolytes, Vol. 6, Consultants Bureau, New York (1968), p. 113.
2.   S. F. Pal'guev, S. V. Karpachev, A. D. Neuimin, and Z. S. Volchenkova, Dokl. Akad. Nauk SSSR, Vol. 134, p. 1138 (1960).

3. S. F. Pal'guev and A. D. Neuimin, in: Electrochemistry of Molten and Solid Electrolytes, Vol. 1, Consultants Bureau, New York (1961), p. 90.
4. S. V. Karpachev and S. F. Pal'guev, in: Electrochemistry of Molten and Solid Electrolytes, Vol. 1, Consultants Bureau, New York (1961), p. 73.
5. S. F. Pal'guev and Z. S. Volchenkova, Tr. Inst. Elektrokhim. Uralsk. Fil. Akad. Nauk SSSR, No. 2, p. 157 (1967).
6. A. D. Neuimin and S. F. Pal'guev, Tr. Inst. Elektrokhim. Uralsk. Fil. Akad. Nauk SSSR, No. 3, p. 133 (1962).
7. V. N. Chebotin and L. M. Solov'eva, Fiz. Tverd. Tela (in press).
8. Z. S. Volchenkova and S. F. Pal'guev, Tr. Inst. Elektrokhim. Uralsk. Fil. Akad. Nauk SSSR, No. 2, p. 173 (1961).
9. A. D. Neuimin and S. F. Pal'guev, Tr. Inst. Elektrokhim. Uralsk. Fil. Akad. Nauk SSSR, No. 5, p. 145 (1964).
10. T. Y. Tien and E. S. Subarao, J. Chem. Phys., Vol. 31, p. 104 (1963); Vol. 39, p. 1041 (1963).
11. D. W. Strickler and W. G. Carlson, J. Am. Ceram. Soc., Vol. 47, p. 122 (1964).

A HIGH-TEMPERATURE X-RAY CAMERA

P. F. Rumyantsev, I. I. Smil'gevich,
and A. S. Shaposhnikov

High-temperature x-ray structure and phase analysis of various materials call for improvements in the construction of high-temperature x-ray cameras, with the aim of increasing their working reliability and precision. The expanding areas of study require construction of cameras which best take into account the diverse (often contradictory) requirements involved in studying the phase composition and structure of materials heated to a specific temperature. These difficulties have not yet been surmounted in any one of the models proposed until now.

A number of cameras are built so that the sample is heated by an electric furnace [1-6]. It is difficult to determine the sample temperature accurately in cameras of this type (having photoregistry of x-ray intensity) due to sizeable temperature gradients (the furnace is exposed to x rays on all sides at the sample level [1]). This deficiency is corrected in cameras with the ionization registry of x-ray intensity, which have the furnace exposed only on one side at the sample level [2-6]. However, here it is impossible to avoid changes in specimen position with a change in temperature, especially with sample transformations involving changes in volume and shape.

The construction of another type of camera is based on the principle of contact sample heating. The sample, in the form of a powder, is applied to the surface of a metal wire, heated by an electric current. Cameras of this type which utilize photoregistry of x-ray intensity permit study of the sample at temperatures even above the material solidus temperature (where the most significant changes of volume and shape occur in the specimen) [7-9]. However, with some cameras it is impossible to achieve qualitative exposures when the heated material develops coarse crystallization, since the sample is not rotated [7]. The absence of a hermetically sealed atmosphere in other cameras of this type results in chemical interaction with air at elevated temperatures.

The principal deficiency in most cameras proposed until now is their inadequately high safe operating temperature (600° [8], 1000° [9], 1200°C [10]).

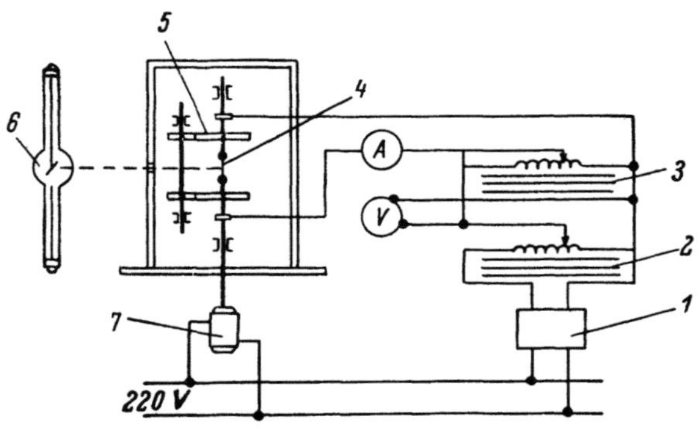

Fig. 1. Electrical layout of a high-temperature x-ray camera.

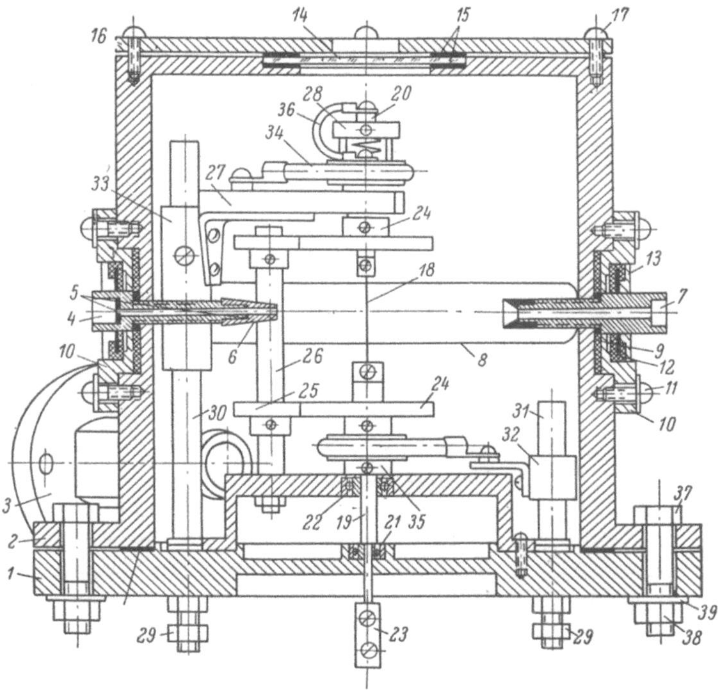

Fig. 2. Construction of a high-temperature x-ray camera.

A high-temperature x-ray camera operating on the above principle of contact sample heating was constructed at the Institute of Silicate Chemistry. The apparatus and components used for operation of the camera were manufactured in the domestic industry. The specifications of the camera construction enable the uppermost temperature of operation to be increased to 1500°C or higher. In the latter case (since the camera is hermetically sealed) an atmosphere can be maintained which is necessary for operation of a molybdenum, tungsten, or rhenium heating element (replacing the platinum—rhodium element specified in the camera described).

Figure 1 shows the electrical layout of the camera, which operates off a 127/220 V ac supply. The voltage stabilizer 1 (type SNÉ-220-0.75) eliminates any voltage fluctuations in the power supply. The autotransformer 2 (type RNO-250-2 and the stepdown transformer 3 (type RNO-250-5) regulate the electric system during operation, which is set at 150 V and 10 A by the voltmeter V and ammeter A. The heating element 4 and attached sample (at which a beam of x rays is directed from the x-ray tube 6) is rotated by the Warren motor 7 (type SD-2) through the drive gear 5. An additional resistance in parallel to the heater eliminates current fluctuations due to resistivity changes in the sliding contacts.

Figure 2 shows the construction of the high-temperature x-ray camera. The frame 1, where all components providing sample rotation and heater power input are mounted, holds the cylindrical body of the camera 2. The sleeve 3 on the camera body permits attainment of a vacuum or any given atmosphere in the camera. The collimator 4 with the two diaphragms 5 and the tip 6, as well as the x-ray trap 7, are secured to the camera body by a ring-shaped turning with two windows 8, the width of which is less than the width of the turning. The middle line of the turning and windows is near the center of the body axis intersecting the collimator—trap axis.

The ring-shaped turning is sealed by a Teflon washer with a 0.2-mm thick beryllium foil across it which transmits x rays, but shields the photographic film from light. The box 10, consisting of two half-rings, is fastened to the turning by the screws 11, and is secured to the

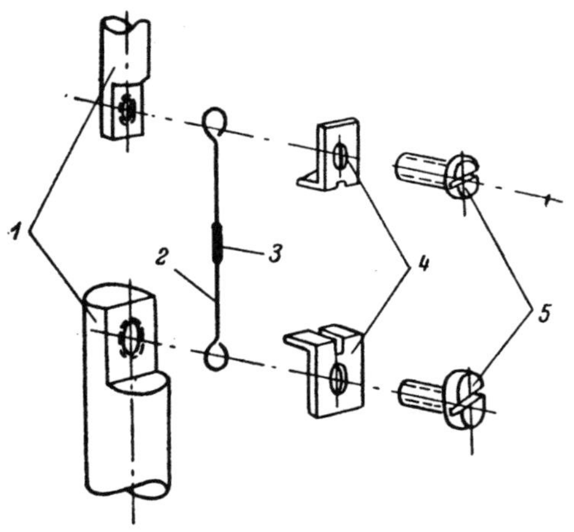

Fig. 3. Order of heater assembly.

collimator and trap from above the light-protective collar 12 with the packings 13. The viewing window in the upper part of the camera body consists of the glass 14, which is sealed into place by the rubber rings 15, the metal gasket 16, and the screws 17. The heater and attached sample are fastened to the split axles 19 and 20 with screws with use of special tools. The split axle 19, held by the ball bearings 21 and 22, is rotated at a rate of 2 rpm by the electric motor connected to the split axle by the coupling 23.

The transmission, consisting of the two gears 25 mounted on the axle 26, ensures the synchronized rotation of both split axles. A tightening device 28 is mounted on the split axle 20, and is riveted to the slip bearing on the support 27. The electric current is brought in through the terminals 29, the leads 30 and 31 (insulated from the camera frame), and the collars 32 and 33 to the sliding contact 34 (the metal rollers of which are riveted to the split axles). The flexible conductor 36 prevents current fluctuations in the tightening device. The camera body is fastened to the frame by the bolts 37, and the nuts and washers 38 and 39. A rubber gasket provides the packing between the body and the frame of the camera.

The precision of work with the camera usually depends on the precision in preparing the details of its assembly and on the care in its installation. Alignment must especially be had between the heater rotation axis and the box axis, in addition to a positioning of the collimator-trap axis perpendicular to the box axis at the middle of the heater. Superposition of the heater and the rotation axis is achieved by careful centering under a microscope. A check on the camera balance is accomplished by two methods. First, by comparison of x-ray patterns for specimen exposures at two camera frame positions 180° apart. The camera is accurately balanced on coincidence of corresponding lines in both x-ray patterns. The second method utilizes comparison of the interplanar spacings computed for several materials [NaCl, SiO_2 (quartz), etc.] with their tabular data. If the interplanar spacings calculated for these materials (after adequately careful study) differ from the tabular data by the same order of magnitude (i.e., can be best approximated by imposing the corresponding corrections), then the camera is exactly balanced. For the camera described, such a correction is for the line shift due to absorption and primary beam divergence (Kurdyumov correction [11]). Its magnitude is determined by the formula

$$x = r \sqrt{1 + \frac{R^2}{a^2} + 2\frac{R}{a}\cos 2\theta} \, , \tag{1}$$

where x is the line shift, R is the camera radius, a is the distance from the point source to the specimen, and r is the specimen radius.

Use of the collimator tip in the described camera protects the film from the diffraction effect at the edges of the diaphragm nearest the specimen.

The shape of the heater 2 is shown in Fig. 3. It is made of 0.35-mm diameter platinum or platinum—rhodium wire. The material being examined, carefully ground to a powder, is applied to the middle portion of the heater, which is coated with asphalt or bakelite lacquer. The sample diameter consequently should not exceed 0.5 mm and its length should be about 5 mm. The heater 2 and attached sample 3 are fastened to the ends of the split axle 1, and are checked for rigidity. The heater is secured by the washers 4 and screws 5. The final sample mounting

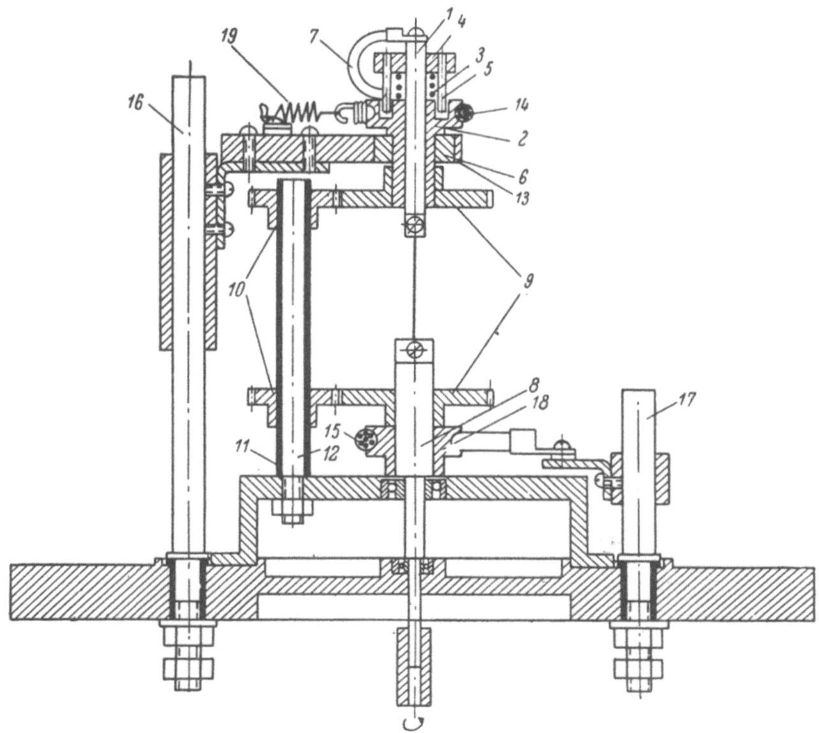

Fig. 4. Details of sample rotation assembly and heater power input.

is checked visually through the collimator. In this procedure, the sample should be situated in the center of the field of vision for each of the four basic camera positions (0, 90, 180, and 270° from the working position).

Figure 4 shows the construction of the tightening mechanism, which eliminates radial sample displacement due to thermal expansion of the heater while its ends are riveted. The upper split axle 1 is shifted in the axial direction on the sleeve 2. The spring 3, compressed on one side by the bushing 2 and on the other by the disk 4, tightens the heater. The pin 5 ensures a coupling between the split axle and the sleeve, which rotates freely on the slip bearing 6 mounted on the bracket 13. The flexible conductor lead 7 eliminates the current fluctuations in the heater due to the reistivity change in parts of the tightening assembly. It is necessary that there are no sharp changes in the resistivity of the conducting contacts when power is applied to the rotating heater, i.e., the strength of the current to the heater should remain constant (and in so doing provide a constant sample temperature for the entire duration of the exposure). This is accomplished with a slide wire and auxiliary resistance in parallel to the heater. The copper conductor cables of the sliding contacts 14 and 15 are joined to the leads 16 and 17. One is situated in the groove of the roller on the sleeve 2 and the other in the groove of the roller 18. The spring 19 presses the leads against the rollers. Synchronized rotation of the split axles 1 and 8 is accomplished by the two gears 9 (affixed to the divided axles) and by the two gears 10 (attached to the electrically insulated sleeve 11, which rotates freely on the axle 12).

The sample temperature can be measured with a precision necessary for the conditions of the experiment by three methods: (1) optical pyrometer; (2) magnitude of the current passing through the heater; and (3) by change in the crystal lattice parameter of platinum.

The feasibility of obtaining an almost unlimited number of x-ray photographs on the same sample at different temperatures assures application of the described high-temperature x-ray camera in determination of the melting temperatures of crystalline compounds (nonmetals), study of their phase composition, and in the study of polymorphic and chemical transformations at different temperatures. In those cases where the crystal unit-cell parameter and structural type can be determined, it can be used to determine the thermal expansion coefficient.

LITERATURE CITED

1. J. R. Johnson, J. Am. Ceram. Soc., Vol. 37, No. 8, p. 360 (1954).
2. P. F. Konovalov, A. I. Efremov, and B. V. Volkonskii, Ionization X-Ray Structure Determination Apparatus for Study of Crystalline Materials at Different Temperatures [in Russian], Leningrad (1958).
3. A. G. Boganov, L. P. Makarov, and V. S. Rudenko, Dokl. Akad. Nauk SSSR, Vol. 161, No. 2, p. 332 (1965).
4. D. K. Smith and C. F. Cline, J. Am. Ceram. Soc., Vol. 45, No. 5, p. 249 (1962).
5. C. R. Hauska and E. J. Keplin, J. Sci. Instr., Vol. 41, No. 1, p. 23 (1964).
6. J. P. Holden, J. Sci. Instr., Vol. 41, No. 11, p. 706 (1964).
7. T. Kubo and H. A. Kobori, J. Phys. Coll. Chem., Vol. 54, No. 8, p. 121 (1950).
8. A. V. Progrushchenko, Zavodsk. Lab., Vol. 24, No. 1, p. 104 (1958).
9. Yu. I. Petrov, Pribory i Tekhn. Éksp., Vol. 4, No. 8, p. 162 (1963).
10. M. M. Umanskii, V. V. Zubenko, S. S. Kvitka, G. V. Bogolyubov, V. K. Rubtsov, and G. I. Shelkovnikov, Apparatus for Study of the Physicomechanical Properties and of the Structure of Metals and Materials. Advanced Scientific-Technical Industrial Practice, 7th ed., [in Russian] (1960).
11. G. V. Kurdyumov, Zh. RFKhO, Ch. Fiz., Vol. 58, p. 745 (1926).

APPARATUS FOR STUDYING THE THERMAL DIFFUSIVITY COEFFICIENTS OF REFRACTORY MATERIALS

Ya. A. Landa and E. Ya. Litovskii

The steady-state cylinder or disk method is now most frequently used in studying the thermophysical parameters of industrial refractory materials, for determining the thermal conductivity coefficient; the mixing calorimeter is used for determining the heat capacity. The thermal diffusivity coefficient of a material is usually calculated from its thermal conductivity and heat capacity. In some cases it is evidently more suitable to determine the thermal diffusivity coefficient directly — the more so since many nonsteady-state methods are sufficiently simple and quick.

The aim of the present work was to build an apparatus for studying the thermal diffusivity coefficient of refractory materials in the 300–1100°C temperature range. The thermal diffusivity coefficient determinations are performed in this apparatus by means of a steady-state heating of cylindrical specimens [1–3]. The essence of the method consists in the following.

Let an infinitely long cylinder be heated radially by a steady, axially symmetrical heat flux. The temperature drop in the cylinder is then related to the thermal diffusivity coefficient through the formula [3]

$$\vartheta\,(r,\tau)=\frac{b_0}{4a_0}\cdot r^2\left[1-\frac{b_0}{16a_0}\,(k_\lambda+k_a-k_{b,\,r})\,r^2\right],\tag{1}$$

where

$$\vartheta\,(r,\,\tau)=t\,(r,\tau)-t\,(0,\tau);$$

t is the temperature; r is a flow coordinate; τ is time; $b = dt/d\tau$ is the rate of temperature rise; a_0 is the coefficient of thermal diffusivity; $k_a=\frac{a-a_0}{\vartheta}$; $k_\lambda=\frac{1}{\lambda}\cdot\frac{\lambda-\lambda_0}{\vartheta}$; $k_{b,\,r}=\frac{1}{b_0}\cdot\frac{b-b_0}{\vartheta}$. The zero subscripts denote that the corresponding parameters relate to a temperature $t_0 = t(0,\tau)$ on the cylinder axis. If the experimental conditions are chosen such that the inequality

$$\left|\frac{b_0}{16a_0}\,(k_\lambda+k_a-k_{b,\,r})\,r^2\right|\leqslant 0.01\tag{2}$$

is maintained, the temperature will obey all the rules of a regular second-order heat cycle, i.e.,

$$a_0=\frac{b_0r^2}{4\vartheta\,(r,\,\tau)}\,.\tag{3}$$

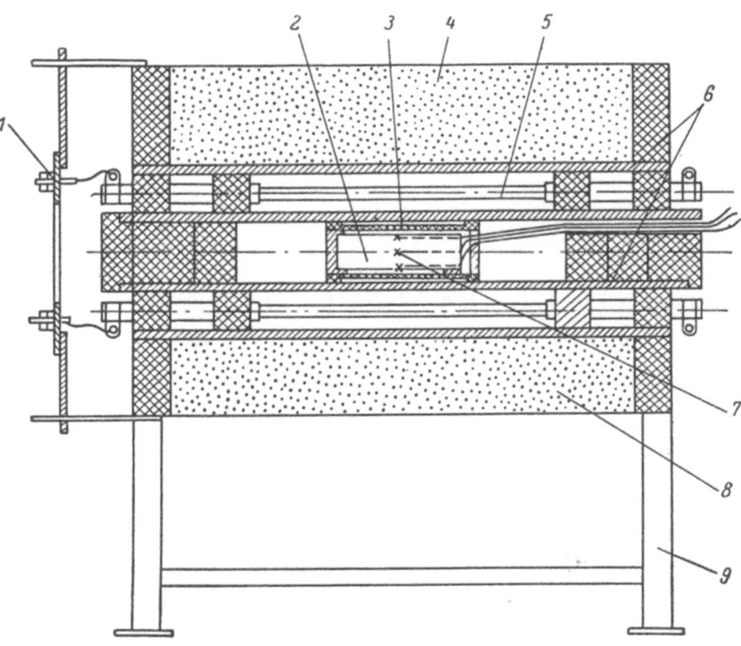

Fig. 1. Heating furnace (in section). 1) Terminal plates; 2) specimen; 3) stainless steel tube for temperature equalization; 4) metal casing; 5) Silit resistors; 6) fireclay tubes; 7) thermocouple; 8) lightweight fireclay filling; 9) support frame.

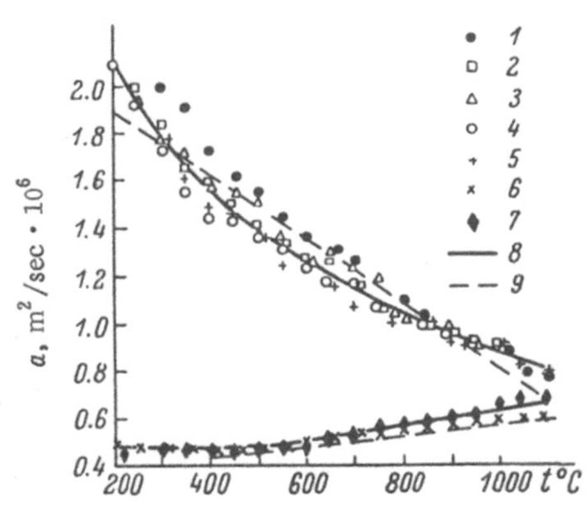

Fig. 2. Experimental curves. 1-5) Magnesite specimens; 6,7) ultralightweight specimens; 8) average experimental values; 9) thermal diffusivity values calculated with the data of [4].

The apparatus consists of a heating furnace, a type AOSKT-25/0.5 regulating transformer (with nominal rating of 220 V, 150 A) a control panel, and a type ÉPP-09 electronic potentiometer. A section of the heating furnace is shown in Fig. 1. The furnace is heated by eight Silit resistors connected in parallel by pairs. The resistors are located between two fireclay tubes which are centered relative to each other and the furnace casing by the fireclay support rings.

Fireclay granules were used for thermal insulation, and were placed in the space between the outer tube and the metal furnace casing. The specimen is inserted in a special, thick-walled stainless steel tube which is intended for temperature equalization along the specimen length and periphery. The use of the tube practically eliminates the effect of furnace convection on the specimen temperature distribution. Grounding the tube permits use of an automatic temperature recorder, since the effect of high-frequency induction on operation of the electronic potentiometer is eliminated.

The specimens to be studied are in the shape of cylinders 50 mm in diameter (d) by 200 mm in length (l). With the l:d ratio used, the specimen under study can be considered an infinite cylinder [1, 2].

Five chromel—alumel thermocouples are attached to the specimen, one at the center and four at the surface. A 3-mm diameter hole is drilled in the specimen to accommodate the central thermocouple. The surface thermocouples are cemented to grooves on the cylindrical specimen. The junction and the thermoelectrodes are covered with a thin layer of the powdered material under study, mixed with a small amount of sodium silicate solution. Averaging the readings of the four surface thermocouples largely excludes chance errors in the temperature measurement. The thermocouple readings are recorded with the electronic potentiometer.

To test the measuring technique, materials were chosen having a sharply different specific gravity and thermal-diffusivity-coefficient temperature dependence, namely, magnesite brick and standard ultralightweight brick. The scatter in the experimental data (Fig. 2) does not exceed the values calculated for the maximum relative error (12-15%). The experimental results agree closely with the thermal diffusivity coefficients computed from known thermal conductivity coefficients and heat capacities for the materials in question.

LITERATURE CITED

1. O. A. Kraev, Teploénergetika, No. 4 (1958).
2. N. Yu. Taits and E. M. Gol'dfarb, Zavosk. Lab., No. 3 (1950).
3. E. S. Platunov, Izv. Vuzov. SSSR, Priborostoenie, No. 5 (1964).
4. A. F. Kolechkova, E. G. Zadvornova, and A. R. Soltan, Tr. Vses. Inst. Ogneuporov, Vol. 35 (1963).

PART III

SYSTEMS CONTAINING REFRACTORY OXIDE COMPOUNDS

A. Systems Containing Rare—Earth Oxides

IRREVERSIBLE POLYMORPHIC TRANSFORMATIONS IN RARE-EARTH OXIDES

V. S. Rudenko and A. G. Boganov

Irreversible polymorphism, rarely if ever found in simple binary compounds of typical ionic bonding and definite valence, occurs quite frequently among silicates and numerous oxide and nonoxide refractory compounds. The complicated polymorphism of silica, aluminum oxide, titanium oxide, and numerous rare-earth oxides is a characteristic example.

Phenomenological thermodynamics describes a monotropic polymorphic transformation on the basis of the concept of metastable phases produced in random fashion, often in a quenching medium. However, it is now clear that this description is not adequate for all cases, since it does not explain the principal reasons for the existence (or more precisely, the stability) of the metastable phases.

An obviously interesting object for the study of irreversible polymorphism is the rare-earth sesquioxide series.

According to conflicting hypotheses [1-6], the rare-earth oxides crystallize in three structural types, usually denoted the C-, A-, and B-modifications (corresponding to the cubix, hexagonal, and monoclinic systems of crystallochemistry). Figure 1 shows the phase relationships for the polymorphism over a broad temperature range up to the melting temperature. Data from the most thorough works characterizing the polymorphic transformations observed are cited below:

Goldschmidt et al. (1925) [1]	All B- to A-form and some C- to B-form transformations reversible.
Shafer and Roy (1959) [3]	Transformation considered as reversible.
Roth and Schneider (1960) [4]	All transformations irreversible. Metastable low-temperature modifications.
Warshaw and Roy (1961) [5]	Both C- to A- and C- to B-form transformations reversible. Each modification stable in a corresponding temperature range.
Glushkova and Boganov (1965) [6]	All transformations irreversible. Metastable low-temperature forms.

As seen from the above data, the literature until quite recently contains a number of contradictory opinions regarding the reversibility or irreversibility of the transformations

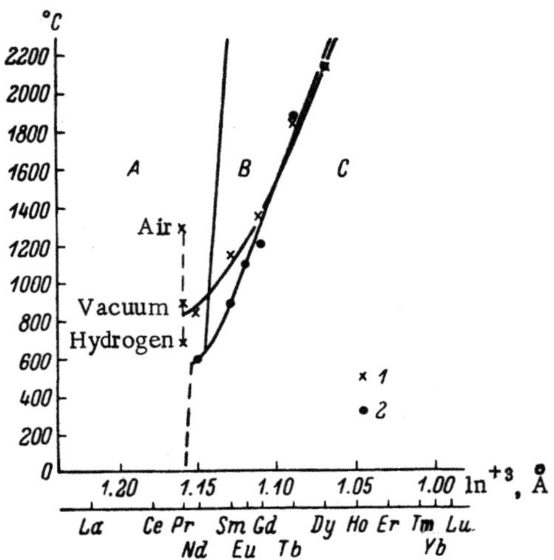

Fig. 1. Phase relations in the polymorphism of the rare-earth sesquioxides. 1) Data of present work; 2) data of [5].

observed. The results of the literature data can be briefly summarized by three points: 1) until recently, all authors without exception viewed phase transformations in rare-earth sesquioxide systems as a manifestation of classic polymorphism; 2) the character of the polymorphic transformations was not conclusively resolved, though most investigators conclude that they are irreversible; 3) the reasons and mechanism for irreversible flow processes during transition were not only not investigated, but in general were practically not even considered.

Only recently has a new approach been proposed for handling the phenomena observed in the rare-earth oxides [6]. These authors [6] noted the effect of gas formation accompanying the irreversible transformations in the oxides of neodymium and samarium. The results of this work led to the supposition that the low-temperature, metastable rare earth oxide phases owe their existence to the presence of extraneous impurities, and that they possess thermodynamic stability in their existing temperature range only as a result of incorporation of extraneous ions into the oxide lattice. The amount of impurity was estimated to be on the order of 1.5-2%. In further development of this idea, the opinion was expressed in [7] that the C-form (for example of Nd_2O_3), consists of a hydrate of composition $3Nd_2O_3 \cdot H_2O$. The latter authors also assume the simultaneous existence of a C-form of the "pure" oxide.

Our work in this connection had as its goal the investigation of the mechanism and the determination of the true nature of the polymorphic transformations in rare-earth oxides.

RESULTS

The clearly irreversible character of the transformations was established by direct x-ray study of the behavior of the oxides under vacuum and at high temperatures (up to 2000-2150°C) [8, 9], as well as by accurate thermoanalysis. Data on the transition temperatures are given in Fig. 1. The cubic C-modification is the low-temperature form in all cases where polymorphism occurs. On heating, it transforms to the hexagonal A- or the monoclinic B-form.

The results of comparative experiments performed in air or vacuum disclosed a strong dependence of the transition temperature on the nature of the surrounding atmosphere. Special experiments were accordingly performed to study the transformation processes in the oxides of praseodymium, neodymium, and samarium under different atmospheres: oxidizing, reducing, and neutral (Fig. 2). As seen from the figure, the data clearly demonstrate the dependence of the transition temperature on the oxidizing—reducing properties of the surrounding environment. The existence of such a dependence definitely indicates that in all three cases the polymorphic transformations are accompanied by chemical reactions leading to a slight change in the chemical composition of the oxide phase.

Weight determinations were conducted to obtain a quantitative evaluation of the composition changes thought to accompany the phase transformations. Repeated reproducible experiments established that a noticeable weight loss (however not exceeding 0.04-0.1 wt.%) occurs in the oxides of neodymium, praseodymium, samarium, and gadolinium in the temperature regions of their transformation.

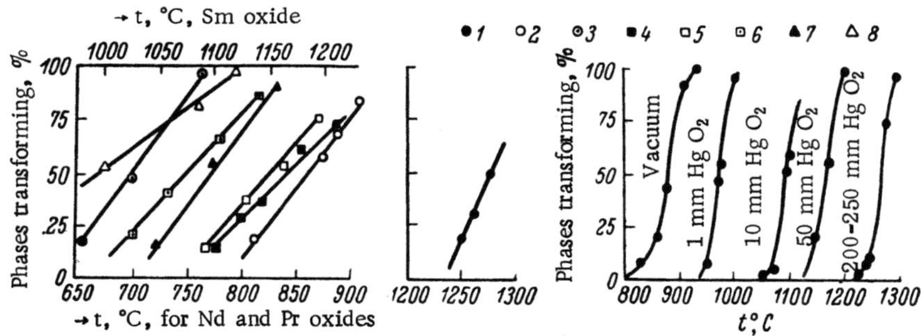

Fig. 2. Temperature dependence of transformations in the oxides of praseodymium, neodymium, and samarium on different atmospheres. Pr oxide: 1) air and various oxygen pressures; 2) argon (5 kg/cm^2); 3) hydrogen (130 kg/cm^2); Nd oxide: 4) air; 5) vacuum (1 · 10^{-5} mm Hg); 6) hydrogen (130 kg/cm^2); Sm oxide: 7) air; 8) hydrogen (130 kg/cm^2).

The results obtained necessitated performing a chemical analysis to determine the slight change of phase composition in the transformation process. Here it is necessary to note that the rare-earth elements have a valence of 3 for the majority of the most stable compounds. However, it is also a well-known fact that a number of rare-earth compounds can display an intermediate valence, varying between the limits of 2 to 4 over a complete series of compounds.

Exact chemical analysis data on the excess oxygen content of praseodymium oxide (relative to the Pr_2O_3 composition) showed that the transition from the cubic to hexagonal form is not a polymorphic transformation in the true sense of the word, i.e., a transformation limited to the very same composition (namely, the composition Pr_2O_3), as it is convenient to consider it. This transition is actually in a limited way associated with the occurrence of reduction processes in the higher praseodymium oxides during transition to the Pr_2O_3 composition (on heating). The quantitative change of PrO_x composition during the transition is approximately 0.05 wt.% oxygen, completely agreeing with weight loss determination data.

Based on the data obtained, the transformation of cubic to hexagonal praseodymium oxide, which is irreversible in vacuum or neutral atmosphere, should become reversible under slightly oxidizing conditions. A special experiment was carried out which completely confirmed this assumption. The reverse transition was effected by incorporation of no more than 0.1 wt.% oxygen into the original sesquioxide lattice. Terbium oxide, to a significant extent, also displays analogous behavior. The difference is only that the additional reduction involved in the transition from the cubic to the high-temperature form occurs in a composition region not above, but somewhat below the sesquioxide. A direct indication of this is the possibility of obtaining the reversible transition at comparatively low temperatures (on the order of 800-900°C), but with oxidation of the B-form. During this oxidation to the sesquioxide composition appears to be sufficient.

It was nevertheless determined that the oxygen content of the low-temperature forms of neodymium, samarium, and gadolinium oxides does not exceed a value corresponding to the sesquioxide composition (to an accuracy of 0.02 wt.%). Furthermore, the phase transformations in these oxides also appear irreversible not only in vacuum and inert environment, but also in an oxidizing atmosphere (in distinction to the oxides of terbium and praseodymium). Nevertheless, as was already indicated, the transition processes here are also accompanied by a noticeable weight loss, and display apparent sensitivity to the atmosphere in which they occur. On the other hand, the conditions for obtaining the low-temperature form of these oxides from aqueous

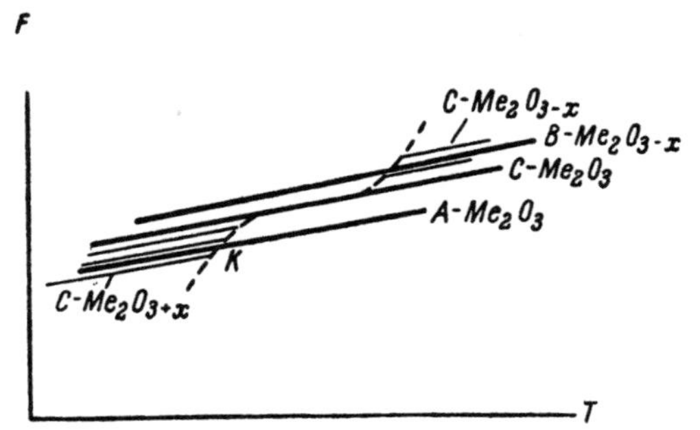

Fig. 3. Schematic representation for thermodynamic equilibrium in the C-, A-, and B-forms of the rare-earth oxides.

solutions (of salts and hydrates) generally lead to retention of residual hydroxide groups in the oxide lattice (even at temperatures considerably above the decomposition temperature of the original compounds).

All of this serves as a basis to confirm that the case under consideration involves the isomorphous substitution of pairs of hydroxyl complexes for specific oxygen atoms in the lattice of the low-temperature C-form. Since the transformation of the sesquioxides of neodymium, samarium, and gadolinium is assumed to be polymorphic, the phase transformation actually appears to accompany the reaction:

$$Me_2O_{3-x}(OH)_{2x} = Me_2O_3 + xH_2O; \quad x \leqslant 0.05 \quad \text{wt.\%}.$$

Explanation of the reversibility of the transformations (observed by Warshaw and Roy in experiments performed in presence of water vapor) is obtained within confines of this conclusion.

The crystallochemical possibility for partial hydration of the low-temperature sesquioxide lattice is naturally justified by its structure, since the lattice of the C-modification of the rare-earth oxides may be considered as derived from an original PrO_2-type lattice, one-third of whose anionic sites are vacant. On hydration of the lattice, two hydroxyl groups are substituted for one oxygen ion and one vacancy. Partial residual hydration of the C-form oxide lattice in the amount observed (~0.05 wt.%), of course, cannot be identified with the formation of compounds of a hydrate nature (which would clearly be erroneous). It is easy to estimate that the above quantity of hydroxyl groups can fill only 1% of the vacant oxygen sites (~0.25% of all the oxygen sites). Therefore, the predominant bonding character of such a solid solution naturally appears the same as for the lattice of the completely dehydrated oxide — it is not at all applicable to each of the stoichiometric hydrates.

The thermodynamic explanation of the stabilization of the low-temperature oxide forms by excess oxygen atoms or residual hydroxyl groups can be illustrated by the schematic representation shown in Fig. 3. As seen from Fig. 3, the high-temperature A-form of the stoichiometric sesquioxide possesses a lower free energy over the entire temperature range and consequently appears the most stable. Correspondingly, the C-form is least stable. It is further proposed that partial substitution for vacancies in the cubic sesquioxide lattice by oxygen or hydrogen lowers the lattice free energy. Under these assumptions, a relative thermodynamic equilibrium (point K) is reached between the A-form of sesquioxide composition and the C-form, whose lattice incorporates a number of atoms which increase its stability. The transition is irreversible when the atoms vacating the lattice during transition cannot return to it by a reverse process (i.e., on cooling), and conversely, the transition becomes reversible in the presence of the component required for the reverse reaction.

The phase relationships of the C—B transitions in the oxides of terbium and dysprosium are explained in an analogous manner. The only difference is that here a lowering of thermodynamic stability (relative to the B-form of the same composition) occurs, as a result of the partial reduction of the C-forms below the sesquioxide composition.

In absence of accurate thermodynamic data for the study of the oxides, we calculated the electrostatic components of the lattice energies of the C- and A-forms of praseodymium oxide, and the energy changes in the cubic form during oxidation. The methods used in the calculations were developed as a result of special studies performed by the authors, with cooperation of I. I. Cheremisin. These are the topics of discussion of separate articles [10, 11].

The values for the C- and A-forms of Pr_2O_3 were

$$U_{C-Pr_2O_3} = -10.1678\,\text{Å}^{-1}e^2; \quad U_{A-Pr_2O_3} = -10.3848\,\text{Å}^{-1}e^2.$$

As seen, the A-form has the lowest Coulomb energy value, and, consequently, within limits of the assumptions of predominantly ionic bonding for the compounds, has the least amount of internal energy.

On the other hand, it was established that the Coulomb energy of the cubic form of PrO_x steadily decreases in the composition interval $1.5 \leq x \leq 2$ (increases in absolute value), according to oxidation, and finally assumes a value of $-17.2493\,\text{Å}^{-1}e^2$ for PrO_2. During this, the amount of oxygen increases to 4.85 wt.% ($PrO_{1.5} \rightarrow PrO_2$).

Thus, the energy difference between the C- and A-forms of Pr_2O_3 amounts to $0.2170\,\text{Å}^{-1}e^2$ (as follows from the calculation) on incorporation of approximately 0.15 wt.% oxygen into the original $C-Pr_2O_3$ lattice. It is seen that the latter computations, in spite of their uncertainty in assuming predominantly ionic bonding for rare-earth oxide lattices, well agree with the results of experimental observations. Furthermore, they also agree with the experimental data of Stubblefield et al. [12] on the heats of solution of the C- and A-forms of praseodymium sesquioxide.

LITERATURE CITED

1. V. M. Goldschmidt, F. Ulrich, and T. Barth, Skr. Norsk. Akad. Oslo, Kl. I, No. 5 (1925).
2. A. Jandelli, Gazz. Chim. Ital., Vol. 77, Nos. 7-8, p. 312 (1947).
3. M. V. Shafer and R. Roy, J. Am. Ceram. Soc., Vol. 42, p. 563 (1959).
4. R. S. Roth and S. J. Schneider, J. Res. Natl. Bur. Stnd., Vol. 64A, p. 309 (1960).
5. J. Warshaw and R. Roy, J. Phys. Chem., Vol. 65, No. 11, p. 2048 (1961).
6. V. B. Glushkova and A. G. Boganov, Izv. Akad. Nauk SSSR, Ser. Khim., No. 7, p. 1131 (1965).
7. V. B. Glushkova, É. K. Keler, and Yu. G. Sokolov, Dokl. Akad. Nauk SSSR, Ser. Khim., Vol. 158, No. 1, p. 151 (1964).
8. A. G. Boganov and V. S. Rudenko, Dokl. Akad. Nauk SSSR, Ser. Khim., Vol. 161, No. 3, p. 590 (1965).
9. A. G. Boganov, L. P. Makarov, and V. S. Rudenko, Dokl. Akad. Nauk SSSR, Ser. Mat. i Fiz., Vol. 161, No. 2, p. 332 (1965).
10. A. G. Boganov, I. I. Cheremisin, and V. S. Rudenko, Fiz. Tverd. Tela, Vol. 8, No. 6, p. 1410 (1966).
11. A. G. Boganov, I. I. Cheremisin, and V. S. Rudenko, this volume, p. 12.
12. C. T. Stubblefield, H. Eick, and L. Eyring, J. Am. Chem. Soc., Vol. 78, No. 12, p. 3018 (1956).

EFFECT OF RARE-EARTH OXIDE IMPURITIES
ON THE POLYMORPHISM OF ZIRCONIUM DIOXIDE

V. B. Glushkova and L. V. Sazonova

Zirconium dioxide, with a melting temperature of 2700°C, is a promising material for high-temperature technology. It is known [1] that ZrO_2 crystallizes at elevated temperatures (above 2300°C) in a face-centered cubic (fluorite-type) lattice. On decrease of temperature the cell parameter and, consequently, also the lattice interatomic spacing decrease. If ZrO_2 were to preserve the cubic structure down to room temperature, the hypothetical oxygen—oxygen distance in such a structure would be 2.5 Å compared to 2.8 Å for typical oxygen compounds. This abnormally low value explains the instability of cubic ZrO_2 at low temperatures, and also explains the existence of the distorted modifications – tetragonal and monoclinic. The transition from the cubic to tetragonal modification at 2300°C is accompanied by a 3% increase in volume (by tentative data), while transition from the tetragonal to monoclinic modification at 1150–1250°C results in further distortion of the fluorite lattice and a significant (7–9%) increase in the unit cell volume of ZrO_2.

The abrupt volume changes during the phase transformations lead to cracking of ZrO_2 pieces. Many publications are thus devoted to the problems of stabilizing ZrO_2 in the cubic or tetragonal modification. Up to the present all such problems have not been conclusively resolved. The present article gives the results of work carried out in our laboratory during the past year and a half.

Explanation of the stabilization mechanism of ZrO_2, which permits defining the limit of stability for the fluorite-type structure, is possible by the principles of crystallochemistry. According to such concepts, the fluorite-type lattice is stable only with a R_c/R_a ratio of ≥ 0.732. Since R_c/R_a actually equals 0.66 for ZrO_2, in order to stabilize the cubic lattice the average cationic radius must be increased by substituting larger metal ions for a portion of the zirconium cations.

The rare-earth oxides, which possess similar chemical properties and differ chiefly by change in cationic radius, are beautiful objects for studying the effect of the stabilizing-impurity cation size on the phase transformations. These oxides have very high melting points and their addition to zirconium dioxide does not diminish its highly refractory qualities. Furthermore, introduction of rare-earth oxide impurities to ZrO_2 enables production of materials which are stable at high temperatures in both oxidizing and reducing atmospheres, and possess several specific properties characteristic of the rare earths.

We studied the effect of rare-earth oxide impurities (oxides of Nd, Sm, Eu, Er, Yb) on the polymorphism of ZrO_2. The classic theory of phase transformations requires the existence of a distinct temperature for the transition of the monoclinic to the tetragonal modification. At this temperature the free energy of the monoclinic phase equals the free energy of the tetragonal

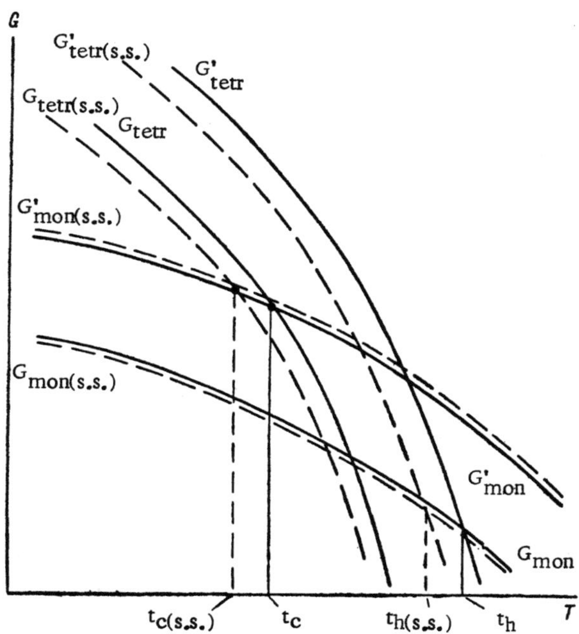

Fig. 1. Change in free energy of zirconium dioxide and its solid solutions as a function of change in temperature.

Table 1. Polymorphic Transformation Temperature of ZrO_2 as Affected by Rare-Earth Impurities

Ln_2O_3 content, mol.%	Nd_2O_3		Sm_2O_3		Eu_2O_3		Er_2O_3		Y_2O_3	
	heat-ing	cool-ing	heat-ing	cool-ing	heat-ing	cool-ing	heat-ing	cool-ing	heat-ing	cool-ing
1	1050	900	870	720	800	650	750	550	850	750
2	1020	850	—	—	750	550	620	450	650	500
3	1000	830	840	650	—	—	450	300	580	450
5	970	750	820	620	700	500	Absent		550	450
10	920	570	Absent		Absent		Absent		Absent	

phase. However, as was shown by French authors [2], the free energy is a function not only of pressure and temperature, but also of the parameters α and β, which characterize the internal strain energy of the phase and its surface energy (dependent on the state of the adjacent phase). These parameters α and β are determined by the physical state of the phases in the system and are essentially different for a nucleating new phase and for the surrounding matrix of an old phase. If the free energy of the monoclinic matrix is denoted as $G_{mon} = f_{mon}(p, T, \alpha_{mon}, \beta_{mon})$, and that of a nucleus of the tetragonal phase in a monoclinic matrix as $G'_{tetr} = f_{tetr}(p, T, \alpha'_{tetr}, \beta'_{tetr})$, then $G_{mon} = G'_{tetr}$ corresponds to the transformation temperature on heating (t_h). Analogously, $G_{tetr} = G'_{mon}$ corresponds to the transformation temperature on cooling (t_c), where $G'_{mon} = f_{mon}(p, T, \alpha'_{mon}, \beta'_{mon})$, the free energy of a nucleus of the monoclinic phase in a tetragonal matrix, and $G_{tetr} = f_{tetr}(p, T, \alpha_{tetr}, \beta_{tetr})$ or the free energy of the tetragonal matrix. Figure 1 shows the change in free energy of ZrO_2 and of a solid solution of composition $Zr^{4+}_{1-x}Me^{3+}_x O^{2-}_{2-\frac{x}{2}} \square_{\frac{x}{2}}$.

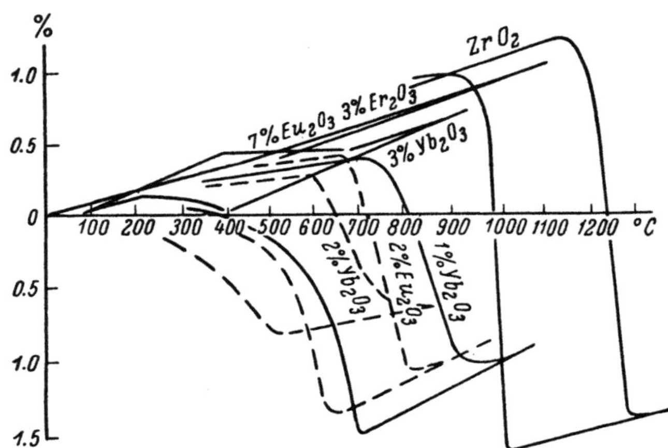

Fig. 2. Change in linear specimen size as a function of
temperature and composition of solid solutions.

This evidently also explains the presence of a hysteresis in the ZrO_2 phase transformations. A dilatometric and x-ray investigation using a high-temperature apparatus gave the transition of the monoclinic to the tetragonal modification of ZrO_2 as being in the interval 1190–1250°C. On cooling, the reverse transition from the tetragonal to monoclinic form begins at 1000°C.

Investigation of ZrO_2—rare-earth oxide mixtures showed that the reactions involve formation of substitutional solid solutions of the general formula $Zr^{4+}_{1-x}Me^{3+}_xO^{2-}_{2-\frac{x}{2}}\square_{\frac{x}{2}}$. During this the x-ray patterns display a shift in the maxima of the ZrO_2 lines, while the thermograms and dilatometric curves show a lowering of the monoclinic $\rightarrow$ tetragonal transformation temperature. It was shown that the solubility of the rare-earth oxides in the monoclinic form of ZrO_2 does not exceed 1-2 mol.% and remains almost constant over the rare earth series. Their solubility in the tetragonal modification is about 4% at 1200°C, and rises with increase in temperature.

It can be surmised that solid solution formation lowers the free energy of the tetragonal form. At the same time the free energy of the monoclinic matrix does not change, and the free energy of a nucleus of the monoclinic phase in the tetragonal matrix can still increase (owing to the increased stability of the tetragonal phase). Thus incorporation of impurities should lead to a shift in the free energy curves, depending on the degree of crystal lattice perfection and individual characteristics of the impurity introduced. It should also result in a lowering of the polymorphic transformation temperature and in an increase in the hysteresis amplitude of ZrO_2-base solid solutions relative to pure ZrO_2.

Table 1 gives the change in polymorphic transformation temperature as a function of rare-earth oxide impurity. The data on yttrium-oxide impurities are also given for comparison. It is seen from the table that the transformation temperature decreases with increase in the amount of rare earth oxide in solid solution, and with decrease in rare-earth ionic radius.

The effect of small amounts of rare-earth oxide impurities appears most evident on the dilatometric curves. Figure 2 shows the dilatometric curves for ZrO_2 and its solid solutions containing different amounts of rare-earth oxides. It is evident from the figure that the tetragonal to monoclinic transformation temperature is considerably lower in the solid solutions than in pure ZrO_2. For a given rare-earth oxide impurity, the lowering of the transformation temperature becomes greater with a smaller rare-earth cationic radius. The volume change

($\Delta V/V_{mon}$) occurring with the phase transformation is 8-9% for ZrO_2, 5.6% for a solid solution containing 5 mol.% Nd_2O_3 [3], and about 4% for a solid solution with 1 mol.% Er_2O_3. Further lowering of the polymorphic transformation temperature and a decrease in the volume changes accompanying the transition occur with increase of impurity. It should be noted that the volume changes on the monoclinic to tetragonal transformation in the solid solutions not only are less in absolute value, but also extend over a broader temperature range. This in particular applies to the cooling curves (Fig. 2).

All of this means that pieces made from solid solutions of rare-earth oxides in ZrO_2 are subjected to fewer structural stresses at temperatures near the polymorphic transformation, increasing their thermal stability. Thus, for example, ZrO_2 specimens containing 2 mol.% Eu_2O_3 or 1 mol.% Er_2O_3, in spite of the hysteresis loop on their dilatometric curves, do not exhibit the cracking typical for pieces made of pure ZrO_2.

Study of the stability of ZrO_2-base solid solutions is of great practical significance. In the course of the investigation, the effect of mode of sample preparation on the composition and crystal structure of the products resulting after anneal was studied. A metastable, cubic, fluorite-type solid solution forms when the specimens are prepared by coprecipitation in the amorphous state, or as intermediate products from easily decomposing salts. The formation of such solid solutions occurs at low temperatures, and the solid solutions are well crystallized at 600-800°C. In spite of their thermodynamic metastability, they possess ample stability. The decay of these solid solutions with formation of the equilibrium phases is observed only at higher temperatures (above 1000°) and occurs in two stages. The first stage is the decay to two very defective cubic solid solutions of an equilibrium phase composition. The second stage is the structural ordering of such phases. The following is cited as an example of the decay scheme and structural transformations in a metastable, ZrO_2-base solid solution containing 10 mol.% Nd_2O_3 [4]:

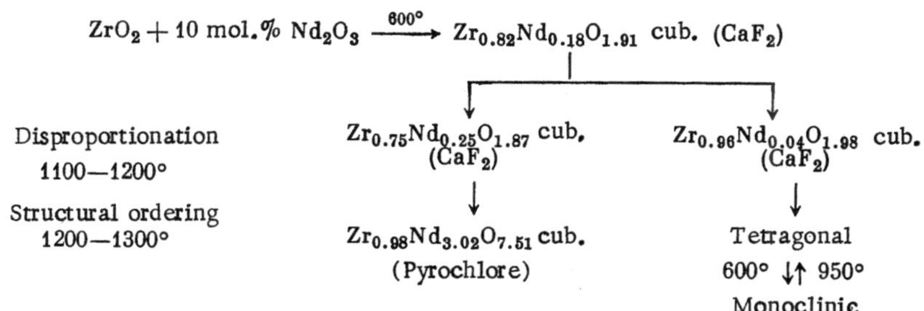

$$ZrO_2 + 10\text{ mol.\% }Nd_2O_3 \xrightarrow{600°} Zr_{0.82}Nd_{0.18}O_{1.91} \text{ cub. }(CaF_2)$$

Disproportionation
1100—1200°

Structural ordering
1200—1300°

$$Zr_{0.75}Nd_{0.25}O_{1.87} \text{ cub. }(CaF_2)$$

$$Zr_{0.96}Nd_{0.04}O_{1.98} \text{ cub. }(CaF_2)$$

$$Zr_{0.98}Nd_{3.02}O_{7.51} \text{ cub. (Pyrochlore)}$$

Tetragonal
600° ↓↑ 950°
Monoclinic

It was shown also that an increase of rare-earth oxide content in the metastable solid solution retards the process of decay to the equilibrium phases. Thus a solid solution with 1 mol.% Eu_2O_3 decays at 1000-1100°C while a 2-mol.% Eu_2O_3 solid solution displays no formation of equilibrium products after heating 5 h at 1200°C [5].

The advantage of these methods (coprecipitation and salt decomposition) is the high uniformity of the resulting product and the relatively low temperature for solid-solution crystallization. The reaction begins at a higher temperature (above 1200°C) on use of oxide mixtures for original constituents, and usually results in specimen disintegration. Nevertheless, here the equilibrium product forms directly.

The equilibrium systems between ZrO_2 and the rare-earth oxides were studied in the temperature range 600-2000°C to produce ZrO_2-base materials which are stable with prolonged heating at high temperatures. High-temperature studies on compositions containing 1 to 33 mol.% rare-earth oxide showed that in all cases a single-phase product forms at elevated

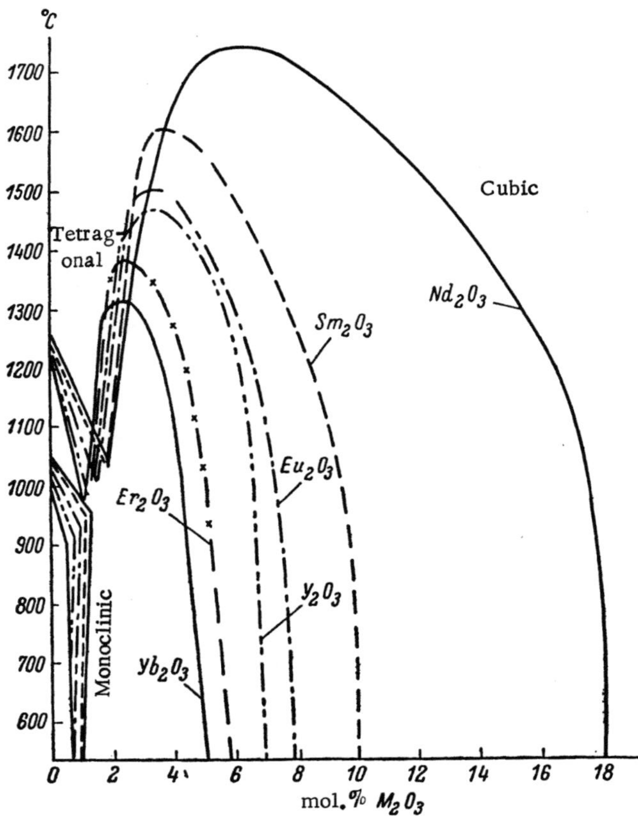

Fig. 3. Phase diagram for ZrO_2—Ln_2O_3 systems in the zirconium dioxide-rich region.

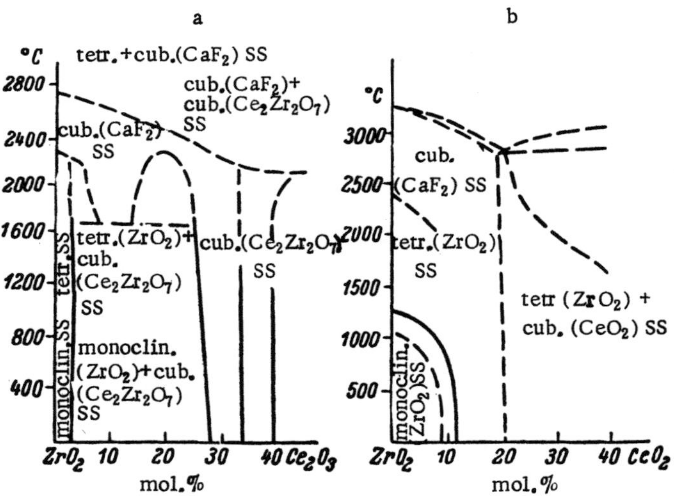

Fig. 4. Phase diagrams for the ZrO_2—CeO_2 and ZrO_2—Ce_2O_3
systems.

temperatures — a solid solution whose zirconium ions are partly substituted by trivalent rare earth ions. A corresponding number of oxygen vacancies form simultaneously in the anionic sublattice. At smaller rare-earth oxide content (down to 3-5%), the solid solution has a tangential structure; with larger content it becomes a cubic, CaF_2-type solid solution. With further increase of Ln_2O_3 content for rare earths of larger ionic radii, the solid solution also becomes cubic, but now having a pyrochlore structure.

The cubic solid solution obtained at high temperature decays on slow cooling or anneal at lower temperatures to a mixture of two solid solutions. One is tetragonal (monoclinic at low temperatures), and contains less than 2-5 mol.% rare-earth oxide. The other is richer in rare-earth oxide, and crystallizes in the pyrochlore (Nd_2O_3) or fluorite structure. The temperature boundary of the two-phase region is lower for a smaller rare-earth ionic radius (Fig. 3). It was also shown that the width of the two-phase region decreases with decrease in rare-earth ionic radius. Thus, for complete stabilization of ZrO_2, 19 mol.% Nd_2O_3 or 7-8 mol.% Er_2O_3 is required as impurity.

The dilatometric curves of these systems do not display the hysteresis loops indicating complete stabilization of ZrO_2. The solid solutions of such composition are stable, and essentially undergo no decay on prolonged heating at elevated temperatures.

Cubic solid solutions of lower rare-earth content than required by the phase diagram decay at temperatures below 1700-1400°C. However, decay of the high-temperature solid solution to two phases occurs at a slower rate and involves a smaller volume change than does the monoclinic → tetragonal transformation in pure ZrO_2. Therefore, there exists (in practice) the possibility for stabilizing ZrO_2 with less oxide impurity than required by the equilibrium phase diagram. This applies particularly to the heavy rare-earth oxides, where addition of 3-4 mol.% prevents decomposition of ZrO_2-base specimens.

Zirconium dioxide-base solid solutions with cerium oxide impurities exhibit special behavior. Figure 4b is a revised phase diagram of the ZrO_2-CeO_2 system according to Duwez and Odell [6], and Fig. 4a shows the phase diagram of the $ZrO_2-Ce_2O_3$ system (in reducing atmosphere) according to the data of Leonov, Keler, and Andreeva [7]. A sharp decrease in the polymorphic transformation temperature occurs at small CeO_2 concentrations in the ZrO_2-CeO_2 system. With CeO_2 contents above 13 mol.%, the tetragonal ZrO_2-base solid solution becomes stable over the entire temperature range from room temperature. However, reduction of the Ce^{4+} ions to the trivalent state occurs on heating the latter solid solution in an atmosphere of low oxygen partial pressure. This process is accompanied by a change in the amounts and composition of the phases and by evolution of oxygen. It consequently produces sample disintegration and cracking.

The $ZrO_2-Ce_2O_3$ system is analogous to the $ZrO_2-Nd_2O_3$ system, determined in detail in [3]. The $ZrO_2-Ce_2O_3$ system has a two-phase region, containing monoclinic (tetragonal) ZrO_2 solid solution and a pyrochlore-type cerium zirconate-base solid solution below 1700°C and at concentrations between 3 and 27 mol.% Ce_2O_3. Zirconium dioxide solid solutions containing Ce_2O_3 undergo intense oxidation on heating in air. Both oxidation and reduction processes are accompanied by sizeable changes in volume which lead to cracking and disintegration of the material.

It was thus shown that the best materials to stabilize ZrO_2 in the cubic form are the oxides of the heavy rare earths and of yttrium. Minimal amounts of oxide impurities (3-7 mol.%) are required to produce stable solid solutions. The resulting solid solutions are stable in various gaseous environments and undergo no phase transformations at elevated temperatures up to 2000°C. A prolonged anneal of the solid solutions at a temperature above 2000°C should lead to volatilization of the stabilizing impurity, with a consequent change in the solid solution composition and a transition to a two-phase region. The literature data of [8] on the vapor pressures of a number of oxides at temperatures above 2000°C are:

$$\begin{array}{lccccc}
\text{Oxide} & MgO & CaO & Ce_2O_3 & Pr_2O_3, & Eu_2O_3 \\
p_i,\ \text{atm} & 10^{-4} & 10^{-5} & 10^{-4} & & 10^{-5}
\end{array}$$

$$\begin{array}{lcccc}
\text{Oxide} & Y_2O_3, & Yb_2O_3, & Tu_2O_3 & Tb_2O_3,\ Dy_2O_3,\ Er_2O_3,\ Sm_2O_3 \\
p_i,\ \text{atm} & & 10^{-6} & & 10^{-7}
\end{array}$$

As seen from the above numbers, the oxides of the heavy rare earths and of yttrium have the lowest vapor pressures. Consequently, ZrO_2 solid solutions having these oxide impurities also will have the greatest high-temperature stability. These data also indicate that it is necessary to introduce a somewhat larger amount of oxide additive than required by the phase diagrams to increase the period of high-temperature service of zirconium refractories.

LITERATURE CITED

1. A. G. Boganov, V. S. Rudenko, and L. P. Makarov, Dokl. Akad. Nauk SSSR, Vol. 160, No. 5, p. 1065 (1965).
2. R. Collongues, M. Perez-y-Jorba, and J. Lefevre, Bull. Soc. Chim. France, Vol. 149 (1962).
3. I. A. Davtyan, V. B. Glushkova, and É. K. Keler, Izv. Akad. Nauk SSSR, Neorg. Mat., Vol. 1, No. 5, p. 743 (1965).
4. L. V. Sazonova, I. A. Davtyan, and V. B. Glushkova, Izv. Akad. Nauk SSSR, Neorg. Mat., Vol. 1, No. 11, p. 1965 (1965).
5. I. A. Davtyan, V. B. Glushkova, and É. K. Keler, Izv. Akad. Nauk SSSR, Neorg. Mat., Vol. 2, No. 5, p. 890 (1966).
6. P. Duwez and F. Odell, J. Am. Ceram. Soc., Vol. 33, No. 9, p. 247 (1950).
7. A. I. Leonov, É. K. Keler, and A. B. Andreeva, Izv. Akad. Nauk SSSR, Neorg. Mat., Vol. 2, No. 1, p. 137 (1966).
8. N. A. Toropov and V. P. Barzakovskii, High-Temperature Chemistry of Silicate and Oxide Systems [in Russian], Izd. Akad. Nauk SSSR, Moscow (1963).

CRYSTAL STRUCTURE OF
GADOLINIUM PYROSILICATE

Yu. I. Smolin and Yu. F. Shepelev

The structure of gadolinium pyrosilicate, $Gd_2Si_2O_7$ [1] (a = 13.87 Å, b = 5.073 Å, c = 8.33 Å; $Pna2_1$; z = 4) was determined by x-ray methods, utilizing a three-dimensional assembly of F(hk$l$).

The intensity measurements were performed on a single-crystal diffractometer with scintillation counter. The perpendicular beam method was used with monochromatic Mo $K\alpha$ radiation. The (hk$l$) reflections were measured with respect to k from 0 to 5 and to $(\sin\theta)/\lambda$ = 1.1 Å^{-1} for each layer. They were measured on a (hk0)-zone crystal in another apparatus. There are 1450 reflections in all. The crystal was in the shape of a perfect sphere 0.3 mm in diameter. The intensities were adjusted to the polarization, kinematic factor, and absorption.

The atomic coordinates were found by a three-dimensional Patterson function analysis and by different electron density projections. They were computed by the method of least squares for all F(hk$l$) measurements up to an R-factor value of 0.073.

The results of the improved determinations of the atomic coordinates, the mean square errors, and the individual temperature factors are presented in Table 1.

The pyrogroups in the cell lie in two layers parallel to the (001) plane; the gadolinium atoms are distributed between these layers. The gadolinium has a sevenfold coordination with oxygen, with the "bridge" oxygen of the pyrogroup entering the sevenfold coordination polyhedron of gadolinium. The Si—O—Si angle is 158°40' ± 30'.

Table 1. Atomic Coordinates, Square Errors, and Temperature Factors

Atom	x/a	σ_x/a	y/b	σ_y/b	z/c	σ_z/c	B Å^2
Gd_1	0.12551	0.00004	0.33730	0.00013	0.99831	0.00007	0.375
Gd_2	0.12564	0.00004	0.33739	0.00013	0.51409	0.00007	0.364
Si_1	0.3205	0.0002	0.3744	0.0007	0.2505	0.0009	0.314
Si_2	0.5390	0.0002	0.6253	0.0006	0.2498	0.0008	0.170
O_1	0.2715	0.0007	0.4769	0.0022	0.0876	0.0013	0.579
O_2	0.2658	0.0006	0.4857	0.0021	0.4130	0.0012	0.337
O_3	0.3457	0.0006	0.0706	0.0017	0.2465	0.0015	0.435
O_4	0.4211	0.0005	0.5557	0.0016	0.2448	0.0011	0.231
O_5	0.5472	0.0008	0.7858	0.0023	0.0866	0.0014	0.608
O_6	0.5456	0.0008	0.7882	0.0025	0.4206	0.0015	0.729
O_7	0.5988	0.0005	0.3526	0.0017	0.2560	0.0019	0.595

The spacings of the Si—O rings in the subgroup (1.604 Å, 1.651 Å, 1.581 Å, 1.588 Å, 1.648 Å, 1.614 Å; average 1.614 Å) are noticeably less than the spacings of the Si—O bridges (1.672 ± 0.008 Å, 1.673 ± 0.007 Å).

The average Gd—O spacing in the Gd coordination polyhedron is 2.40 Å, and nearly equals the sum of the ionic radii of gadolinium and silicon.

In closing, the authors express thanks to I. A. Bondar' who supplied the crystals studied, and to N. A. Toropov for assistance and support in the work.

LITERATURE CITED

1. N. A. Bondar', L. N. Koroleva, and N. A. Toropov, Proceedings of the Sixth Conference on Experimental and Technical Mineralogy and Petrography [in Russian], Izd. Akad. Nauk SSSR, Moscow (1961), pp. 303–309.

THE CHEMISTRY OF CERIUM IN OXIDE SYSTEMS: INVESTIGATION OF THE EFFECT OF TEMPERATURE AND GASEOUS ENVIRONMENT ON SOLID-PHASE REACTIONS

A. I. Leonov, A. B. Andreeva, V. E. Shvaiko-Shvaikovskii, and É. K. Keler

Cerium and its compounds are used or are familiar in more than 20 branches of industry as refractories, radioceramics, phosphors, catalysts, modifiers, and for other special purposes. The number of experimental studies of the synthesis of cerium compounds and studies of their high-temperature properties is limited because of the experimental difficulties associated with the variable valence of cerium. Inadequate knowledge of the properties of cerium results in the frequent assignment of an incorrect valence to cerium in its compounds. This brings about the urgent necessity for a systematic study of the physicochemical properties of cerium and its compounds.

It is known that cerium forms two principal oxides, CeO_2 and Ce_2O_3, with the higher oxide CeO_2 stable in air. The heat of formation of CeO_2 is $\Delta H^0_{298} = -260.18$ kcal/mole [1]. Cerium dioxide has a fluorite-type crystal structure, a lattice constant $a = 5.41$ Å [2], a specific gravity of 7.3 g/cm^3, and a melting point of $2725 \pm 25°C$.

We obtained the trivalent cerium oxide (Ce_2O_3) in a current of dry hydrogen at an elevated temperature. The heat of formation of Ce_2O_3 is $\Delta H^0_{298} = -435.00$ kcal/mole [1]. Ce_2O_3 has a hexagonal, La_2O_3-type structure with the lattice parameters $a = 3.889$ Å, $c = 6.054$ Å, $c/a = 1.556$ [3]. It has a calculated density of 6.87 g/cm^3, and a yellow-green color. The melting temperature of Ce_2O_3 (in hydrogen) is $2160 \pm 30°C$; it oxidizes on heating in air to form CeO_2.

A distinguishing feature of cerium is the nature of its transition from the tetra- to trivalent state. For example, dissociation of CeO_2 to $CeO_{1.99}$ begins at 800°C under an oxygen pressure of 10^{-18} atm, and reaches the final limit ($CeO_{1.5}$) at 10^{-27} atm [4]. This corresponds to change in thermodynamic potential of about 30 kcal/mole CeO_2. Such a property of the cerium oxides facilitates defect formation in the anionic sublattice and strongly increases the ease of compound reaction.

Cerium dioxide interacts with several alkaline earth oxides. It forms a limited solid solution (to 15 mol.% CaO) with calcium oxide [5, 6]. It forms a strontium cerate ($SrCeO_3$) and a limited solid solution (to 9 mol.% SrO) in the CeO_2-SrO system [5, 7, 8]. According to the data of [9], CeO_2 forms a simple eutectic with BeO and MgO. Solid solutions of three types form in the CeO_2-ZrO_2 system: monoclinic, tetragonal, and cubic [10]. Cerium oxide forms limited solid solutions with the lanthanide oxides [11]. Continuous series of solid solutions were defined in the CeO_2-UO_2 [12, 13] and CeO_2-PuO_2 systems [13, 14].

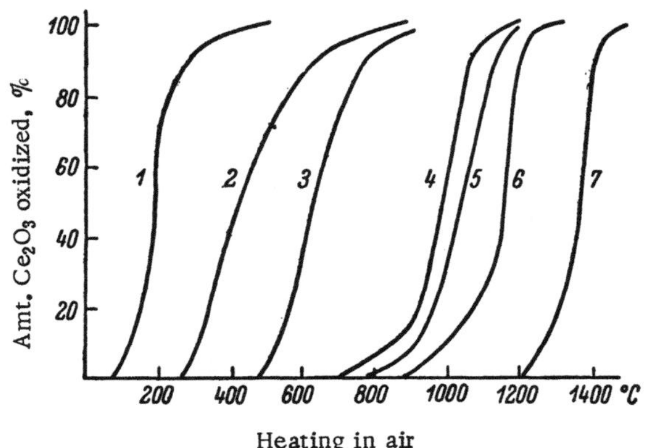

Fig. 1. Decomposition of cerium compounds on heating in air. 1) $Ce_2Zr_2O_7$; 2) $Ce_2Ti_3O_{8.4}$; 3) $CeGaO_3$; 4) $Ce_2Si_2O_7$; 5) $CeCrO_3$; 6) $CeAlO_3$; 7) $Ce_2O_3 \cdot 11Al_2O_3$.

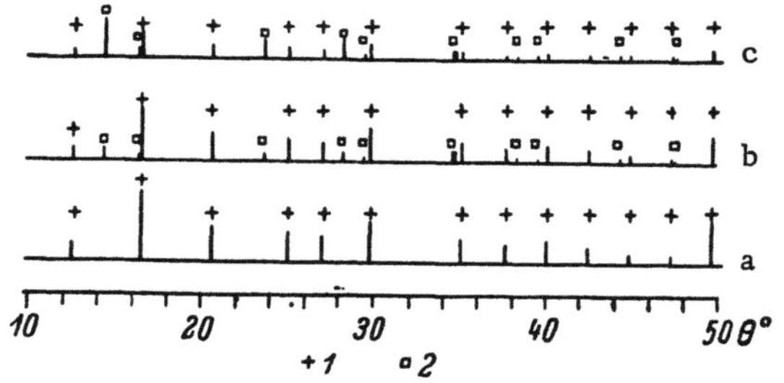

Fig. 2. X-ray patterns of solid solutions heated for one hour in air at 1400°C. a) $(CeAlO_3—3LaAlO_3)$ solid solution; b) $(CeAlO_3—LaAlO_3)$ solid solution; c) $(3CeAlO_3—LaAlO_3)$ solid solution. 1) $(xCeAlO_3—yLaAlO_3)$ solid solution; 2) CeO_2.

Trivalent cerium behaves as an analog of the lanthanides in compounds with other oxides, forming numerous compounds both as natural minerals and synthetics. These include the aluminates, the chromite, the gallate, titanates, silicates, zirconates, niobates, tantalates, and many others. The chemistry of cerium is therefore primarily represented by the compounds of trivalent cerium.

Compounds of perovskite structure form by the reaction

$$2CeO_2 + Me_2O_3 \rightleftarrows 2CeMeO_3 + \tfrac{1}{2}O_2,$$

where Me = Al, Cr, or Ga. The reactions begin to occur in air from left to right at about 1700°C. At a lower temperature, the reactions occur in the reverse direction. The synthesis of the pure compounds is therefore performed in a reducing atmosphere, or in vacuum-sealed quartz ampoules.

Two compounds, $CeAlO_3$ and $Ce_2O_3 \cdot 11Al_2O_3$, form in the $Ce_2O_3—Al_2O_3$ system [15]. $CeAlO_3$ has a perovskite type structure with a lattice constant indexed in the cubic system as $a_0 = 3.77$ Å. It has a specific gravity of 6.17 g/cm³, an index of refraction of Nm = 2.02, a birefringence of 0.01, and a melting point (under hydrogen) of 2030 ± 25°C. $CeAlO_3$ oxidizes in air over the temperature interval 850-1200°C, to form CeO_2 and Al_2O_3, as shown in Fig. 1 (curve 6). At oxygen partial pressures of 10^{-3} mm Hg and lower, there is no decomposition of the compound over the entire temperature interval from room temperature to its melting point. At oxygen pressures of 0.2, 2.0, and 20.00 mm Hg, $CeAlO_3$ decomposes in the respective temperature ranges 1000-1360, 1000-1460, and 1000-1560°C. Synthesis of the compound again proceeds in the temperature region above the decomposition range.

The compound $Ce_2O_3 \cdot 11Al_2O_3$ has a β-Al_2O_3 structure with a specific gravity of 4.07 g/cm³. It has an index of refraction of Nm = 1.80, a weak birefringence, and a melting temperature (in hydrogen) of 1950 ± 25°C. It decomposes in air over the temperature interval 1200-1400°C, with formation of CeO_2 and Al_2O_3 (Fig. 1, curve 7); above 1650°C the compound $Ce_2O_3 \cdot 11Al_2O_3$ again forms.

The properties of solid solutions between $CeAlO_3$ and other rare-earth monoaluminates were studied in [16]. Continuous series of solid solutions form under reducing atmospheres in the $CeAlO_3—LaAlO_3$ and $CeAlO_3—SmAlO_3$ systems. Here the trivalent state of cerium is stabilized with the increasing content of the second solid-solution component. As x-ray phase analysis has shown (Fig. 2), solid solutions with cerium contents up to $Ce_{0.25}La_{0.75}AlO_3$ do not decompose on heating in an oxidizing atmosphere, and can be synthesized in air from the oxides.

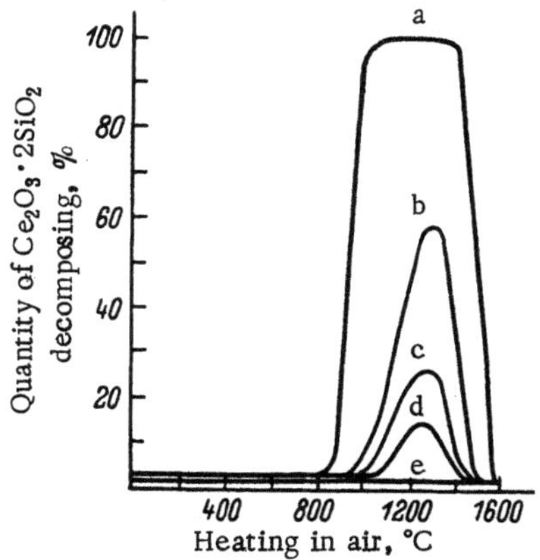

Fig. 3. Decomposition of Ce$_2$O$_3$ · 2SiO$_2$—La$_2$O$_3$ · 2SiO$_2$ solid solutions on heating in air. Ce$_2$O$_3$ · 2SiO$_2$ content in solid solution (mol.%): a) 100; b) 90; c) 80; d) 75; e) 50.

Cerium chromite, synthesized in hydrogen, decomposes on heating in air at 800-1100°C (Fig. 1, curve 5), but again forms at 1660°C and higher temperatures. The melting temperature of CeCrO$_3$ is 2300 ± 30°C in air and 2220 ± 30°C in hydrogen. The lower melting temperature in hydrogen is explained by the reduction of chromium oxide and formation of a nonstoichiometric compound. The lattice constant of CeCrO$_3$ is a = 3.85 Å; its specific gravity is 6.92 g/cm^3 and its color is yellow-green.

The synthesis of cerium gallate was performed from a mixture of chemically pure cerium dioxide and metallic gallium in a sealed, evacuated, quartz ampoule held 5 h at 1300°C. The lattice constant a of CeGaO$_3$ is 3.87 Å. The compound decomposes on heating in air over the temperature interval 470-800°C, forming CeO$_2$ and Ga$_2$O$_3$ (Fig. 1, curve 3).

Cerium dioxide does not form any compounds with TiO$_2$. A compound of composition CeO$_{1.6}$ · 2TiO$_2$ forms at a partial oxygen pressure of 1 mm Hg and a temperature of 1400°C. A compound of variable composition, which can be expressed by the formula $(Ce_2O_3)_{1+x}$ · 3TiO$_{2-y}$, forms in hydrogen. The compound of Ce$_2$O$_3$ · 3TiO$_{1.8}$ composition has a perovskite-type structure with a lattice constant a_0 = 3.89 Å. It has a specific gravity of 5.39 g/cm^3, an index of refraction N_{av} = 2.10, and a dark brown color. The compound decomposes on heating in air over the temperature interval 260-800°C (Fig. 1, curve 2). The electrical properties of the compounds CeAlO$_3$, CeCrO$_3$, and Ce$_2$O$_3$ · 3TiO$_{1.8}$ are described in [17-20].

Cerium dioxide does not interact chemically with silica. Cerium sesquioxide (Ce$_2$O$_3$) in a hydrogen atmosphere forms silicates of composition Ce$_2$O$_3$ · SiO$_2$, 2Ce$_2$O$_3$ · 3SiO$_2$, and Ce$_2$O$_3$ · 2SiO$_2$ [21]. Cerium diorthosilicate begins to oxidize in air at 700°C (Fig. 1, curve 4).

It was determined in [22] that continuous series of solid solutions form in the Ce$_2$O$_3$ · 2SiO$_2$—La$_2$O$_3$ · 2SiO$_2$ and Ce$_2$O$_3$ · 2SiO$_2$—Nd$_2$O$_3$ · 2SiO$_2$ systems when the synthesis is conducted in a reducing atmosphere. Only limited solid solutions form when the synthesis is performed in air. Figure 3 shows the stability, on heating in air, of Ce$_2$O$_3$ · 2SiO$_2$—La$_2$O$_3$ · 2SiO$_2$ solid solutions as a function of composition. Solid solutions containing to 50 mol.% Ce$_2$O$_3$ · 2SiO$_2$ are stable on heating in air from room temperature to 1600°C (Fig. 3e). At 75 and 80 mol.% Ce$_2$O$_3$ · 2SiO$_2$ content, the solid solutions partially decompose over the temperature interval 1000-1300°C, with, respectively, 16 and 25 mol.% of the diorthosilicate decomposing. The reverse reaction, solid solution formation, occurs at 1400-1500°C with reduction of their original composition. The compound Ce$_2$O$_3$ · 2SiO$_2$ is unstable from 700 to 1600°C (Fig. 3a).

Reactions in the Ce$_2$O$_3$—ZrO$_2$ system were studied in a reducing atmosphere [23]. It established the presence of a compound Ce$_2$Zr$_2$O$_7$ of pyrochlore type; monoclinic ZrO$_2$-base solid solutions (below 1000°C) and tetragonal (above 1000°C); a cubic solid solution (from 5 to 17 mol.% Ce$_2$O$_3$) stable at high temperatures; a metastable cubic solid solution from 65 to 77 mol.% Ce$_2$O$_3$; a Ce$_2$O$_3$-base solid solution of hexagonal structure; and lastly of immiscibility regions among the indicated phases. A polymorphic transformation is observed in ZrO$_2$ from 0 to 27 mol.% Ce$_2$O$_3$. The transformation is completely suppressed at larger concentrations. The compound Ce$_2$Zr$_2$O$_7$ has a pyrochlore-type structure with a lattice constant a_0 = 10.70 Å, a specific gravity of 6.18 g/cm^3, and a melting temperature in hydrogen of 2300 ± 30°C. It decomposes in air over the temperature interval 80-400°C (Fig. 1, curve 1), forming a metastable, cubic solid solution of CeO_2 and ZrO_2.

The CeO_2 present in the ZrO_2 solid solution transforms to the trivalent state at elevated temperatures in a reducing atmosphere (H_2, CO, NH_3), a vacuum of 10^{-3}-10^{-4} mm Hg, a current of inert gas (Ar, He), or in the atmosphere of a gas furnace having a low oxygen partial pressure (P_{O_2} = 1.4 · 10^{-5} atm at 1400°C). The alternating oxidation—reduction of cerium-containing zirconium refractories causes disintegration and cracking of the material, as a result of the volume changes accompanying the oxidation—reduction processes.

CONCLUSIONS

The chemistry of cerium is determined by its valence. Trivalent cerium exists in compounds with other oxides as an analog of the lanthanides and forms numerous compounds: silicates, titanates, aluminates, the chromite, the gallate, niobates, tantalates, and many others. It was established that the thermodynamic stability of cerium increases when the above compounds form solid solutions with analogous compounds of other rare earths. Because of these properties, cerium occurs in the trivalent state in many natural minerals.

Cerium oxides of both the tri- and tetravalent state form solid solutions with ZrO_2, depending on the oxygen partial pressure of the gaseous phase.

Tetravalent cerium usually interacts with oxides of large ionic radius — as, for example, the lanthanide and alkaline-earth oxides (CaO, SrO, BaO), where CeO_2 forms limited solid solutions as well as a compound of the type $A^{2+}B^{4+}O_3$.

LITERATURE CITED

1. F. A. Kuznetsov, Thermodynamic Study of the Oxides of Cerium [in Russian], Abstract of Candidate's Dissertation, Moscow (1961), p. 15.
2. A. M. Cherepanov and S. G. Tresvyatskii, Highly Refractory Materials and Components from Oxides [in Russian], Izd. Metallurgiya, Moscow (1964), p. 391.
3. D. J. M. Bevan, J. Inorg. Nucl. Chem., Vol. 1, Nos. 1-2, p. 49 (1955).
4. D. J. M. Bevan and J. Kordis, J. Inorg. Nucl. Chem., Vol. 26, No. 9, p. 1509 (1964).
5. É. K. Keler, N. A. Godina, and A. M. Kalinina, Zh. Neorg. Khim., Vol. 1, No. 11, p. 2556 (1956).
6. M. Keith and R. Roy, Am. Mineral., Vol. 39, No. 1, p. 1 (1954).
7. É. K. Keler and N. A. Godina, Zh. Neorg. Khim., Vol. 2, No. 1, p. 209 (1957).
8. R. S. Roth, J. Res. Natl. Bur. Stnd., Vol. 58, No. 1, p. 75 (1957).
9. H. Wartenberg and K. Eckhard, Z. Anorg. and Allgem. Chem., Vol. 232, No. 2, p. 179 (1937).
10. P. Duwez and F. Odell, J. Am. Ceram. Soc., Vol. 33, No. 9, p. 274 (1950).
11. J. D. McCullough and J. D. Britton, J. Am. Chem. Soc., Vol. 72, p. 1386 (1950); Vol. 74, p. 5225 (1952).
12. A. Magneli and L. Kihlborg, Acta Chem. Scand., Vol. 5, p. 578 (1951).
13. N. A. Toropov, V. P. Barzakovskii, V. V. Lapin, and N. N. Kurtseva, Handbook of Phase Diagrams of Silicate Systems [in Russian], Izd. Nauka, Moscow (1965), p. 414.
14. R. N. R. Mulford and F. H. Ellinger, J. Phys. Chem., Vol. 62, No. 11, p. 1466 (1958).
15. A. I. Leonov and É. K. Keler, Izv. Akad. Nauk SSSR, Otdel. Khim. Nauk No. 11, p. 1905 (1962).
16. A. I. Leonov, Izv. Akad. Nauk SSSR, Ser. Khim., No. 1, p. 8 (1963).
17. V. A. Ioffe, A. I. Leonov, and I. S. Yanchevskaya, Fiz. Tverd. Tela, Vol. 4, No. 7, p. 1788 (1962).

18. V. A. Ioffe, A. I. Leonov, and M. V. Razumeenko, Fiz. Tverd. Tela, Vol. 6, No. 8, p. 2314 (1964).
19. V. A. Ioffe, A. I. Leonov, and M. V. Razumeenko, Fiz. Tverd. Tela, Vol. 6, No. 8, p. 2405 (1964).
20. V. E. Shvaiko-Shvaikovskii, A. I. Leonov, and A. I. Shelykh, Izv. Akad. Nauk SSSR, Neorg. Mat., Vol. 1, No. 5, p. 537 (1965).
21. A. I. Leonov, V. S. Rudenko, and É. K. Keler, Proceedings of the Sixth Conference on Experimental and Technical Mineralogy and Petrography [in Russian], Izd. Akad. Nauk SSSR, Moscow (1962), pp. 326-331.
22. A. I. Leonov, Izv. Akad. Nauk SSSR, Ser. Khim., No. 12, p. 2084 (1963).
23. A. I. Leonov, É. K. Keler, and A. B. Andreeva, Izv. Akad. Nauk SSSR, Neorg. Mat., Vol. 2, No. 1, p. 137 (1966).

INTERACTION OF SCANDIUM OXIDE WITH OXIDES OF ELEMENTS OF THE TITANIUM SUBGROUP

L. N. Komissarova, B. I. Pokrovskii, and F. M. Spiridonov

The present work is devoted to the high-temperature interaction of scandium oxide with oxides of elements of the titanium subgroup.

THE Sc_2O_3—TiO_2 SYSTEM

Relative to the interaction of these oxides, Brixner [1] indicates the existence of a compound of $2:1$ composition ($Sc_2Ti_2O_7$) (pyrochlore structure, $a = 9.802$ Å). He evidently obtained this result by analogy to other rare-earth oxides. According to our data, two types of compounds form in this system (Fig. 1): the first of $3:2$ composition ($3TiO_2 \cdot 2Sc_2O_3$); and the second of $2:3$ composition ($2TiO_2 \cdot 3Sc_2O_3$).

A primary, fluorite-structure solid solution forms first during the synthesis from a coprecipitated hydroxide mixture. This decomposes to the above compounds during final heating at 1000°C. Both compounds transform reversibly (at respective temperatures of 1150 and 1350°C) to a cubic, fluorite-type phase. The $3:2$ compound ($3TiO_2 \cdot 2Sc_2O_3$) is an ordered phase based on a rhombohedrally distorted fluorite structure with lattice parameters $a = 9.094$ and $c = 17.07$ Å. While we did not determine the structure of this compound, it nevertheless can be assumed that it has a strongly distorted fluorite structure. This is because comparison of x-ray patterns of the compound and those of the CaF_2-type phase give evidence that the compound has a part of its lines regularly grouped near fluorite structure lines — evidently formed by their splitting. The assumption of the link between the CaF_2 structures and the $2:3$ compound is also justified by the fact that this compound is formed from a cubic solid-solution with fluorite structure which completely reverts to itself at a temperature above 1350°C.

The $3:2$ compound is analogous in structure to a compound of the same composition in the ZrO_2—Yb_2O_3 system [2]. However, it contradicts the opinion of the authors [2] on the possibility of its existence in the Sc_2O_3—TiO_2 system. Repeated attempts to synthesize a compound of $2:1$ composition ($Sc_2Ti_2O_7$) proved unsuccessful. We obtained two-phase samples at all temperatures, i.e., the compound $Sc_2Ti_2O_7$ does not exist.

THE Sc_2O_3—ZrO_2 SYSTEM

The phase diagrams in this system are described most completely in works of the French school (Lefevre, Collongue, et al. [2, 3]). These authors discovered several phases based on a CaF_2-type structure.

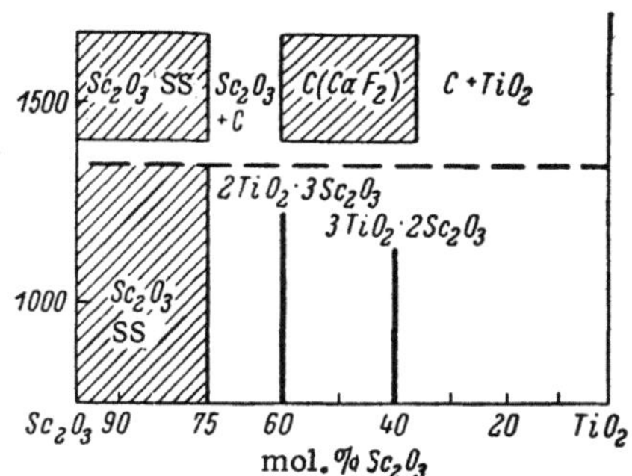

Fig. 1. Phase relations in the Sc_2O_3—TiO_2 system.

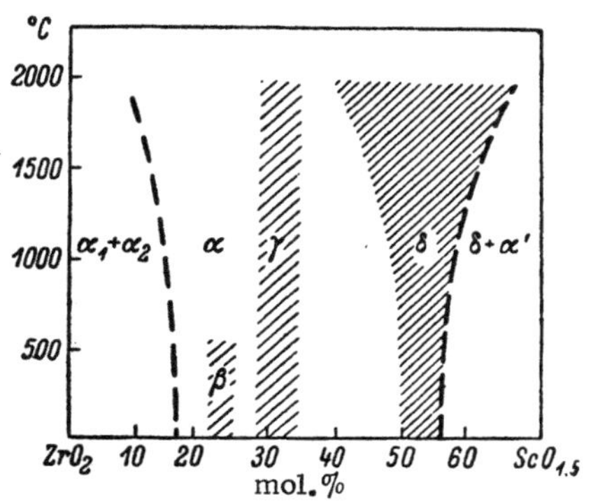

Fig. 2. Phase diagram of the Sc_2O_3—ZrO_2 system.

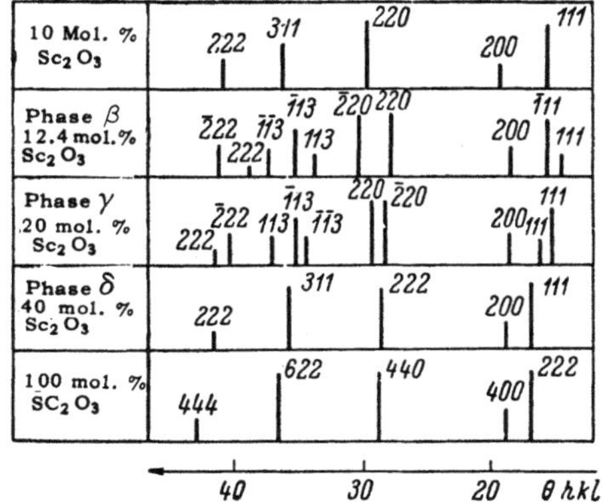

Fig. 3. X-ray patterns of compositions in the Sc_2O_3—ZrO_2 system.

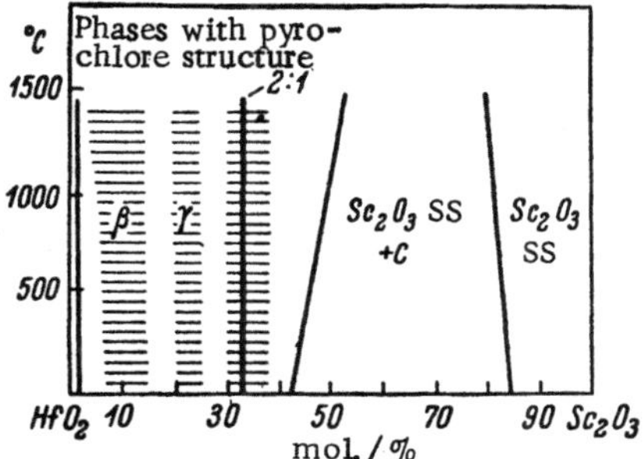

Fig. 4. Phase relations in the Sc_2O_3—HfO_2 system.

The following phases appear with increase of scandium-oxide concentration (Fig. 2):

1) α-Phase: a tetragonally distorted fluorite phase which decomposes on cooling to two other tetragonal phases.

2) β-Phase: a rhombohedrally distorted fluorite structure transforming reversibly to the α-phase with increasing temperature. In the given case an order—disorder transition occurs. As American researchers have shown [4], the approximate composition $Sc_2O_3 \cdot 7ZrO_2$ (or $Sc_2Zr_7O_{17}$) can be ascribed to this phase.

3) γ-Phase: also a rhombohedrally distorted fluorite structure. However, it is distinguished from the β-phase by the reversible character of the distortion. The γ-phase has a rhombohedral angle of >90°, the β-phase has one of <90° (Fig. 3).

4) δ-Phase: has the same indices as the β- and γ-phases and also appears to be a rhombohedrally distorted CaF_2 structure. This phase can be considered as a solid solution based on a compound of 3:2 composition ($3ZrO_2 \cdot 2Sc_2O_3$).

At high temperatures all these phases are characterized by a disordering process and a transition to phases based on the CaF_2-type structure [4], as was already found in the system with TiO_2.

However, study of the above system did not give a conclusive picture of the phase equilibria. There is no indication as to whether here there occurs (for the group between the ordered phases) a two-phase region or a continuous transition from one form to the other.

Sc_2O_3—HfO_2 SYSTEM

We encountered the same problem in the study of the Sc_2O_3—HfO_2 system. Here we found, analogous to the zirconium system, phases characterized by distortion, similar to the β- and γ-phases. The parameters are, respectively: a_R = 10.104 Å, β = 89°13', and a_R = 10.086 Å, β = 90°21'.

However, they have a somewhat different location in the system. It is important to note that in this case these phases have a higher degree of ordering, and possess a rhombohedrally distorted pyrochlore structure rather than a fluorite structure. The rhombohedral angle is < 90° for the β-phase and >90° for the γ-phase. A continuous transition is observed between these phases. We further discovered in the system a phase corresponding to the 2:1 composition ($Sc_2Hf_2O_7$), whose stoichiometry corresponds to the composition of a pyrochlore-type compound. This compound has a monoclinic, distorted pyrochlore structure with the parameters $a = c = $ 10.100 Å, b = 10.049 Å, β = 89°20' (Fig. 4).

There is a continuous transition between the γ-phase and the pyrochlore one. There are observed: (a) decrease in superstructure lines, and (b) decrease and disappearance of splitting of the intense lines.

A gradual transition is observed from the monoclinic, distorted pyrochlore to the fluorite structure on further increase of Sc_2O_3 concentration above the stoichiometry of $Sc_2Hf_2O_7$.

Thus, at 1500°C the fluorite-like phase in the Sc_2O_3—HfO_2 system is located in the region from 5 to 40 mol.% Sc_2O_3. A continuous transition is observed within it between compositions of different degrees of ordering. The least degree of ordering is observed at a composition of about 30% Sc_2O_3. On both sides of this there are compositions possessing a higher degree of order. The order is the greatest in a composition 2:1 with respect to β- and γ-phases.

THE Sc_2O_3—ThO_2 SYSTEM

We know of two works by German authors [5, 6] on the interaction of Sc_2O_3 with ThO_2, according to which an insignificant solubility of Sc_2O_3 is observed in ThO_2. This amounts to approximately 1% at 1750°C. However, this system requires further exact study.

Therefore, leading up to the results, the following can be noted:

1) In study of systems of this type it is necessary to carefully consider the specimen thermal history. It is evident that in most works the authors do not achieve a completely equilibrium composition, producing contradictory results. This frequently applies to the work of the German researchers [5, 6], who noted only cubic, fluorite-type solid solutions in the Sc_2O_3—ZrO_2 system.

2) Further, systems of Sc_2O_3 with the oxides in question are characterized by order—disorder type transformations which are beginning to play a greater role in inorganic chemistry, since they occur both with rising temperature and with change of concentration, as was shown in the Sc_2O_3—HfO_2 system.

3) The systems formed by Sc_2O_3 and oxides of elements of the titanium subgroup are analogous in their own way to systems formed by the rare earth oxides and ZrO_2 and HfO_2. A fluorite-structure phase appears as the basic structural unit.

However, a considerably more ordered phase is found in the systems with the oxides of Zr and Hf. They have another composition, and the corresponding phase diagrams can be sooner compared with those formed by oxides of a rare earth existing in different degrees of oxidation (e.g., CeO_2—Ce_2O_3).

LITERATURE CITED

1. L. H. Brixner, J. Inorg. Chem., Vol. 3, No. 7, p. 1065 (1964).
2. R. Collongues, F. Queuroux, and M. Perez-y-Jorba, Bull. Soc. Chim. France., No. 4, p. 1141 (1965).
3. J. Lefevre, Ann. Chim., Vol. 8, p. 117 (1963).
4. D. W. Strickler and W. G. Carlson, USAEC Conf., Vol. 45, No. 2, p. 21 (1963).
5. H. H. Moebius, H. W. Witzmann, and F. Zimmer, Z. Chemie, Vol. 4, No. 5, p. 194 (1964).
6. H. H. Moebius, Z. Chemie, Vol. 4, No. 3, p. 81 (1964).

SOLID-PHASE REACTIONS BETWEEN PENTOXIDES OF NIOBIUM AND TANTALUM AND OXIDES OF THE RARE-EARTH ELEMENTS

E. P. Savchenko, N. A. Godina, and É. K. Keler

Great interest has recently developed in studying the properties and formation conditions of rare-earth compounds, which serve as base materials in the production of substances with valuable technological properties.

The rare-earth niobates and tantalates are among a number of little-studied compounds. At present only the rare-earth orthoniobates of general formula $LnNbO_4$ have been well studied. An x-ray investigation of compounds of this type was conducted by Komkov, who established that such compounds crystallize in the monoclinic system and form an isostructural series [1].

Krylov, Keller, and Stubican [2-5] synthesized the rare-earth orthoniobates and ortho-tantalates. The works of Rooksby and White [6, 7] established, besides the orthoniobates, still two other compound types of general formulas Ln_3AO_7 and LnA_3O_9 (where A = Nb, Ta), in the $Ln_2O_3—Nb_2O_5$ and $Ln_2O_3—Ta_2O_5$ systems.

However, both these compounds and the orthotitanates (general formula $LnTaO_4$) were little studied. Furthermore, none of the above authors performed a detailed investigation on solid phase reactions in the $Ln_2O_3—Nb_2O_5$ and $Ln_2O_3—Ta_2O_5$ systems. The present work is devoted to this problem.

EXPERIMENTAL

Solid phase reactions between the pentoxides of niobium and tantalum and the rare-earth oxides (La_2O_3, Pr_6O_{11}, CeO_2, and Nd_2O_3) were studied.

X-ray, phase-chemistry, and thermal analyses were used as methods of study. The x-ray analysis was performed by the powder method, using copper radiation and ionization registry.

The chemical analysis of phases in the $Ln_2O_3—Nb_2O_5$ and $Ln_2O_3—Ta_2O_5$ systems (where Ln = La, Nd) was based on the solubility of unreacted lanthanum and neodymium oxides in solutions of ammonium salts, in which the rare earth niobates and tantalates are insoluble.

To find the amount of reacted praseodymium oxide, the active oxygen content was determined in annealed compositions of $Pr_6O_{11} + Nb_2O_5$ and $Pr_6O_{11} + Ta_2O_5$ [8]. Oxygen generated during the reaction process was determined directly for the $CeO_2 + Nb_2O_5$ compositions.

Niobium pentoxide (99.6% Nb_2O_5), tantalum pentoxide (99.0% Ta_2O_5), and 99.8% pure rare-earth oxides were used as starting reagents.

1. Interaction of Niobium Pentoxide with the Oxides of Lanthanum, Cerium, Praseodymium, and Neodymium

The study of the interaction between niobium pentoxide and the rare-earth oxides was performed in the 900-1650°C temperature range. The La_2O_3—Nb_2O_5 system was the one most thoroughly studied, on compositions ranging from 5 to 95 mol.% La_2O_3.

It was established that four compound types form in the La_2O_3—Nb_2O_5 system. These have La_2O_3 : Nb_2O_5 ratios of 1:1 ($LaNbO_4$), 3:1 (La_3NbO_7), 1:3 ($LaNb_3O_9$), and 1:6 ($La_2Nb_{12}O_{33}$).

Lanthanum orthoniobate ($LaNbO_4$) crystallizes in a monoclinic form and is isostructural to the mineral fergusonite. Lanthanum orthoniobate undergoes a reversible polymorphic transformation in the temperature range 550-600°C, changing from the monoclinic to the tetragonal modification. It melts congruently at 1620 ± 30°C.

Lanthanum metaniobate ($LaNb_3O_9$) has a distorted perovskite structure and crystallizes in the tetragonal form. This compound melts incongruently at 1420 ± 30°C, forming orthoniobate and liquid. However, if the heating and cooling is performed quickly (over several minutes), the lanthanum metaniobate can be obtained in the metastable state.

The compound La_3NbO_7 crystallizes in the orthorhombic form and is isostructural to the mineral bergite. This compound melts congruently at 1780 ± 30°C.

The conditions for formation of the lanthanum niobates were studied. X-ray and chemical analyses of $La_2O_3 + Nb_2O_5$ mixtures, annealed at 900-1200°C, showed that reactions occur very energetically in this temperature range. Orthoniobate forms in the amount of 98.5% in a mixture of $1La_2O_3 + 1Nb_2O_5$ annealed 2 h at 1100°C.

X-ray analysis established that there is a stepwise formation process for this compound. The compound La_3NbO_7 appears in the first intermediate reaction stage at 900-1000°C. On increase of the temperature and duration of anneal, the La_3NbO_7 reacts with Nb_2O_5 to form lanthanum orthoniobate.

The formation process of lanthanum metaniobate is also stepwise. It was discovered that La_3NbO_7 and $LaNbO_4$ form as primary intermediate reaction products in mixtures of the composition $1La_2O_3 + 3Nb_2O_5$ (corresponding to the metaniobate). La_3NbO_7 was the predominant phase in specimens annealed at 900°C, while $LaNbO_4$ was predominant at 1000°C. Further increase in temperature to 1200°C and in duration to 12 h resulted in formation of the metaniobate ($LaNb_3O_9$).

Investigation of the formation conditions and properties of the cerium niobates established that they are identical in structural types and properties to the lanthanum niobates. Only three compounds, of the general formulas $LnNbO_4$, Ln_3NbO_7, and $LnNb_3O_9$ and analogous to the niobates of lanthanum and cerium, were produced in the Pr_2O_3—Nb_2O_5 and Nd_2O_3—Nb_2O_5 systems. Compounds of the type $Ln_2Nb_{12}O_{33}$ were not found in these systems.

During the synthesis condition study, a stepwise formation character was established for the ortho- and metaniobates in all the systems studied. Such can be represented by the following schemes: for the orthoniobates,

$$Ln_2O_3 + Nb_2O_5 (1:1) \rightarrow Ln_3NbO_7 + Nb_2O_5 \rightarrow LnNbO_4;$$

for the metaniobates,

$$Ln_2O_3 + Nb_2O_5 (1:3) \rightarrow Ln_3NbO_7 + Nb_2O_5 \rightarrow LnNbO_4 + Nb_2O_5 \rightarrow LnNb_3O_9.$$

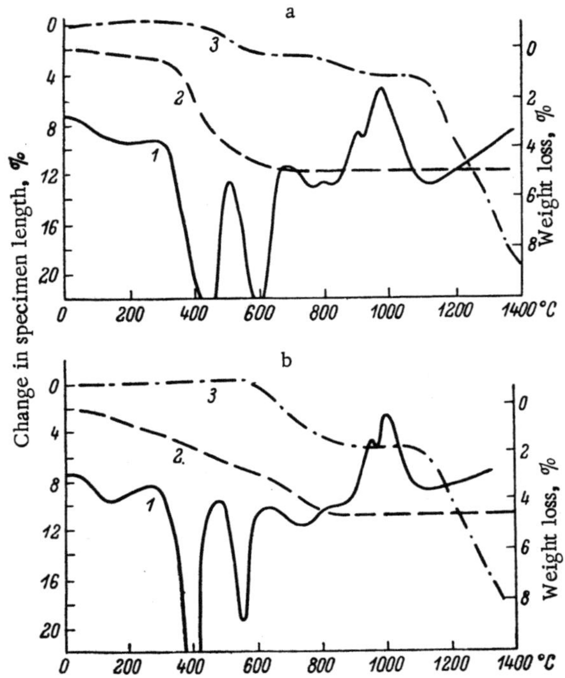

Fig. 1. Heating curves for formation of (a) lanthanum orthoniobate and (b) neodymium orthoniobate. 1) Differential thermal curve; 2) weight loss; 3) change in linear dimensions.

We observed analogous reactions (a definite succession of compound formation during solid phase reaction) in the Ln_2O_3—SiO_2 and Ln_2O_3—Al_2O_3 systems [9, 10]. Compounds richer in the rare-earth oxides (relative to other compounds in these systems) formed as primary reaction products.

The lanthanum and neodymium niobates formed under the same conditions at 1100-1200°C. A temperature of 1300°C was required for synthesis of the praseodymium niobates, while synthesis of the cerium niobates was effected at 1350-1400°C (with prolonged annealing of about 15-20 h).

The heating curves obtained during formation of the lanthanum and neodymium niobates are shown in Fig. 1. It is seen from Fig. 1 that endothermic effects are recorded on the differential curves. These correspond to decomposition of the hydrates and carbonates of lanthanum and neodymium, which form during storage of these oxides in air. The exothermic effects correspond to the processes of compound formation.

The change in linear dimension curves shows that the shrinking process, starting at a temperature corresponding to the exothermic effect, is slowed because of specimen disintegration in the process of synthesis.

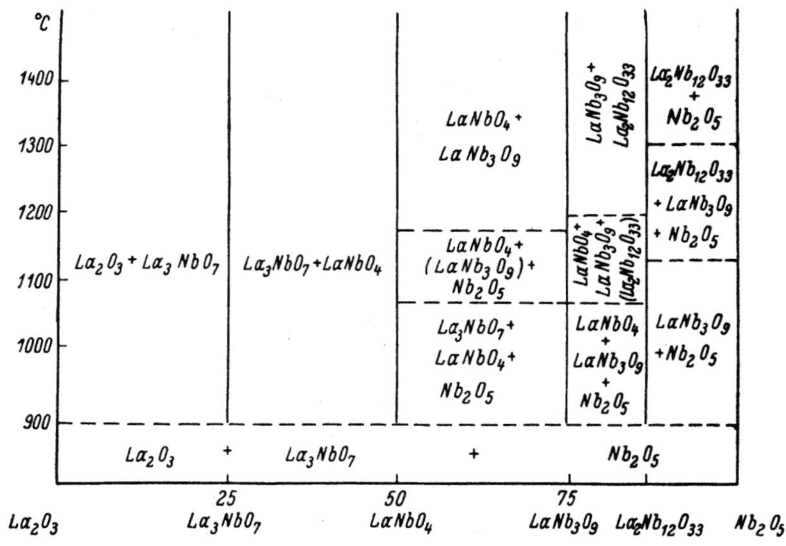

Fig. 2. Phase composition of annealed mixtures in the La_2O_3—Nb_2O_5 system. (Phases found in small amounts are in parentheses.)

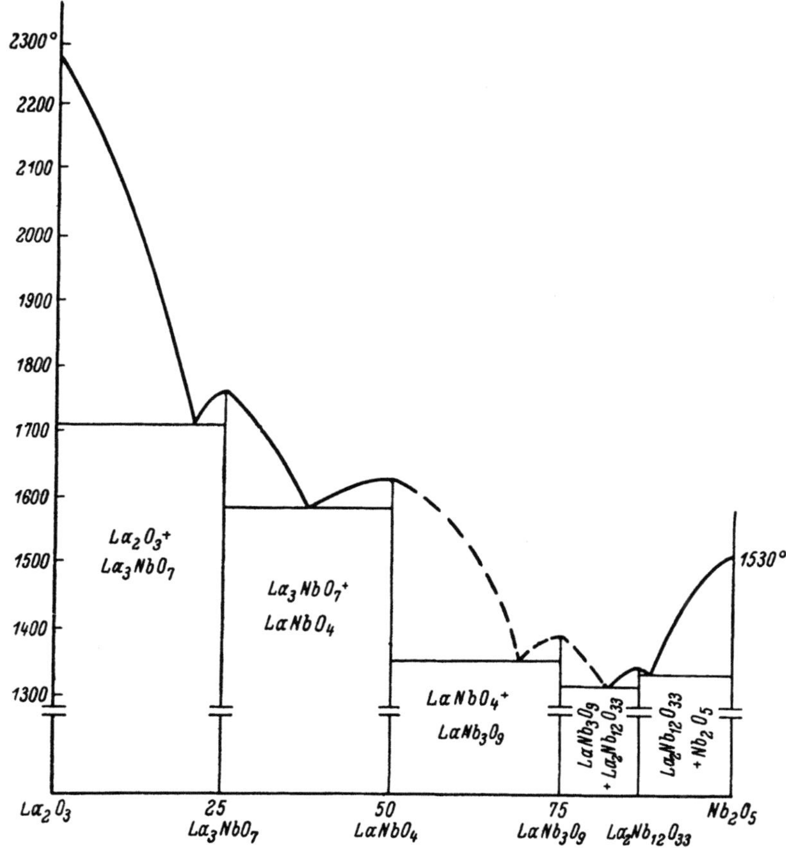

Fig. 3. Tentative phase diagram of the La_2O_3—Nb_2O_5 system.

We studied in detail similar phenomena occurring on formation of the zirconates, hafnates, and cerates of the alkaline-earth elements [11, 12]. For these compounds it was shown that sample disintegration during synthesis results from the breaking away of reaction products at the surface of the reacting component grains. The high rate of solid-phase transformation experimentally observed can be explained by the lesser role of the slowly occurring volume diffusion processes.

The entire series of experiments enabled us to construct a scheme for the phase relationships in annealed La_2O_3 + Nb_2O_5 mixtures, and to construct a tentative phase diagram for the La_2O_3—Nb_2O_5 system (Figs. 2 and 3).

2. Interaction of Tantalum Pentoxide with the Oxides of Lanthanum, Cerium, Praseodymium, and Neodymium

A study of the interaction between tantalum pentoxide and the oxides La_2O_3, CeO_2, Pr_6O_{11}, and Nd_2O_3 was performed in the temperature range 900-1800°C. Compositions from 80 to 20% Ta_2O_5 were studied. It was established that three compound types form, of the general formulas Ln_3TaO_7, $LnTaO_4$, and $LnTa_3O_9$.

Based on the equal ionic radii of Nb^{5+} and Ta^{5+} (0.66 Å), it can be anticipated that all resulting tantalum compounds will be isostructural to the niobium compounds. However, the experiments showed that the lanthanum, cerium, and praseodymium orthotantalates are not isostructural to the corresponding niobates, and that they form a new structural series. The structure of these compounds has not yet been determined.

The results we obtained contradict the work of Krylov and Pinaeva-Strelina [3], but confirm the study of Stubican.

The lanthanum and cerium orthotantalates melt congruently in the temperature range 1880-1900°C. It was discovered that praseodymium orthotantalate can also be obtained in the monoclinic modification typical of praseodymium orthoniobate, but that it is not in the pure state and contains an impurity phase isostructural to $LaTaO_4$ and $CeTaO_4$. Monoclinic $PrTaO_4$ was obtained on annealing specimens of composition $1Pr_6O_{11} + 1Ta_2O_5$ at 1400-1550°C. A final anneal of such specimens at 1600-1750°C led to the irreversible transformation of monoclinic $PrTaO_4$ to a crystal modification isostructural to $LaTaO_4$ and $CeTaO_4$, and stable up to a congruent melting point of 1900 ± 30°C.

The metatantalates $LnTa_3O_9$ (Ln = La, Ce, Pr) and the compounds synthesized of general formula Ln_3TaO_7 proved isostructural to the corresponding niobates.

An analogy of the structural types for the resulting compounds was only established for the niobates and tantalates of neodymium. Neodymium orthotantalate is isostructural to neodymium orthoniobate, and also crystallizes in the monoclinic form. This compound was observed to melt congruently at 1920 ± 30°C.

The compounds Ln_3TaO_7 (where Ln = La, Pr, Nd) melt without decomposition in the temperature range 1990-2030°C. The metatantalates $LaTa_3O_9$, $CeTa_3O_9$, $PrTa_3O_9$, and $NdTa_3O_9$ melt in the temperature range 1780-1810°C, during which incongruent melting was only observed for neodymium metatantalate.

Thus there is an analogy between the $Nd_2O_3-Ta_2O_5$ and $Nd_2O_3-Nb_2O_5$ systems not only with respect to the structural types of the compounds forming, but also with respect to compound properties.

On examination of the formation conditions of the tantalates, it is evident that synthesis of the tantalates is effected at higher temperatures relative to the niobates. For example, a temperature of 1400-1450°C is required to produce lanthanum orthotantalate, while the lanthanum orthoniobate (as already noted) can be prepared at 1100°C.

The definite succession we established for formation of the niobates was also noted in solid phase transformations in the $Ln_2O_3-Ta_2O_5$ systems. It is quite evident that this would be observed in the $Nd_2O_3-Ta_2O_5$ system, where structures of the resulting compounds are noticeably different.

On interaction of Nd_2O_3 with Ta_2O_5 (as also in systems with Nb_2O_5), Nd_3TaO_7 forms as a product, independent of the ratio of oxides in the original mixture. On further anneal, Nd_3TaO_7 reacts with Ta_2O_5 to form $NdTaO_4$ and $NdTa_3O_9$.

A study of the spectroscopic characteristics of doped rare-earth niobate and tantalate specimens indicated them to be prospective laser materials.

CONCLUSIONS

1. Solid phase reactions were studied in the $Ln_2O_3-Nb_2O_5$ and $Ln_2O_3-Ta_2O_5$ systems (Ln = La, Ce, Pr, Nd) over the temperature range 800-1800°C.

2. Three types of compounds of general formulas $LnAO_4$, Ln_3AO_7, and LnA_3O_9 (where A = Nb, Ta) were observed. Still another compound type, with the general formula $Ln_2Nb_{12}O_{33}$, was noted in the $La_2O_3-Nb_2O_5$ and $Ce_2O_3-Nb_2O_5$ systems.

3. A definite succession for compound formation during the synthesis process was established.

4. It was shown that a complete analogy between the structural types of the resulting compounds is observed only in the Nd_2O_3—Nb_2O_5 and Nd_2O_3—Ta_2O_5 systems. The compounds synthesized are of interest for modern technology.

LITERATURE CITED

1. A. I. Komkov, Dokl. Akad. Nauk SSSR, Vol. 4, p. 126 (1959).
2. E. I. Krylov, V. N. Sanatina, and A. K. Shtol'ts, Zh. Neorg. Khim., Vol. 6, No. 5, p. 1135 (1961).
3. E. I. Krylov and M. M. Pinaeva-Strelina, Zh. Neorg. Khim., Vol. 8, No. 10, p. 2254 (1963).
4. C. Keller, Z. Anorg. Allgem. Chem., Vol. 318, No. 2, p. 89 (1962).
5. V. S. Stubican, J. Am. Ceram. Soc., Vol. 47, No. 2, p. 94 (1964).
6. H. P. Rooksby and E. A. D. White, J. Am. Ceram. Soc., Vol. 47, No. 2, p. 94 (1964).
7. H. P. Rooksby and E. A. D. White, J. Am. Ceram. Soc., Vol. 48, No. 2, p. 447 (1965).
8. É. K. Keler, N. A. Godina, and E. P. Savchenko, Izv. Akad. Nauk SSSR, Otd. Khim. Nauk, No. 10, p. 1735 (1961).
9. É. K. Keler, N. A. Godina, and E. P. Savchenko, Izv. Akad. Nauk SSSR, Otd. Khim. Nauk, No. 10, p. 1728 (1961).
10. N. A. Godina and É. K. Keler, Izv. Akad. Nauk SSSR, Ser. Khim., No. 1, p. 24 (1966).
11. A. K. Kuznetsov, Zh. Neorg. Khim., Vol. 2, No. 10, p. 2327 (1957).
12. É. K. Keler, N. A. Godina, and M. G. Degen, Zh. Prikl. Khim., Vol. 34, p. 1764 (1961).

SYSTEMS CONTAINING REFRACTORY OXIDE COMPOUNDS

B. Refractory Oxide Systems Not Containing Rare—Earth Elements

PHASE EQUILIBRIA IN SEVERAL SECTIONS OF THE $CaO-BaO-GeO_2-SiO_2$ SYSTEM

R.G. Grebenshchikov, A. K. Shirvinskaya, V. I. Shitova, and N. A. Toropov

The germanates are the nearest crystallochemical analogs to the silicates. The literature contains rather incomplete information on the conditions of synthesis and equilibrium relationships in germanate systems. Studies of various germanosilicate systems are being performed at the Institute of Silicate Chemistry, devoted particularly to the occurrence of isomorphous substitution in the anionic sublattice. The present article discusses a number of germanate and germanosilicate systems which comprise the $CaO-BaO-GeO_2-SiO_2$ system.

Thermal, x-ray, microscopic (slides), and crystallo-optical analysis methods were used for determining all phase diagrams described in the present article. The high-temperature study of the specimens (determination of the liquidus curves of the systems) was performed in a microfurnace using a technique developed at the above Institute. The differential thermal analysis curves were obtained on apparatus for complex and microthermal analysis. The x-ray study was performed at room temperature and above (to 1400°C) on a diffractometer utilizing ionization registry (Cu $K\alpha$ radiation, Ni filter). For an additional check on phase composition, all samples from the systems studied were examined under immersion oils. Highly refractive phosphoric liquids were used for determining the high indices of refraction. The density of specimens was determined by pycnometer with distilled kerosene at 25°C. The specific germanates and their intermediate compositions were obtained by sintering corresponding equimolar amounts of the starting materials (analytical grades $CaCO_3$ and $BaCO_3$, chemically pure GeO_2, and 99.9% pure SiO_2). Homogeneity of the synthesized germanates was ensured in some cases by repeated anneal and intermediate pulverizing.

THE $CaO-GeO_2$ SYSTEM

Previous researchers [1-5] established the presence of five individual compounds in this system: Ca_3GeO_5, Ca_2GeO_4, $CaGeO_3$, $CaGe_2O_5$, and $Ca_3Ge_2O_5$, We were the first to study the phase diagram of the $CaO-GeO_2$ system [6], which is shown in Fig. 1. Six chemical compounds were defined in this system: Ca_3GeO_5, Ca_2GeO_4, $Ca_3Ge_2O_5$, $CaGeO_3$, $CaGe_2O_5$, and $CaGe_4O_9$. Table 1 gives values of the interplanar spacings of these compounds. Their fundamental physical properties are given in Table 2, and the heating curves of the compounds are shown in Fig.2.

Tricalcium Germanate Ca_3GeO_5

As follows from Fig. 1, Ca_3GeO_5 is thermodynamically stable in the 1320-1880°C temperature range, and can usually be quenched from high temperatures. Below 1320°C, CaO and

Table 1. Interplanar Spacings of the Calcium Germanates

Ca_3GeO_5		$\alpha\text{-}Ca_2GeO_4$		$\gamma\text{-}Ca_2GeO_4$		$Ca_3Ge_2O_7$		$CaGeO_3$		$CaGe_2O_5$		$CaGe_4O_9$	
d/n	I/I_0	d/n	I/I_0	d/n	I/I_0	d/n	I/I_0	d/n	I/I_0	d/n	I/I_0	d/n	I_0I_0
4.38	12	3.89	35	5.56	10	4.40	30	5.12	20	4.97	21	5.15	25
3.91	26	3.50	23	4.84	7	3.96	25	4.12	10	3.40	16	4.07	35
3.58	22	2.82	80	4.30	48	3.67	13	3.91	61	3.20	56	3.64	32
3.09	52	2.72	100	4.10	13	3.42	33	3.60	35	3.13	70	2.88	100
2.81	100	2.17	15	3.83	13	3.34	40	3.38	73	3.01	100	2.79	35
2.64	68	1.97	41	3.39	41	3.21	26	3.27	39	2.79	17	2.71	27
2.36	15	1.77	22	3.06	60	3.09	30	3.18	52	2.63	88	2.63	60
2.22	30	1.60	25	2.79	90	3.00	100	3.05	100	2.53	42	2.44	30
2.00	7	1.58	19	2.76	100	2.81	70	2.78	25	2.38	26	2.39	30
1.96	20	1.49	18	2.63	35	2.76	66	2.65	61	2.32	20	2.34	27
1.79	46	1.36	11	2.52	13	2.63	24	2.56	68	2.18	14	2.25	23
1.65	26	1.24	5	2.38	24	2.58	16	2.39	30	2.05	20	2.12	13
1.56	24	1.19	4	2.16	10	2.49	10	2.29	14	2.03	31	1.80	20
1.51	35	1.15	5	2.09	21	2.23	21	2.21	36	1.95	11	1.78	27
1.49	11			1.95	37	2.19	20	2.14	19	1.82	10	1.76	42
1.41	8			1.85	19	2.15	14	2.05	32	1.77	15	1.68	30
				1.82	48	2.10	5	1.93	30	1.69	56	1.61	18
				1.77	18	2.05	14	1.88	34	1.65	45	1.57	19
				1.69	38	1.99	6	1.82	18	1.58	31	1.52	18
				1.65	38	1.93	10	1.76	20	1.53	30	1.49	18
				1.54	16	1.89	10	1.66	30	1.50	26	1.42	26
				1.488	17	1.87	10	1.58	25	1.49	21	1.39	20
				1.477	16	1.82	15	1.57	25	1.45	17	1.36	14
				1.448	15	1.74	20	1.53	25	1.436	24	1.32	20
				1.318	10	1.70	17	1.50	25	1.405	20		
				1.236	10	1.66	14	1.41	18	1.390	26		
				1.188	7	1.58	17	1.37	15	1.366	20		
				1.153	11	1.53	15	1.34	11	1.319	16		
				1.098	10	1.50	16	1.27	15	1.112	15		
						1.46	30			1.057	10		
										1.047	10		

$\gamma\text{-}Ca_2GeO_4$ are the equilibrium phases in the composition range 0 to 33.3 mol.% GeO_2. This upper limit coincides with the incongruent melting temperature of tricalcium germanate (by the reaction $Ca_3GeO_5 \rightleftharpoons CaO + liq.$).

According to thermal analysis data (Fig. 2), Ca_3GeO_5 is characterized by three phase transformations, occurring at 754°, 1034°, and 1163°C. It was shown by high-temperature x-ray analysis that these transformations relate to the following structural transitions in Ca_3GeO_5:

$$\text{Triclinic I} \xrightarrow{754°} \text{Triclinic III} \xrightarrow{1034°} \text{Monoclinic} \xrightarrow{1163°} \text{Rhombohedral.}$$

Thus tricalcium germanate has four polymorphs. The unit-cell parameters were determined for the triclinic I and monoclinic forms of Ca_3GeO_5, and their isostructurality to the corresponding tricalcium silicate (Ca_3SiO_5) was established [3].

Calcium Orthogermanate Ca_2GeO_4

This compound is stable over the entire temperature range up to its congruent melting at 1900°C. It exists in two polymorphs [5, 7]. The low-temperature $\gamma\text{-}Ca_2GeO_4$ modification has an olivine-type structure and is isostructural to $\gamma\text{-}Ca_2SiO_4$. The hexagonal form of Ca_2GeO_4 is stable above 1490°C (Fig. 2) and is isostructural to $\alpha\text{-}Ca_2SiO_4$.

In view of the polymorphism of calcium orthogermanate, the diagram shows the boundary of the equilibria between the γ- and α-forms at a temperature of 1490°C. The high-temperature modification of $\alpha\text{-}Ca_2GeO_4$ is unstable at room temperature, since a reversible transformation to $\gamma\text{-}Ca_2GeO_4$ occurs on cooling below 1450°C. However, the synthesis of calcium

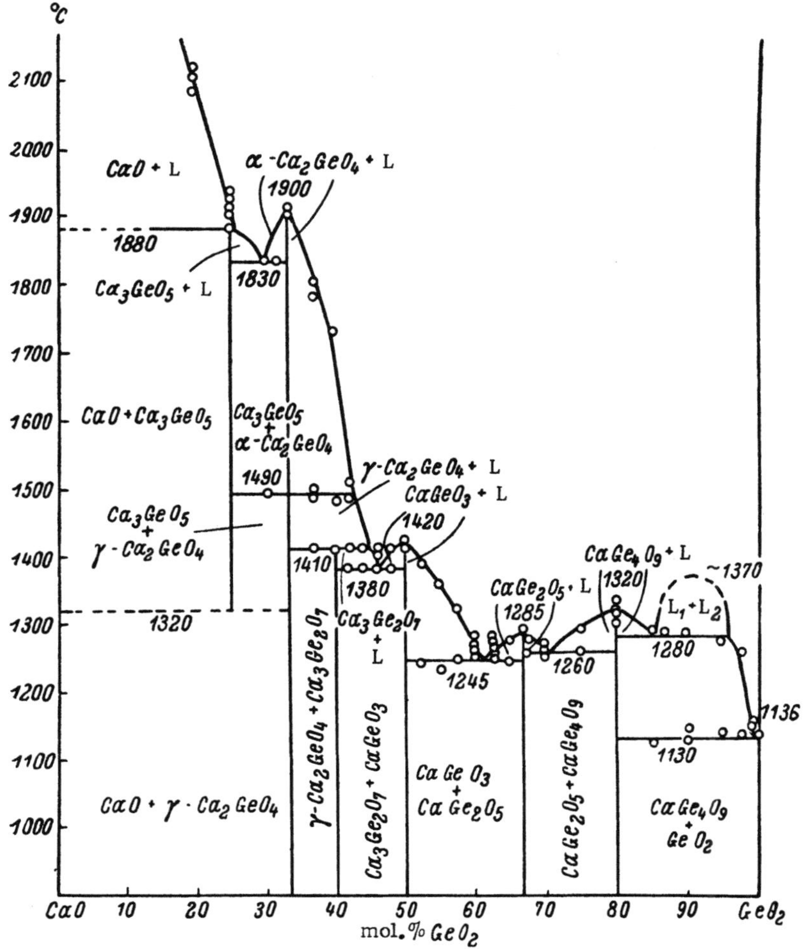

Fig. 1. Phase diagram of the $CaO-GeO_2$ system.

orthogermanate above 1490°C in the presence of 1.0–1.5 wt.% B_2O_3 with final air quenching guarantees the stabilization of α-Ca_2GeO_4.

Calcium Diorthogermanate $Ca_3Ge_2O_7$

We were the first to observe this compound. It melts with decomposition via a peritectic reaction at a temperature of 1410°C, according to the scheme $Ca_3Ge_2O_7 \rightleftharpoons \gamma$-$Ca_2GeO_4$ + liquid. The properties of the compound Ca_3GeO_7 are determined by x-ray and infrared spectroscopic investigations, and $[Ge_2O_7]^{6-}$ diortho groups were observed in this particular germanate. The presence of anionic radicals in $Ca_3Ge_2O_7$ generally analogous to the F_2X_7 diortho groups (F = Si, B, P, S, Ge, etc.) attests to the crystallochemical similarity of calcium diorthogermanate and the corresponding silicate, $Ca_3Si_2O_7$ (rankinite).

Calcium Metagermanate $CaGeO_3$

This compound melts congruently at 1420°C. It crystallizes in only one structure (wollastonite-type). When $CaGeO_3$ is subjected to high temperatures and pressures (700°, 40–70 atm), it transforms from the wollastonite form to a new modification of garnet structure.

Table 2. Basic Physical Properties of the Calcium and Barium Germanates

Compound	Index of refraction		Density, g/cm^3	Symmetry and unit cell parameters, Å	Structural analog
	Ng	Np			
Ca_3GeO_5	1.754	1.750	3.55	Triclinic; $a=12.43$, $b=7.24$, $c=25.50$; $\alpha=90°0'$, $\beta=89°48'$, $\gamma=89°54'$ [3].	Ca_3SiO_5
α-Ca_2GeO_4....	1.745		3.62	Hexagonal; $a==5.49$, $c=7.09$ [4].	α-Ca_2SiO_4
γ-Ca_2GeO_4....	1.730	1.700	3.53	Orthorhombic; $a=6.82$, $b=5.25$, $c=11.4$ [4].	γ-Ca_2SiO_4
$Ca_3Ge_2O_7$	1.750	1.737	3.63	—	$Ca_3Si_2O_7$ (?)
$CaGeO_3$	1.714	1.702	3.75	Triclinic; $a=8.15$, $b=7.58$, $c=7.31$, $\alpha=90°01'$, $\beta=94°28'$, $\gamma=103°27'$ [2].	β-$CaSiO_3$
$CaGe_2O_5$	1.88	1.84	4.86	—	$CaTiSiO_5$
$CaGe_4O_9$	1.78		4.61	Hexagonal; $a==11.12$, $c=4.73$ [2].	$BaTiSi_3O_9$
Ba_3GeO_5	1.91	1.90	5.79	Tetragonal.	Ba_3SiO_5
Ba_2GeO_4	1.87	1.83	5.71	Orthorhombic; $a=7.74$, $b=5.88$, $c=10.35$ [13].	Ba_2SiO_4
$Ba_3Ge_2O_7$	1.767		—	—	—
$BaGeO_3$	1.726	1.678	4.73	Hexagonal; $a==7.64$, $c=10.80$ [14].	α-$CaSiO_3$
$BaGe_4O_9$	1.805	1.780	5.09	Hexagonal; $a==11.61$, $c=4.74$ [2].	$BaTiSi_3O_9$
$BaGe_{19}O_{39}$ ...	1.734	1.725	—	Quartz-like crystal structure	—
GeO_2 quartz ...	1.724	1.705	4.28	Hexagonal $a==4.972$, $c=5.648$.	SiO_2 (quartz)
GeO_2 rutile ...	1.985	1.945	6.28	Tetragonal; $a==4.396$, $c=2.863$.	TiO_2

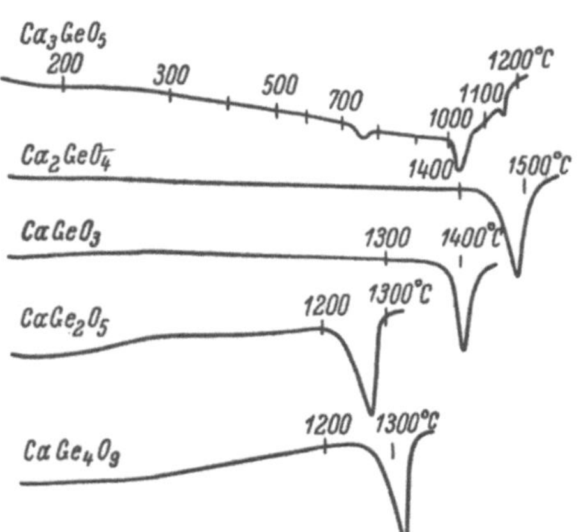

Fig. 2. Heating curves of the calcium germanates.

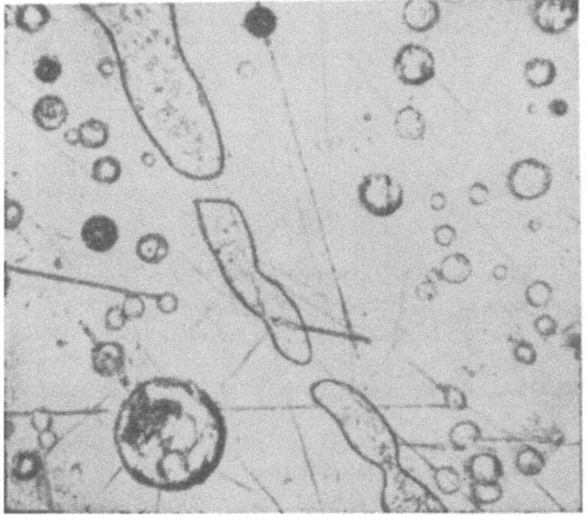

Fig. 3. Region of two immiscible liquids (6 mol.% CaO and 94 mol.% GeO_2), 1350°C, 200×.

Calcium Digermanate $CaGe_2O_5$

This compound melts congruently at 1280°C. It is known in one structural form. One germanium atom of the $CaGe_2O_5$ molecule is situated in a sixfold coordination, conforming to its crystallochemical analogy with sphene ($CaTiSiO_5$).

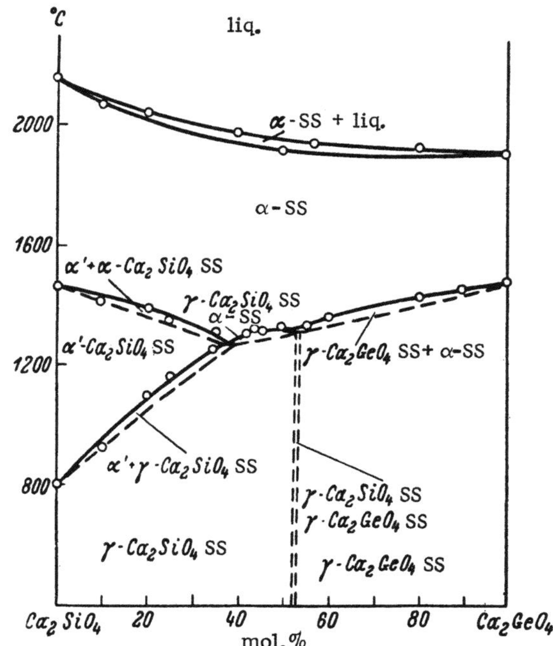

Fig. 4. Phase diagram of the Ca_2SiO_4–Ca_2GeO_4 system.

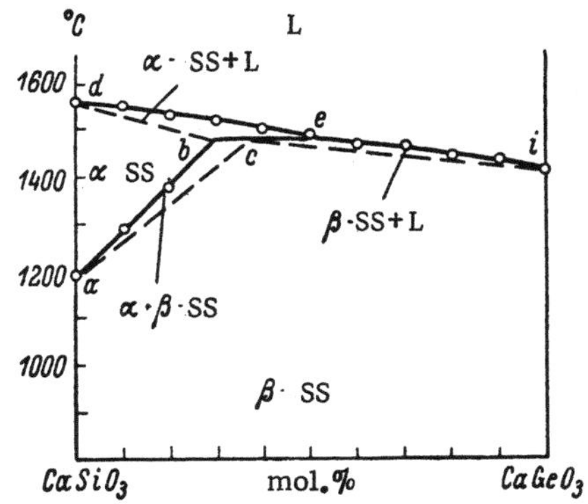

Fig. 5. Phase diagram of the $CaSiO_3$–$CaGeO_3$ system.

Calcium Tetragermanate $CaGe_4O_9$

This compound melts congruently at 1320°C, but has a rather sloping melting maximum, in distinction to the previous compounds. $CaGe_4O_9$ also is the crystallochemical analog to the titanosilicate — in the given case to benitoite ($BaTiSi_3O_9$).

The $CaGe_4O_9$–GeO_2 join is a binary eutectic system with liquid immiscibility. The region of liquid immiscibility extends from 86 to 96 mol.% GeO_2. The upper limit for stratification was approximately determined and is denoted in Fig. 2 by the crossed lines. The critical point for stratification is at approximately 1370°C. The photomicrograph in Fig. 3 visibly demonstrates the character of the liquid immiscibility in this system.

THE Ca_2SiO_4–Ca_2GeO_4 SYSTEM

We first studied this system in 1963 [7]. We determined the latest phase diagram of the Ca_2SiO_4–Ca_2GeO_4 system and it is shown in Fig. 4. Solid solutions of olivine structure and a narrow miscibility gap were defined in the low-temperature portion of the system. The α'-base Ca_2SiO_4 solid solution contains up to 38 mol.% of the germanate component. The γ-Ca_2SiO_4 solid-solution region transforms directly to a solid solution in the 38–48 mol.% Ca_2GeO_4 range, the α' form of the solid solution. A continuous solid solution forms between α-Ca_2SiO_4 and α-Ca_2GeO_4 in the subsolidus region of the diagram. The α'-Ca_2SiO_4 and α solid solutions are unstable under normal conditions and transform to γ-Ca_2SiO_4- and γ-Ca_2GeO_4-base solid solutions. The metastable β-Ca_2SiO_4 solid solution contains up to 15 mol.% Ca_2GeO_4.

THE $CaSiO_3$–$CaGeO_3$ SYSTEM

We were the first to study the phase diagram of this system [8], which is shown in Fig. 5. There are two single-phase fields in the subsolidus portion of the diagram: a continuous solid solution of wollastonite structure, and a limited pseudowollastonite-base solid solution. The solid solutions join with a peritectic reaction at 1480°C.

THE BaO–GeO_2 SYSTEM

Six compounds were established in this system; we were the first to prepare three of them: Ba_3GeO_5, $Ba_3Ge_2O_7$, and $BaGe_{19}O_{39}$ [9, 10]. The germanates Ba_3GeO_5, Ba_2GeO_4, $Ba_3Ge_2O_7$,

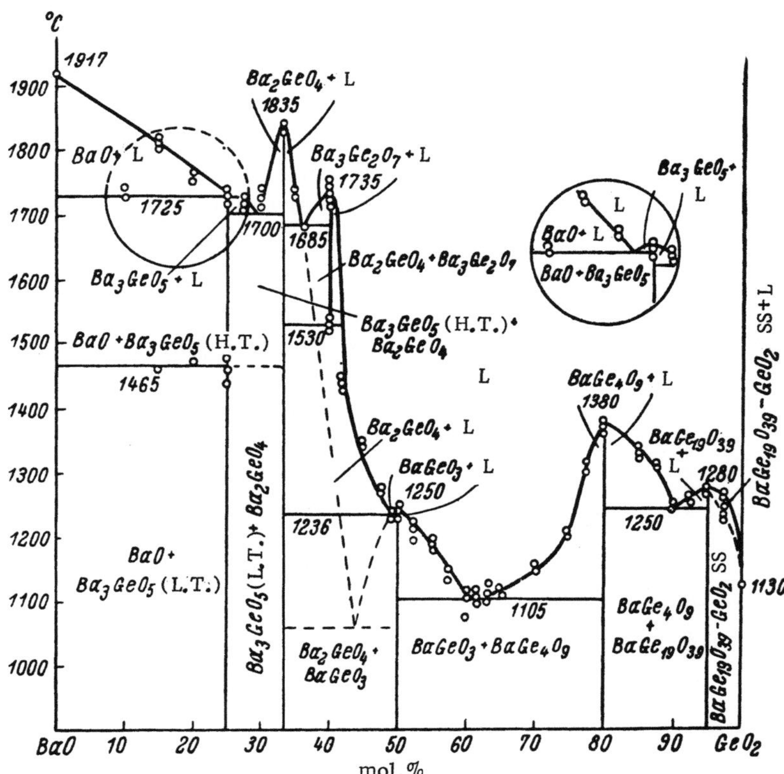

Fig. 6. Phase diagram of the BaO—GeO₂ system. H.T. = high-temperature form; L.T. = low-temperature form, of Ba₃GeO₅.

and BaGeO₃ are crystallochemical analogs to the corresponding barium silicates. Barium tetragermanate is the crystallochemical analog to benitoite, BaTiSi₃O₉. The compound BaGe₁₉O₃₉ is at present the only representative having a similar stoichiometry for both the alkali and alkaline-earth elements.

The phase diagram of the BaO—GeO₂ system is shown in Fig. 6 [10]. As seen from the diagram, Ba₃GeO₅ is thermodynamically stable under normal conditions. In view of the polymorphism of tribarium germanate, invariant inversion lines are shown on the diagram at 1465°C. It was not possible to determine precisely the melting character of Ba₃GeO₅. The choice of incongruent melting at 1725°C is based on use of the analogy between the melting character of the silicates and fluoroberyllates of composition A₃BX₅. The diagram shows areas which point to the possibility of congruent melting of Ba₃GeO₅.

In the absence of mineralizers, barium metagermanate, BaGeO₃, crystallizes only in the pseudowollastonite form, which is stable even after prolonged heating. A chain-like variety of BaGeO₃ forms on adding BaSiO₃ to it (even in as small amounts as 5 mol.%). BaGeO₃ melts congruently at 1250°C.

Parts of the BaGeO₃—BaGe₄O₉ and BaGe₄O₉—BaGe₁₉O₃₉ binary systems are represented as simple eutectic systems. The eutectic of the first is at 1105°C; of the second at 1250°C.

Part of the BaGe₁₉O₃₉—GeO₂ system displays continuous solid solutions. As for structural character, they evidently represent a case of defect lattices, forming with inclusion of large Ba cations in the quartz-like frame of GeO₂.

The values for the indices of refraction and densities of the barium germanates are quoted in Table 2.

Table 3. Compressive Strength of 1:3 Specimens of Calcium Orthogermanate and Its Solid Solutions (kg/cm^2)

Specimen	Storing time		
	7 days	28 days	3 months
β-Ca_2SiO_4	24	28	51
γ-Ca_2GeO_4	39	50	100
γ-$Ca_2Si_{0.8}Ge_{0.2}O_4$	20	30	40
γ-$Ca_2Si_{0.6}Ge_{0.4}O_4$	22	45	70
γ-$Ca_2Si_{0.4}Ge_{0.6}O_4$	37	49	78
β-$Ca_2Si_{0.9}Ge_{0.1}O_4$	45	88	105
α'-$Ca_2Si_{0.8}Ge_{0.2}O_4$	78	118	157
α'-$Ca_2Si_{0.65}Ge_{0.35}O_4$	90	128	176
α-Ca_2GeO_4	176	196	170

$BaSiO_3$—$BaGeO_3$ SOLID SOLUTIONS

As is known, the low-temperature forms of $BaSiO_3$ and $BaGeO_3$ have a ring-like structural type (pseudowollastonite lattice). Both compounds have crystal structures of hexagonal symmetry with similar values for their lattice constants — $BaSiO_3$: $a = 7.50$ Å, $c = 10.58$ Å; $BaGeO_3$: $a = 7.64$ Å, $c = 10.80$ Å.

The high-temperature forms of $BaSiO_3$ and $BaGeO_3$ are compounds of pyroxenoid structure, characterized by a new type of silicon—oxygen chain with two tetrahedra in a like interval along the chain axis. The unit-cell parameters of the chain-like variety of $BaSiO_3$ are $a = 4.54$ Å, $b = 5.56$ Å, $c = 12.27$ Å; for $BaGeO_3$ they are $a = 4.58$ Å, $b = 5.68$ Å, $c = 12.76$ Å.

The pseudowollastonite form was obtained during this study during solid-phase synthesis of barium metagermanate. The chain-like variety of $BaGeO_3$ forms on addition of 2 wt.% Fe_2O_5 to the sample. The ring-like form of $BaSiO_3$ was synthesized by hydrothermal means: $Ba(OH)_2 \cdot 8H_2O$ and amorphous SiO_2 starting materials in the mole ratio required were subjected to autoclave treatment for 48 h at 350°C in the presence of excess H_2O.

We defined the continuous solid solutions of pyroxenoid structure in the $BaSiO_3$—$BaGeO_3$ system on the basis of the monotonic change of interplanar spacings (according to x-ray studies), and also on the basis of the linear change in the indices of refraction.

By our preliminary data, the chain-like varieties of $BaSiO_3$ and $BaGeO_3$ also form continuous mutual solid solutions.

THE Ba_2SiO_4—Ba_2GeO_4 SYSTEM [11, 12]

In this system solid solutions of the β-K_2SO_4 structural type were discovered, with a narrow region of miscibility gap. The phase diagram is characterized by a maximum on the liquidus curve and by peritectic melting of the saturated solid solutions.

CEMENTING PROPERTIES OF CALCIUM GERMANATE

The first evidence of cementing properties of calcium germanate was discovered by V. F. Zhuravlev, as exemplified in dicalcium germanate. The cementing properties of the latter compound were confirmed in Ca_2GeO_4 and discovered in Ca_3GeO_5 [1].

According to [1], the hydraulic activities of γ-Ca_2GeO_4 and β-Ca_2SiO_4 are of the same order, since Ca_3GeO_5 yields in strength to Ca_3SiO_5. Nevertheless, the latter data indicate that the strength on compression of samples composed of mixtures of Ca_3SiO_5 and Ca_3GeO_5 is considerably above that of pure tricalcium silicate.

According to our data [5], the rate of hydration of dicalcium germanate is many times larger than that of dicalcium silicate. It can thus be anticipated that cementing materials based on Ca_2GeO_4 will have high-strength characteristics. We studied the cementing properties of the α- and γ-forms of pure Ca_2GeO_4 and of solid solutions based on γ-Ca_2GeO_4 and γ-, β-, and α'-Ca_2SiO_4. Stabilization of the β- and α'-form Ca_2SiO_4 solid solutions is effected by addition of 1 wt.% B_2O_3 as mineralizer. To study the cementing properties of dicalcium germanate and its solid solutions, tests were performed on the strength of cubic specimens of the 1:3 solid solution (size $1.41 \times 1.41 \times 1.41$ cm). The normal consistency of the setting was determined by dew drop. The cubes were stored under moist conditions. As seen from Table 3, the dicalcium germanate (especially in the α-form), and its solid solutions possess obvious hydraulic activity. This is of great interest in the further study of the cementing properties of the germanosilicates.

In conclusion, we note that study of the germanosilicate systems is especially of practical interest, since it opens additional possibilities for development of new materials having given sets of physicochemical properties.

LITERATURE CITED

1. S. M. Royak and I. A. Prokhvatilova, Dokl. Akad. Nauk SSSR, Vol. 141, No. 4, p. 880 (1961); Trudy NIItsement, No. 17 (1962).
2. G. Eulenberger, A. Wittman, and H. Nowotny, Monatsh. Chem., Vol. 93, p. 1046 (1962).
3. T. Hahn and W. Eysel, Naturwiss., Vol. 50, p. 471 (1963).
4. W. Eysel and T. Hahn, Neues Jahrb. Mineral. Abhandl., No. 6, p. 137 (1963).
5. N. A. Toropov and A. K. Shirvinskaya, Zh. Prikl. Khim., Vol. 36, No. 4, p. 717 (1963).
6. A. K. Shirvinskaya, R. G. Grebenshchikov, and N. A. Toropov, Izv. Akad. Nauk SSSR, Neorg. Mat., Vol. 2, No. 2, p. 332 (1965).
7. N. A. Toropov and A. K. Shirvinskaya, Dokl. Akad. Nauk SSSR, Vol. 153, No. 5, p. 1081 (1963).
8. A. K. Shirvinskaya, R. G. Grebenshchikov, V. I. Shitova, and N. A. Toropov, Izv. Akad. Nauk SSSR, Neorg. Mat., Vol. 2, No. 10, p. 1900 (1966).
9. R. G. Grebenshikov, N. A. Toropov, and V. I. Shitova, Dokl. Akad. Nauk SSSR, Vol. 153, p. 842 (1963).
10. R. G. Grebenshchikov, N. A. Toropov, and V. I. Shitova, Izv. Akad. Nauk SSSR, Neorg. Mat., Vol. 1, No. 7, p. 1130 (1965).
11. R. G. Grebenshchikov, N. A. Toropov, and V. I. Shitova, Izv. Akad. Nauk SSSR, Neorg. Mat., Vol. 1, No. 1, p. 121 (1965).
12. R. G. Grebenshchikov, Proceedings of the Seventh Conference on Experimental and Technical Mineralogy and Petrography [in Russian], Izd. Nauka, Moscow (1966).
13. H. Strunz and P. Jacob, Neues Jahrb. Mineral., Vol. 73, No. 4, p. 73 (1960).
14. F. Liebau, Neues Jahrb. Mineral. Abhandl., Vol. 94, p. 1209 (1960).

PHASE TRANSITIONS AND CHANGES IN SEVERAL ENGINEERING PROPERTIES OF HIGHLY REFRACTORY MATERIALS FROM THE $MgO-Al_2O_3-Cr_2O_3$ SYSTEM UNDER ACTION OF CaO AND Fe_2O_3

Ya. V. Klyucharov

For some time considerable attention has been very routinely paid to studying the properties of highly refractory oxides and their combinations. Interest has recently increased in those refractories which (for varying reasons) still have no practical uses, but which may have specialized applications in the future, both in the pure state and as additive mineralizers, stabilizers, or regulators of a desired trend in certain component properties.

At the same time, great significance has been attributed to studying the ultrapure refractory oxides, the properties of which display marked improvement over the same materials of normal purity. In both instances we are primarily discussing materials which will in part be used for specialized tasks.

Moreover, many problems of prime national significance await scientific answers which will decide the proper direction of development for very important branches of industry involving vast amounts of output (metallurgy, power generation, the cement industry, etc.).

The production of the highly stable spinellide—periclase refractories, which provide high refractoriness and thermal and chemical stability in contact with various reagents (chiefly the oxides of iron, calcium, silicon, and several others) in industrial furnaces and power plants, is foremost among the above problems.

To date a few highly refractory compositions have been found by an empirical approach. These are chiefly of the chromite—periclase and spinellide—periclase type. To some extent they fill the requirements for refractoriness and thermal stability. However, there has been altogether inadequate coverage of their chemistry. The nature and order of the phase transitions have not been precisely defined, and neither has the effect of these transitions on the engineering properties of components during slag abrasion. Lacking, for example, are data on the optimum composition for the spinel portion of the refractory. This prevents the scientific characterization and proper choice of the technology (and, consequently, direction of development) for production of this nature.

Solution of these very important problems is possible only by systematic studies on the corresponding compositions in the pure oxides, i.e., by simply knowing the true molecular chemical affinity for the processes of slag abrasion on refractories. These compositions occur in the $MgO-Al_2O_3-Cr_2O_3$ system on action of Fe_2O_3, CaO, and other oxides as reagent-contaminants.

Over the past years, the refractory technology staff of the Lensoveta Leningrad Technical Institute has devoted a number of publications to studies on this system. This was done according to one plan, using (when possible) the same methods of investigation. These publications have established the nature and order of the phase transitions in the $MgO-Cr_2O_3-Fe_2O_3$, $MgO-Cr_2O_3-CaO$, $MgO-Cr_2O_3-ZrO_2$, and $MgO-Cr_2O_3-SiO_2$ systems. They also clarified the relationship between the phase composition and engineering properties of corresponding compositions. Recommendations of considerable practical value have been made on the basis of these physicochemical and engineering characteristics [1].

Work of this nature is now leaning toward studies on the $MgO-Al_2O_3-Cr_2O_3$ ternary system. Later the studies will consider compositions of this system which involve the action of Fe_2O_3 and CaO. This will allow the chemical affinity for the processes to be described in still closer approximation.

As before, the new studies are predominantly dissertations by degree candidates. However, they include also the research theses of instructors and faculty members. These all follow the same general unified plan and have the same ultimate goal formulated above.

The present paper reports the principal results, to present, of this cycle of investigations.

Several compositions in the $MgO-Al_2O_3-Cr_2O_3$ system (with and without Fe_2O_3 additive) were studied with great detail in the dissertation of S. A. Suvorov. It was determined that at first the individual spinels $MgAl_2O_4$ and $MgCr_2O_4$ both form in this system. Their solid solutions then form at the places of contact. The $Al_2O_3-Cr_2O_3$ solid solutions (which exist in presence of sufficient MgO to form both the individual spinels) form at higher temperatures, and are not observed.

Formation of solid solutions between the two spinels already begins at 800°C: the x-ray pattern of an annealed sample of composition $MgO : Al_2O_3 : Cr_2O_3 = 2 : 1 : 1$ displays the magnesiochromite and spinel solid-solution lines. The $MgAl_2O_4$ lines are somewhat displaced toward the side of higher d/n values.

Three cubic spinel phases are also observed under the microscope. Of these, two have nearly the same indices of refraction, approximating those of $MgAl_2O_4$ and $MgCr_2O_4$.

Intense solid-solution formation begins at 1200°C, when the structure of the spinels and solid solutions is still not normalized. Solid-solution formation is practically complete at 1400°C. Only one solid solution with 0.5% free magnesium oxide exists in the products of annealing the compositions $MgO : Al_2O_3 : Cr_2O_3 = 5 : 4 : 1$ and $5 : 1 : 4$. Earlier completion of the process is evidence of the effective mineralizing action of Cr_2O_3 in the low-temperature spinel synthesis.

The described phase transitions have a large effect on the engineering properties of the anneal products. Thus, the solid solutions in the Al_2O_3-rich portion of the system have considerably more mechanical strength (Table 1, 1700° anneal).

In the presence of a ferruginous melt, as with the solid-phase reactions, the phase transitions terminate in formation of solid solutions between the spinellide phase and the ferruginous reagent.

This is confirmed by the presence of only spinel lines in the x-ray patterns. The disintegration of the spinellide phases of the $MgO-Al_2O_3-Cr_2O_3$ system under action of ferruginous melts is the result of a layer-wise spalling of the solid solutions, especially of those rich in $MgCr_2O_4$. This is to the greatest extent displayed in $MgCr_2O_4$ itself. The complete chemical analysis of the material comprising the spalling portion suggests also that the ratio $MgO : Cr_2O_3$ (and in presence of $MgAl_2O_4$, the ratio $MgO : Al_2O_3 : Cr_2O_3$) remains practically constant (Table 2).

Table 1. Engineering Properties of Spinels

Phase	Composition	Annealing temp., °C	Compres. strength, kg/cm^2	Apparent porosity, %	Change in linear dimensions, %	
					by height	by diameter
I	$5MgO : 4Al_2O_3 : 1Cr_2O_3$ ($MgAl_{1.6}Cr_{0.4}O_4$)	800	75	54.5	+ 0.16	+ 0.14
		1000	42	57.7	+ 0.42	+ 1.00
		1200	19	55.4	+ 0.58	+ 0.58
		1400	13	51.7	− 2.14	− 2.23
		1700	395	28.6	−16.2	−19.4
II	$2MgO : 1Al_2O_3 : 1Cr_2O_3$ ($MgAlCrO_4$)	800	152	52.6	+ 0.07	+ 0.067
		1000	106	56.6	+ 0.35	+ 0.39
		1200	54	53.4	+ 0.26	+ 0.10
		1400	43	52.0	− 0.55	− 0.66
		1700	144	42.1	−10.3	−10.5
III	$5MgO : 1Al_2O_3 : 4Cr_2O_3$ ($MgAl_{0.4}Cr_{1.6}O_4$)	800	426	44.8	+ 0.39	+ 0.27
		1000	429	47.3	+ 0.39	+ 0.34
		1200	181	50.6	+ 0.23	+ 0.26
		1400	18	49.7	+ 1.45	+ 0.46
		1700	105	46.6	− 2.50	− 2.40

The spalling of zones in accordance with their iron oxide saturation is the result of differences in the thermal expansion coefficients; it also causes a decrease in mechanical strength, especially for the $MgCr_2O_4$-rich compositions, which display no plastic flow when cooled under oxidizing conditions.

An increase in $MgCr_2O_4$ content causes formation of a second phase, Fe_2O_3 (or more exactly R_2O_3). This is evidently the result of the poorer wetting by ferruginous magnesiochromite melts. The iron concentration in the solid solution abruptly changes in accordance with the thickness of the reaction layer. This, in turn, relates to the more intense spalling. It is the reason for formation of the thickly lamellar outgrowths on a magnesiochromite specimen exposed to ferruginous reagent.

Thus the $MgCr_2O_4$-rich compositions in the $MgO-Al_2O_3-Cr_2O_3$ system have decreased stability in contact with ferruginous melts. This is a conclusion with a significance difficult to overestimate. This result is confirmed by the data in Table 3.

Table 3 clearly demonstrates the considerable advantages of $MgAl_2O_4$ compositions as compared to predominantly $MgCr_2O_4$ compositions with respect to reaction-zone thickness and, particularly, linear expansion of the reacted layer.

The extremely vigorous action of both Fe_3O_4 and $MgFe_2O_4$ also deserves mention. A $MgFe_2O_4$ melt thus brings about considerable change in the volume of the reacting zones, and to some extent causes the same disintegration of the refractory as does a Fe_3O_4 melt.

The question as to how properties change is of prime significance. This applies especially to the expansion occurring on formation of solid solutions between the spinellide compositions and iron oxides, and during the transition to spinellide—periclase solid solutions with predominant periclase content. The latter can actually serve as models for the phase composition of industrial, heat-resistant superrefractories of the chromite—periclase and spinel—periclase type.

Tests on these models exposed to $MgFe_2O_4$ and Fe_3O_4 melts (crucible method) produced the results in Table 4.

Table 2. Interaction of Ferruginous Melts and Spinellides

Sample comp.	Reagent	Test temp., °C	RO:R₂O₃ ratio in sample	Cr₂O₃:Al₂O₃ ratio in sample	Complete sample analysis, %					Cr₂O₃:Al₂O₃ ratio after test	RO:R₂O₃ ratio after test
					MgO	Cr₂O₃	Al₂O₃	Fe₂O₃	Σ		
MgCr₂O₄	Fe₃O₄*	1750±20	1:3.75	—	10.97	41.76	—	46.51	99.24	—	1:3.81
MgCr₂O₄	MgFe₂O₄**	1750±20	1:3.75	—	14.15	42.08	—	43.43	99.66	—	1:3.75
MgAlCrO₄	Fe₃O₄*	1750±20	1:3.14	1.49:1	13.67	28.51	18.34	39.48	100.00	1.55:1	1:3.42
MgAlCrO₄	MgFe₂O₄**	1750±20	1:3.14	1.49:1	16.90	28.49	18.17	36.39	99.95	1.57:1	1:3.14

*Reagent was introduced as 5 g Fe_2O_3 pellets.

**Results correspond to conditions where the $MgFe_2O_4$ melt contained about 8% excess MgO.

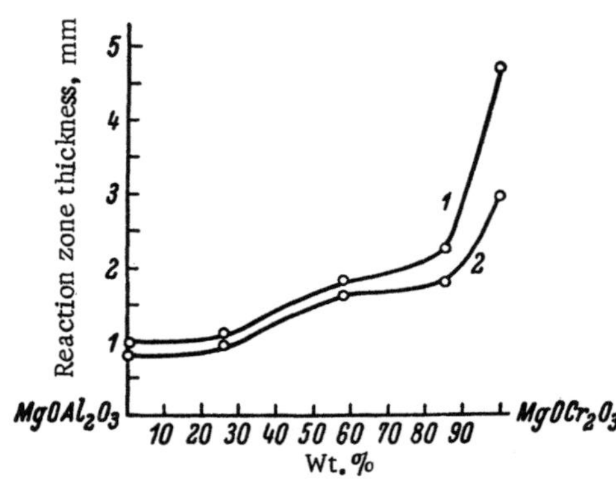

Fig. 1. Change of reaction zone thickness as a function of composition of the spinel phase. 1) During action of a MgO · Fe₂O₃ melt; 2) during action of a Fe₂O₃ melt.

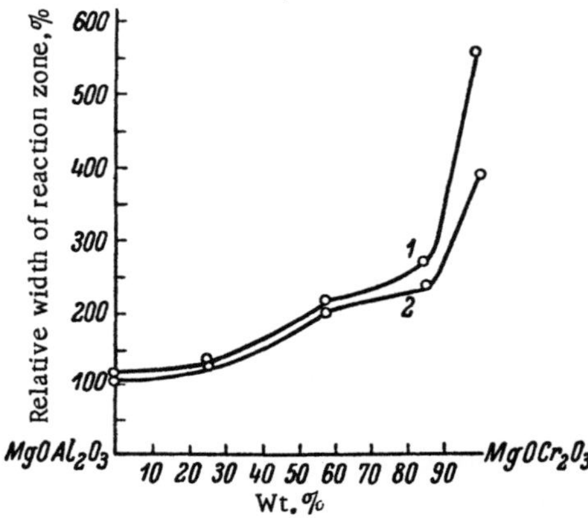

Fig. 2. Relative change of reaction zone thickness as a function of spinel phase composition. 1) During action of a MgO · Fe₂O₃ melt; 2) during action of a Fe₂O₃ melt.

The significant advantages of spinellide—periclase compositions with predominant periclase content become evident on comparison of the data of Tables 3 and 4 on the change of the linear dimensions of the product of annealing. Instead of the considerable (to 55%) expansion of the spinellides there is the moderate (5-7%) shrinkage of spinellide—periclase compositions.

Figures 1 and 2 show the results of measurements on the reaction zone thickness as a function of composition of the spinellide phase in spinellide—periclase samples (30% spinellide and 70% periclase), after action by ferruginous reagents.

Table 3. Action of $MgFe_2O_4$ and Fe_3O_4 on Spinellides

Reagent	Crucible comp.	Initial crucible wall thickness, mm	Thickness of unreacted part of crucible, mm	Wall thickness after annealing, mm	Reaction zone thickness, mm	Linear change of reacted portion, %
$MgFe_2O_4$	$MgCr_2O_4$	5.0	1.1	7.15	6.05	+55
	$MgAl_{0.4}Cr_{1.6}O_4$	5.0	1.2	7.00	5.80	+52
	$MgAlCrO_4$	4.9	0	6.3	6.3	+28
	$MgAl_{1.6}Cr_{0.4}O_4$	4.7	0	5.7	5.7	+21
	$MgAl_2O_4$	4.8	0	5.5	5.5	+11
	MgO	4.8	4.0	4.8	0.8	0
Fe_3O_4	$MgCr_2O_4$	5.0	0	6.0	6.0	+20
	$MgAl_{0.4}Cr_{1.6}O_4$	5.0	0	6.0	6.0	+20
	$MgAlCrO_4$	4.9	0	5.9	5.9	+20
	$MgAl_{1.6}Cr_{0.4}O_4$	4.7	0	5.3	5.3	+12
	$MgAl_2O_4$	4.8	0	5.1	5.1	+ 6
	MgO	4.8	3.9	4.8	0.9	0

Table 4. Results of Tests with the Crucible Method

Crucible comp., wt.%	Pretest properties		Shrinkage after test	
	apparent poros., %	shrinkage, %	$MgFe_2O_4$ reagent	Fe_3O_4 reagent
70% MgO + 30% $MgCr_2O_4$	27.05	5.72	5.0	5.2
70% MgO + 30% $MgCr_{1.6}Al_{0.4}O_4$	27.00	5.90	5.2	5.6
70% MgO + 30% $MgAlCrO_4$	25.51	6.05	6.7	6.5
70% MgO + 30% $MgCr_{0.4}Al_{1.6}O_4$	25.06	6.30	7.1	7.1
70% MgO + 30% $MgAl_2O_4$	24.62	6.45	7.3	7.5
100% MgO	21.71	8.2	9.0	9.0

Table 5. Thickness of Reaction Zone

Crucible comp., wt.%	Relative increase of decomposed zone, %	
	Fe_3O_4 reagent	$MgFe_2O_4$ reagent
30% $MgAl_2O_4$ + 70% MgO . . .	100	110
30% $MgCr_2O_4$ + 70% MgO . . .	420	580

We note that with rising $MgCr_2O_4$ content in the spinellide phase, the reaction zone thickness occurring under action of Fe_3O_4 noticeably increases, from 1 mm for $MgAl_2O_4$ to 3 mm for $MgCr_2O_4$.

$MgFe_2O_4$ causes an analogous but even larger effect. Here the reaction zone thickness in $MgAl_2O_4$ is 1.4 mm, and that in $MgCr_2O_4$ is 4.8 mm.

Finally, the calculated data of Table 5 are convincing, which illustrate the relative (compared to periclase) increase in thickness of the reaction zone thickness as a function of the composition of the spinellide phase.

Briefly formulating the basic results of this portion of the studies, it can thus be stated:

(a) The spinellide phase composition is highly important during its exposure to ferruginous reagents.

(b) The $MgAl_2O_4$ phase has an essential advantage over $MgCr_2O_3$ in this case.

Table 6. Phase Composition After Anneal (wt.%)

Temp., °C	CaO	MgO	Al_2O_3	Cr_2O_3	$CaO\cdot CrO_3$	$9CaO\cdot 4CrO_3\cdot Cr_2O_3$	$CaO\cdot Cr_2O_3$	CA	C_5A_3	$MgCr_2O_4$	$MgAl_2O_4$	Reaction No.	Proposed reactions for interacting phases
Up to anneal	29	8	28	35	—	—	—	—	—	—	—	—	—
700	19	6	28	8	23	—	7	—	—	9	—	1 2 3	$CaO + Cr_2O_3 \rightarrow CaOCr_2O_3$. $2Cr_2O_3 + 4CaO + 3O_2 \rightarrow 4(CaOCrO_3)$. $MgO + Cr_2O_3 \rightarrow MgOCr_2O_3$.
														Reaction No. 2
1000	—	5	8	—	54	—	8	4	13	—	8	4 5 6 7	$4CaO + 2(MgOCr_2O_3) + 3O_2 \rightarrow 4(CaOCr_2O_3) + 2MgO$. $CaO + Al_2O_3 \rightarrow CaOAl_2O_3$. $3(CaOAl_2O_3) + 2CaO \rightarrow 5CaO3Al_2O_3$. $MgO + Al_2O_3 \rightarrow MgOAl_2O_3$.
1200	—	—	—	—	25	13	8	20	9	10	15 Solid solution	8 9 10 11 12	$12(CaOCrO_3) \rightarrow 3(CaOCr_2O_3) + 9CaO4CrO_3Cr_2O_3 + O_2$. $CaOCr_2O_3 + MgO \rightarrow MgOCr_2O_3 + CaO$. $2(MgOCr_2O_3) + 2Al_2O_3 \rightarrow (Al_2O_3Cr_2O_3)_{тв.р.} + (MgOAl_2O_3 \cdot MoOCr_2O_3)_{тв.р.}$ $3(Al_2O_3Cr_2O_3)_{тв.р.} + 12CaO + 3O_2 \rightarrow 3(CaOAl_2O_3) + 9CaO4CrO_3Cr_2O_3$. $5CaO3Al_2O_3 + 2Al_2O_3 \rightarrow 5(CaOAl_2O_3)$.
														Reaction No. 8
1400	—	—	—	—	10	34	6	25	—	11	16 Solid solution	13 14	$3(CaOCr_2O_3) + 3(5CaO3Al_2O_3) + 3O_2 \rightarrow 9CaO4CrO_3Cr_2O_3 + 9(CaOAl_2O_3)$. $9CaO4CrO_3Cr_2O_3 \rightarrow 2(CaOCr_2O_3) + 7CaO + 3O_2$.
1600	—	4	—	—	—	15	35	16	5	8	17 Solid solution	15a 156	$CaO + MgOCr_2O_3 \rightarrow CaOCr_2O_3 + MgO$. $6CaO + 4(MgOAl_2O_3) \rightarrow 5CaO3Al_2O_3 + CaOAl_2O_3 + 4MgO$.

(c) The spinellide—periclase compositions with a high predominance of periclase have just as significant advantages.

(d) There is very intense chemical disintegration of the spinels, not only due to the action of Fe_2O_3 in the solid phase and Fe_3O_4 in melt, but particularly due to the action of $MgFe_2O_4$ melts.

Returning to the same $MgO—Al_2O_3—Cr_2O_3$ system, but now with CaO as the actor, we shall proceed to the experimental data obtained in the thesis work of N. S. Gaenko.

The data of Table 6 illustrate the nature and order of the phase transitions resulting on heating a sample from this system of the composition $CaO:MgO:Al_2O_3:Cr_2O_3 = 2:1:1:1$. Here the phase compositions of the products of annealing are given for the temperatures 700, 1000, 1200, 1400, 1600, and 1750°C. Moreover, these tabular data serve as a basis for the tentative formulation of the formation reactions for the phases which correspond to the phase transitions established experimentally for each temperature.

As seen from the table, heating to 700°C results in about 35% of the calcium oxide forming compounds with chromium oxide: the chromite and chromate of calcium. Some magnesiochromite also forms. On further heating, the chromate continues to form both as a result of the reaction between the calcium and chromium oxides (reaction 2), as well as from decomposition of the magnesiochromite which had formed up to that time (reaction 4). Above 1000°C, the calcium chromate begins to decompose, forming calcium oxychromite.

Formation of the aluminates begins at 1000°C in the presence of sufficient CaO. Evidently, calcium monoaluminate first forms, and then changes to the more stable $5CaO \cdot 3Al_2O_3$ at lower temperatures.

A spinel solid-solution is observed in the 1000–1200°C temperature range. Its occurrence is possible both because of the interaction of magnesiochromite and aluminum oxide (reaction 10), and because of the interaction of both spinels.

Above 1400°C, the CaO forming on decomposition of the oxychromate (reaction 14) vigorously reacts with the spinel solid-solution, transforming the magnesiochromite to calcium chromite and the magnesium aluminum spinel to $5CaO \cdot 3Al_2O_3$ and $CaO \cdot Al_2O_3$ (which transform in the melt and exit from the system, reactions 15a, b).

Magnesium oxide forms as a result of the processes described. At sufficiently high temperatures, substitution of the magnesium by the calcium from the chromium—calcium compounds becomes possible, coupled with the formation of secondary magnesiochromite. However, secondary magnesium aluminum spinel does not result as a consequence of the calcium aluminates being tied up.

The above is the basis for proposing a reversible character for reaction 15a and an irreversible one for reaction 15b.

A very important conclusion can be drawn from this. At high temperatures and with sufficient CaO, the compositions having predominantly magnesium aluminum spinel solid solution produce a large amount of melt, leading to deterioration in the engineering properties of the products of annealing. The high-temperature calcium chromite and magnesiochromite form with the predominance of magnesiochromite solid solution. This has the consequence of improving the engineering properties of such compositions.

We finally note the possibility of reducing the calcium oxychromite content, and thereby lessening its detrimental action on the spinel phase at high temperatures. This is achieved by increasing the amount of chromium oxide in the original oxide mixture. With a bare surplus of chromium oxide, the calcium chromate (having formed at low temperatures) decomposes into the highly stable calcium chromite, with only an insignificant amount of the oxychromite.

This conclusion is apparently only of theoretical significance, since in industrial practice the spinel—periclase refractories are hardly ever planned on the basis of the artificial enrichment of a charge with free chromium oxide.

Briefly formulating the fundamental conclusions for this portion of the studies, we note chiefly that there is a great complexity of phase transitions in the $MgO-Al_2O_3-Cr_2O_3 + CaO$ system. This complexity is the result, particularly, of the variable valence of chromium and the presence of a large number of chromium—calcium compounds, both at low and high temperatures.

A second conclusion, which is of great practical significance, relates to the chemical stability of the spinellide phase (composed of the spinels $MgCr_2O_4$ and $MgAl_2O_4$) in the presence of CaO at high temperatures.

Earlier [2, 3] we showed experimentally that the first of the above spinels is inadequately stable under these conditions, and by studying the respective phase transitions we established the reasons for this fact.

Now, in contrast to the above $MgO-Al_2O_3-Cr_2O_3 + Fe_2O_3$ $(Fe_3O_4 + MgFeO_4)$ system, it is necessary to give a chemical reason of preference for $MgAl_2O_4$ from the standpoint of CaO compatibility; under these conditions, this spinel evidently does not have the advantage of $MgCr_2O_4$. A final discussion on this problem will follow the conclusion of the present studies.

However, it will soon be necessary to consider the simultaneous action of calcium and iron oxides on spinellide—periclase refractories during actual conditions of service. The results of an analogous investigation of the $MgO-Al_2O_3-Cr_2O_3 + CaO, Fe_2O_3$ system (conducted by N. V. Meshalkina and A. S. Grigor'ev) have a decisive significance.

The phase transitions in the magnesium chromium and magnesium aluminum spinels and their solid solutions under action of iron and calcium oxides (coexisting in the combined state) were studied.

The following starting materials were used, which had earlier been synthesized from analytical grade oxides: (a) the spinels $MgAl_2O_4$ and $MgCr_2O_4$; (b) spinel solid-solutions of compositions $MgAl_2O_4 : MgCr_2O_4 = 2 : 1$ and $1 : 2$; and, (c) the calcium ferrites $2CaO \cdot Fe_2O_3$ and $CaO \cdot 2Fe_2O_3$.

Both the equilibrium and nonequilibrium compositions were studied. The latter were exemplified by combinations of predominant amounts of granular periclase and the corresponding spinel solid solutions.

The phase composition of mixtures annealed to 1750°C was studied using microscopic, x-ray, and also thermal analyses on a Kurnakov apparatus. Because of the complexity of this multicomponent system, besides computational methods for defining the phase composition of the products of anneal, a special procedure was used in the chemical phase analysis of compositions which contained simultaneously the solid solutions of both spinels, magnesioferrite, periclase, braunmillerite, as well as the ferrite, aluminate, and chromite of calcium. The scheme of the phase analysis is given on the facing page.

The results obtained so far in this part of the studies can be formulated in the following way:

1. Magnesiochromite is the spinel which is chemically more stable against action of the calcium ferrites $2CaO \cdot Fe_2O_3$ and $CaO \cdot 2Fe_2O_3$. Samples based on it maintain (even after

Phase Analysis Scheme

Batch I

Treatment with 100 ml of boiling 0.075 N HCl for 1 h

Filtrate $\quad\downarrow\quad$ Residue

| MgO_{free} $2CaO \cdot Fe_2O_3$, $\underline{CaO \cdot Al_2O_3}$
 part | Spinel solid solutions, $MgO \cdot Fe_2O_3$,
 $CaO \cdot Cr_2O_3$, $4CaO \cdot Al_2O_3 \cdot Fe_2O_3$, etc. |

Determination of free MgO

Batch II

Treatment with a Ditmar mixture, decomposition in HCl and weight determination of Fe_2O_3

Batch III

Treatment with 100 ml of 10% boiling sodium tartrate for 1 h

Filtrate $\quad\downarrow\quad$ Residue

| Fe_2O_3 in C_2F | $MgO \cdot Fe_2O_3$, etc. |

$\downarrow$

Treatment with 100 ml boiling 7% HCl for 1 h

Filtrate $\quad\downarrow\quad$ Residue

| $MgO \cdot Fe_2O_3$, CaO in C_2F, etc. | Spinel solid solutions, $CaO \cdot Cr_2O_3$ |

Determination of Fe_2O_3, calculation of $MgO \cdot Fe_2O_3$

Treated in a Ditmar mixture, decomposed in HCl

1) Determination of Fe_2O_3 in solid solutions
2) Determination of CaO and calculation of $CaO \cdot Cr_2O_3$
3) Calculation of C_2F content: $Fe_{2}O_{3\,grav} - Fe_2O_{3\,(in\ MF)} - Fe_2O_{3\ (in\ sol.\ solution)} = Fe_2O_{3\ (in\ C_2F)}$
4) Calculation of content of solid solutions: solid solution = (undissolved residue $-$ $CaO \cdot Cr_2O_3$)

Batch IV

Treatment (cold) for 1 h, 15 min in 100 ml 1 N acetic acid

Filtrate $\quad\downarrow\quad$ Residue

| $CaO \cdot Al_2O_3$ and others (incomplete) | C_4AF, spinel solid solutions, etc. |

Determination of Al_2O_3, Calculation of $CaO \cdot Al_2O_3$

Treatment with 100 ml boiling 5% HCl for 1 h

Filtrate $\quad\downarrow\quad$ Residue

| $4CaO \cdot Al_2O_3 \cdot Fe_2O_3$, $MgO \cdot Fe_2O_3$,
 $2CaO \cdot Fe_2O_3$ | Spinel solid solutions, $CaO \cdot Cr_2O_3$ |

Determination of Al_2O_3, calculation of $4CaO \cdot Al_2O_3 \cdot Fe_2O_3$

anneal at 1750°C) high values of the engineering property indicators even with reagent contents up to 50%. This conclusion completely agrees with phase transitions occurring in the magnesio-chromite—calcium ferrite system during heating. Phase-chemical and petrographic analyses of materials annealed at 1600-1750°C revealed only the high-temperature compounds — solid solutions of magnesiochromite with iron oxide, calcium chromite (2170°), and magnesioferrite (1750°C).

2. Magnesium aluminum spinel is less stable against the calcium ferrites. Even with 30% reagent content, after anneal at 1600-1750°C the samples fuse and lose their original shape. Their porosity sharply decreases due to the formation of such high-temperature compounds as the solid solutions of the spinels with iron oxides and magnesioferrite, calcium mono-aluminate (1600°C), and a considerable amount of braunmillerite (1415°C).

3. For the same reasons the spinel solid-solution having predominantly magnesiochromite proved more stable against the calcium ferrites.

4. The dicalcium ferrite is the more active reagent with respect to its action on magnesium aluminum spinel and solid solutions of the compositions $MgAl_2O_4 : MgCr_2O_4 = 2:1$ and $1:2$.

The phase transitions are analogous to the interaction of the above solid solutions with the calcium ferrites. A larger amount of braunmillerite and calcium aluminate forms when there is a prevalence of $MgAl_2O_4$ spinel in the solid solutions, or of CaO in the reagent.

CONCLUSIONS

Completely summarizing the present discussion, we note:

1) the high complexity of phase transitions in the systems studied, due to a) the variable valence of some of their components, b) a considerable number of newly forming materials of sharply differing composition and properties, as well as c) the presence of solid solutions of variable concentration;

2) the very large importance of the spinel phase composition, as well as the spinel content, in the spinellide—periclase compositions;

3) the outstanding and very apparent differences in the chemical stability of specific spinels and their combinations in contact with ferruginous, calciferous, and calciferous—ferruginous reagents;

4) the possibility of using the data obtained to judge the chemical affinity during the slag abrasion of different type spinellide—periclase refractories for corresponding conditions of service.

LITERATURE CITED

1.   Ya. V. Klyucharov, Phase Transformations and Properties of High-Temperature Compositions Based on the Systems $MgO—Fe_2O_3—Cr_2O_3$, $MgO—CaO—Cr_2O_3$, $MgO—Cr_2O_3—SiO_2$, and $MgO—Cr_2O_3—ZrO_2$, in: Silicates and Oxides in High-Temperature Chemistry [in Russian], Izd. Vses. Khim. Obshch. im. D. I. Mendeleeva, Moscow (1963).
2.   Ya. V. Klyucharov and V. G. Eger, Synthesis and Properties of Chromium—Calcium Compounds [in Russian], Trudy Leningr. Tekh. Inst. im Lensoveta, Vol. II (1961).
3.   Ya. V. Klyucharov and V. G. Eger, Zh. Prikl. Khim., Vol. 35, p. 1916 (1962).

CRYSTALLIZATION OF MELTS IN
THE BeO–Al$_2$O$_3$–SiO$_2$ SYSTEM

I. A. Dmitriev and E. G. Semin

Study of the compounds forming in ternary silicate systems and of the questions and characteristics concerning their crystallization is of great scientific and practical interest. Of like interest is the effect of various catalyzing impurities on the course of crystallization and the elemental states in these systems. This particularly applies to the little-studied BeO—Al$_2$O$_3$—SiO$_2$ system. This system must be investigated to define the synthesis conditions and properties of new glass—crystalline materials which are of possible use in the ceramic industry.

The present work is devoted to studying the crystallization and component state problems in quenched melts of the BeO—Al$_2$O$_3$—SiO$_2$ system. These melts have a composition approximating that of the natural mineral beryl (3BeO · Al$_2$O$_3$ · 6SiO$_2$).

In selecting as the catalyzing additives Fe, Ni, Mn, and Ti, we considered the fact that such are among those transition elements often found as impurities in the mineral beryl [1].

The melting, quenching, and secondary anneal of the fused samples were performed according to a technique described earlier [2]. The transition-metal additives were introduced as their chemically pure oxides. The resulting beryl glass is a glass—crystalline material composed of glassy and finely crystalline phases in the respective amounts 95-98 and 2-5%. The introduction of catalyzing-additive mineralizers results in a more thorough crystallization of the beryl melt during the heating and quenching process. After this the finely crystalline phase is uniformly distributed throughout the material volume, which then contains 30-50% of this phase.

Behavior of Silicon Dioxide. Careful choice of x-ray exposure technique enabled identification of a portion of the crystal phase in an unmineralized beryl melt as the more symmetrical, high-temperature modification of SiO$_2$ (α-tridymite), with the interplanar spacings 4.3007, 4.0000, 3.7806, and 1.4600 Å.

Introduction of the transition-metal oxides to the melt promoted recrystallization of α-crystobalite (primarily), α-tridymite, or the two together. This established fact indicates that the phases primarily crystallizing in a pure beryl melt, and in a melt catalyzed by transition metals, are the more symmetrical, high-temperature modifications of SiO$_2$, whose structures approximate the structure of the glass melt [3].

The heat-treatment of quenched beryl melt samples at 900°C results in the recrystallization of the metastable and stable forms of SiO$_2$. All of these phase transitions occur according to a stepwise rule [4].

It should be noted that the presence of various catalyzing additives in the melt to some degree stabilizes certain metastable modifications of SiO_2, affecting the kinetic factors and lowering or increasing the polymorphic transformation temperature. This is apparently associated with the crystallochemical nature and interaction of the additive introduced to the SiO_2.

Behavior of Beryllium Oxide. Study of the formation conditions of the metastable phases resulting with high supercooling is of great significance in the crystallochemistry of the silicates, as it relates to the course of crystallization and to the properties of glass—crystalline materials. It was found while studying the properties of BeO-containing glasses that BeO is an intermediate oxide (BeO, ThO_2, etc.) between the oxides which are typical glass formers (SiO_2, P_2O_5) and those which do not form the glassy state (Nb_2O_5, Ta_2O_5) [5-7]. Under certain conditions (temperature, composition, component ratio), BeO can act as a glass former, joining together the silicon—oxygen radicals.

Morgan and Hummel, in studying the BeO—SiO_2 system, showed that a homogeneous liquid forms in the SiO_2-rich portion of this system at temperatures above 1700°C. Slow-cooling results in decrease of the liquid and in crystallization of the constituent oxide melts.

Silicon dioxide enrichment is also observed in melts of the composition $3BeO \cdot Al_2O_3 \cdot 6SiO_2$ — a homogeneous liquid forms at temperatures above 1700°C. When the melt is cooled, the liquid separates into its constituent oxide melts. These interact with each other, and crystallization of both the oxides and the binary compounds occurs. Rapid cooling of the melt (quenching) promotes formation of the beryl glass. Part of the BeO enters the glass structure and acts as glass former, while the other part, as a result of the high supercooling, separates from the melt structure and is uniformly distributed throughout as a finely crystalline phase — evidently free BeO.

American researchers believed that BeO forms a metastable solid solution with SiO_2, which on heat treatment decomposes to BeO and SiO_2 [9], however, we did not observe the change of SiO_2 crystal lattice parameters which would relate to solid solutions. Study of the properties of quenched beryl melts showed that the phase which BeO enters is metastable. On heating, this metastable glassy phase decomposes to form BeO and SiO_2.

Interpretation of the x-ray patterns of a melt not containing additive showed that part of the BeO enters in the composition as the binary compound $BeO \cdot Al_2O_3$ (chrysoberyl). Increasing the fusion temperature above 1900°C did not retard formation of this compound — even on heating and quenching the beryl glass. Introduction of Ti, Fe, Ni, and Mn additives prevents formation of chrysoberyl during heating and quenching. Chrysoberyl formed on heat treatment at 900-1000°C (with Ti, Mn, Ni additives). Formation of chrysoberyl was not observed in melts catalyzed by iron, even after prolonged high-temperature heat treatment (1140°C, 20 h).

Formation of phenacite ($2BeO \cdot SiO_2$) was also observed in heat-treated melts catalyzed by all the above additives. With manganese additives it even formed during heating and quenching [2]. Formation of phenacite did not occur in an uncatalyzed melt. This completely agrees with the conclusions of other authors regarding the impossibility of synthesizing phenacite without mineralizers [8, 10, 11].

Behavior of Aluminum Oxide. The Al_2O_3 phases first crystallizing are also the more symmetrical, high-temperature modifications. The low-temperature phase state is reached stepwise via intermediate metastable states. Besides crystallization of Al_2O_3 from the melt and recrystallization of the high- into the low-temperature form, the formation and crystallization of the binary compounds sillimanite and mullite were observed after a 900°C heat treatment of the melt. Noticeable interaction of Al_2O_3 and BeO is observed above 1100°C.

It is interesting to note that introduction of Ti^{4+} ions to the melt causes no apparent effect on mullite formation. Introduction of Mn^{4+} (which is reduced to Mn^{2+} during the course of heating the glass) accelerated mullite formation [2]. Additives of Fe^{3+} and Ni^{2+} prevent mullite formation; mullite did not form (even with heat treatment above 1100°C), but formation of sillimanite ($Al_2O_3 \cdot SiO_2$) was noted. An analogous impression was obtained by Budnikov et al. [12] while studying the effect on mullite formation of small amounts of transition-metal additives. They showed that the degree of mullite formation increases with the presence of Group II ions in the glass. The effect of tetravalent metal ions (Ti^{4+}, Mn^{4+}, etc.) on the mullite formation process is nonexistent, though Fe^{3+} and Ni^{2+} ions actually slow mullite formation.

CONCLUSIONS

1. Melts of the $BeO-Al_2O_3-SiO_2$ system corresponding in composition to the mineral beryl were studied.

2. New glass—crystalline materials based on this system were obtained.

3. The behavior of oxides in the system during crystallization of beryl glass was shown.

4. Crystallization of beryl melts catalyzed by transition metal oxides was studied.

LITERATURE CITED

1. V. G. Feklichev, Beryl, Izd. Nauka, Moscow (1964).
2. E. G. Semin, I. A. Dmitriev, V. N. Strekalovskii, and V. S. Bykovskii, Izv. Akad. Nauk SSSR, Neorg. Mat., Vol. 1, No. 11, p. 2026 (1965).
3. V. N. Filipovich, in: The Structure of Glass, Vol. 3, Consultants Bureau, New York (1964), p. 9.
4. W. Eitel, Physical Chemistry of the Silicates, University of Chicago Press (1954).
5. A. A. Appen, Directory of Dissertations for the Degree of Doctor of Chemical Sciences, Institute of Silicate Chemistry, Academy of Sciences of the USSR, Moscow (1953).
6. I. Naray-Szabó, in: The Structure of Glass, Vol. 6, Consultants Bureau, New York (1966), p. 67.
7. I. A. Bondar', in: Structural Transformations in Glasses at Elevated Temperatures, Izd. Akad. Nauk SSSR, Moscow (1965), p. 120.
8. R. A. Morgan and F. Hummel, J. Am. Ceram. Soc., Vol. 32, No. 8, p. 250 (1949).
9. C. W. Schwenzfeier, Beryllium [Russian translation], IL, Moscow (1963).
10. P. P. Budnikov and A. M. Cherepanov, Dokl. Akad. Nauk SSSR, Vol. 74, No. 5, p. 1011 (1950).
11. E. N. Isupova and É. K. Keler, Zh. Neorg. Khim., Vol. 5, No. 5, p. 1126 (1960).
12. P. P. Budnikov, T. N. Keshishyan, and A. V. Volkova, in: Silicates and Oxides in High-Temperature Chemistry, Izd. Vses. Khim. Obshch. im. D. I. Mendeleeva, Moscow (1963), p. 233.

SYNTHESIS AND PROPERTIES OF
CORDIERITE REFRACTORY MATERIALS FROM
AGALMATOLITE

N. A. Sirazhiddinov

Cordierite-base materials, in view of their low coefficient of thermal expansion, find wide application in various branches of technology, particularly in structures operating under conditions of an abrupt drop in temperature [1-5].

The small thermal expansion of cordierite is due to the nature of its crystal structure, which lends very high thermal stability to this material. It is known from literature data that cordierite is surpassed in thermal stability only by quartz glass and several forms of lithium ceramics [6].

The contemporary studies on preparing new cordierite-base materials proceed along two paths: (1) catalyzed crystallization from glass, and, (2) solid-state reactions.

The greatest portion of the industrial bulk obtained by solid-state reaction consists of talc, clay and kaolin, sierrolite, sepiolite, kyanite, and magnesium carbonate. Their compositions occur in the primary crystallization field of cordierite in the $MgO-Al_2O_3-SiO_2$ system [5-16].

The goal of the present work was to synthesize cordierite from a domestic mineral raw material (agalmatolite of the Aktashkii deposit of the UzbekSSR) and to study the mechanism of its formation in the solid state.

A chemical analysis of agalmatolite, given below, showed that its Al_2O_3 and SiO_2 content corresponds exactly to a point symbolic of cordierite in the $MgO-Al_2O_3-SiO_2$ system.

Chemical Composition of Agalmatolite

SiO_2	TiO_2	Al_2O_3	Fe_2O_3	CaO	MgO	C.L.* (1000°)	W (110°)	Σ
51.86	0.04	34.36	0.11	0.30	0.27	12.68	0.43	100.05

*Calcination loss.

Consequently, it is necessary to add 13.7% MgO to agalmatolite to produce cordierite.

Agalmatolite is a cryptolaminated, dense variety of pyrophyllite, and is difficult to pulverize because of its tendency to cleave. Thus the agalmatolite starting material was reduced to a grain size of less than 1 mm, after which it was subjected to wet grinding with MgO for a period of 25 h. After dehydration and drying the cordierite mass was investigated by the DTA method. Then the mass, pressed into the form of pellets, was subjected to heat treatment in

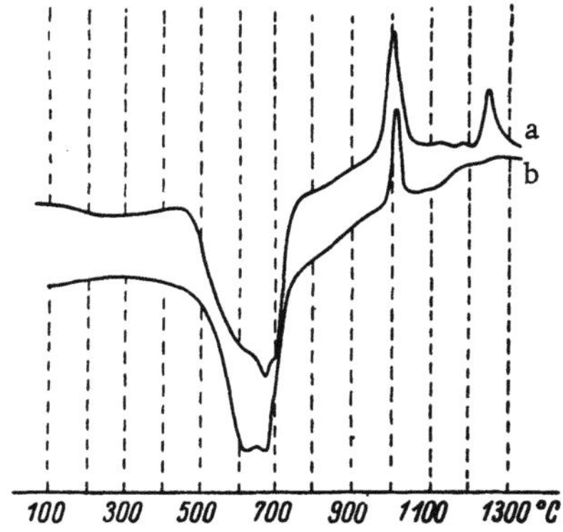

Fig. 1. Heating curves. a) Mixture of cordierite composition; b) agalmatolite from the Aktashkii deposit.

the temperature range 1000-1350°C. The specimens were investigated by x-ray and microscopic methods at 100° intervals.

Figure 1 shows the heating curves of agalmatolite and the original mixture of the cordierite composition, obtained at the Solid-Phase Reactions Laboratory of the Institute of Silicate Chemistry. The agalmatolite heating curve has two endothermic effects (at 620 and 685°C) and one exothermic effect (at 1010°C). The first endothermic effect at 620° corresponds to the separation of chemically bound water with the formation of a metastable phase — dehydrated agalmatolite. We first observed and considered the second endothermic effect at 685° as the decomposition to a metastable phase of finely dispersed γ-Al_2O_3 and amorphous SiO_2.

The exothermic effect at 1010°C is caused by reactions forming mullite on interaction of the products of dissociation of γ-Al_2O_3 and SiO_2. Besides the above endo- and exothermic effects, the heating curve of a mixture of cordierite composition displays still another exothermic effect at 1250°C — a result of the vigorous reaction of cordierite formation. Our concept of the nature of the high-temperature exothermic reaction contradicts the data of Sorrel [10], who believes that mullite formation (mullitization) of kaolin occurs at about this temperature (1267°C) rather than formation of cordierite. According to his views, cordierite forms via an endothermic reaction at a temperature above 1000°C. Similar interpretation of the thermal effects appears unfounded and is not consistent with the results of his investigations. The temperature of mullitization determined by numerous studies, is in the temperature range 1000-1100°C. The cordierite formation reaction occurs just after mullitization, and the nature of the reaction appears to be exothermic.

Persuasive facts, confirming the above, were also obtained in the x-ray study. Figure 2 shows the x-ray patterns of a mixture of agalmatolite and MgO, subjected to heat treatment at various temperatures. It is seen from Fig. 2b that the diffraction maxima at 1000°C, corresponding to the first exothermic effect on the heating curve, basically correspond to formation of mullite. Furthermore, the presence of weak and diffuse maxima of this x-ray pattern allow visualization of the presence of crystobalite, formed as a result of the crystallization of amorphous silica.

With increase of temperature above 1000°C, an interaction occurs between the resulting mullite, silica, and MgO, leading to formation of cordierite. The intensities of the mullite x-ray lines gradually attenuate and their number decreases. However, the cordierite lines gradually become more prominent.

At a temperature of 1200°C, near the second exothermic effect, the vigorous process of cordierite formation occurs. The product at 1350°C consists chiefly of crystals of cordierite containing a very slight amount of mullite (Fig. 2b). If the three very weak mullite lines (at d = 2.71, 2.22, and 1.52) are neglected, this x-ray pattern will be identical to the x-ray pattern of the cordierite obtained from the pure oxides at the same temperature but with a more prolonged duration (Fig. 2a).

X-ray analysis, in conjunction with microscopic and DTA data, led to an explanation of the dissociation mechanism of agalmatolite and of the solid-state formation reactions for cordierite. On the basis of the above studies, the processes in the temperature range 600-1350°C can be represented by the following equations:

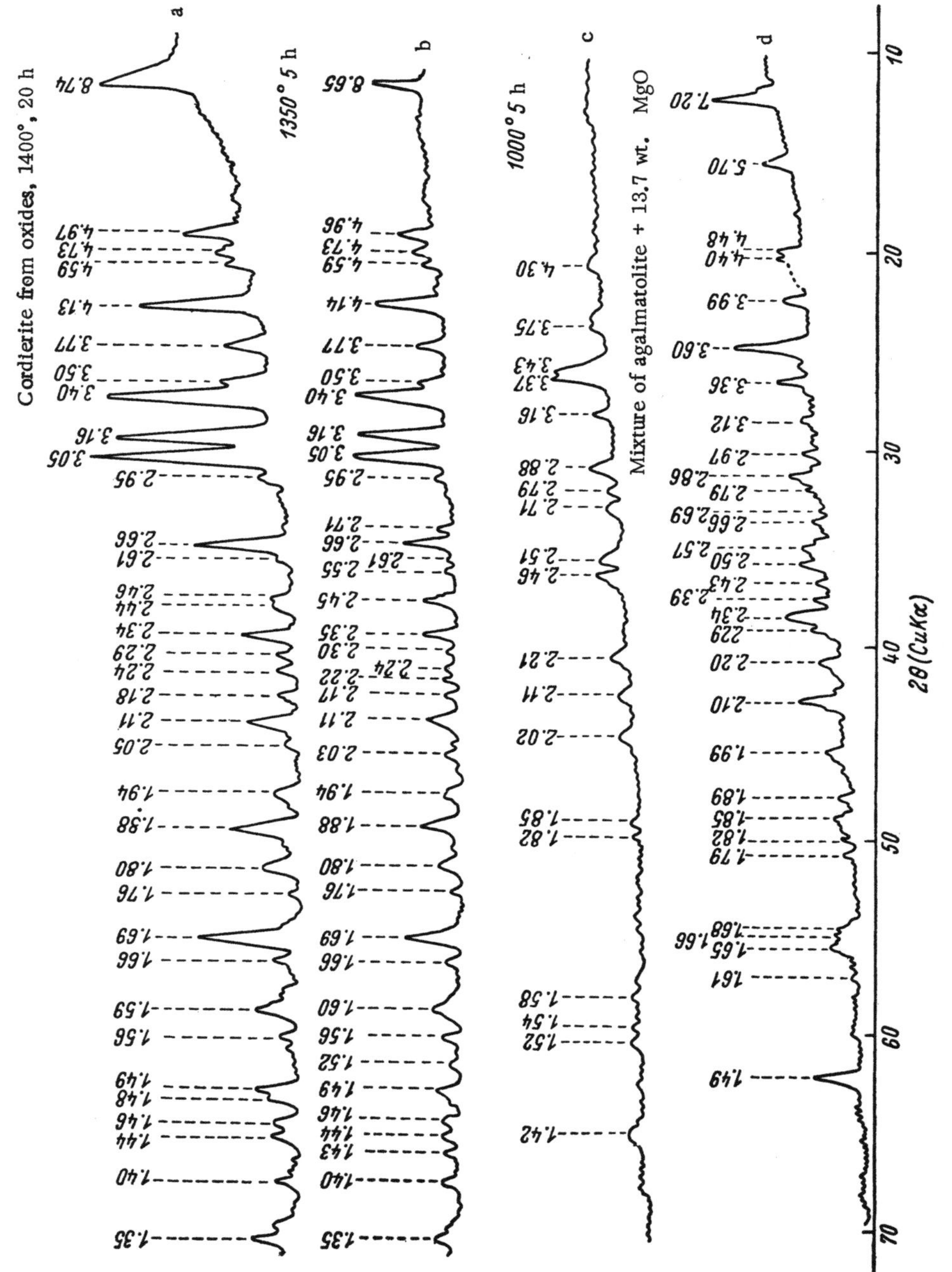

Fig. 2. X-ray patterns of the reaction products of a mixture of agalmatolite + MgO.

Table 1. Several Properties of Cordierite Synthesized Using Agalmatolite

Material	Elastic constants			Bending strength σ_b, kg/cm^2	Coeff. of thermal expansion $\alpha \cdot 10^{-7}$ (20-600°C)
	Poisson's coeff., μ	Shear modulus G, kg/cm^2	Modulus of elasticity E, kg/cm^2		
Cordierite synthesized at 1350° and 5 h........	0.204	$1.17 \cdot 10^5$	$2.82 \cdot 10^5$	127	13
Cordierite synthesized at 1350°, 15 h, and 1200°, 280 h	0.321	$1.24 \cdot 10^5$	$3.28 \cdot 10^5$	285	13.8

Table 2. Change in Shrinkage, Water Absorption, and Specific Gravity of Cordierite as a Function of Temperature (duration of heat treatment 15 min)

Property	1200°	1300°	1380°	1420°
Shrinkage, %	5.4	5.4	6	8
Water absorption, %........	17.4	14.5	7	1.8
Specific gravity, g/cm^3	1.84	1.89	1.97	2.08

$$6(Al_2O_3 \cdot 2.5SiO_2 \cdot 2H_2O) \overset{620°}{=} 6(Al_2O_3 \cdot 2.5SiO_2) + 12H_2O, \tag{1}$$

$$6(Al_2O_3 \cdot 2.5SiO_2) \overset{665°}{=} 6\gamma\text{-}Al_2O_3 + 15SiO_2 \overset{1000°}{=} 2(3Al_2O_3 \cdot 2SiO_2) + 11SiO_2, \tag{2}$$

$$2(3Al_2O_3 \cdot 2SiO_2) + 11SiO_2 + 6MgO \overset{>1000°}{=} 3(2MgO \cdot 2Al_2O_3 \cdot 5SiO_2), \tag{3}$$

$$6(Al_2O_3 \cdot 2.5SiO_2 \cdot 2H_2O) + 6MgO = 3(2MgO \cdot 2Al_2O_3 \cdot 5SiO_2) + 12H_2O. \tag{4}$$

The total reaction (4) includes the dehydration of agalmatolite (1), its dissociation to the free oxides and mullite formation (2), and the final formation of cordierite (3).

It was thus determined that formation of cordierite from a mixture of agalmatolite and magnesium oxide occurs through an intermediate phase − mullite. We did not observe cordierite formation through a spinel phase by the reaction $2MgAl_2O_4 + 5SiO_2 \rightarrow Mg_2Al_4Si_5O_{18}$, as established in a number of works [2, 7]. However, spinel was present in small quantities in the final reaction product as impurity. This only occurred at elevated temperatures.

The polymorphic transformations and structural changes in the synthesized cordierite were also studied. It was established that the primary phase forming is a metastable, high-symmetry (hexagonal) cordierite with disordered structure, which (only after prolonged heat treatment, 250 h at 1200°C) transforms through an intermediate structural state to a stable, ordered, low-symmetry (orthorhombic) cordierite.

The following properties of the synthesized cordierite were studied: elasticity, coefficient of linear thermal expansion, mechanical strength on bending, water absorption, and specific gravity. The data obtained are given in Tables 1 and 2.

The most interesting property of cordierite is the coefficient of thermal expansion. As seen from Table 1, it is quite low. A small increase with duration of synthesis is observed on comparison of the coefficients of linear thermal expansion of cordierite synthesized at 1350° over a 5-h period and cordierite synthesized at 1350° for 15 h and 1200° for 280 h. This is evidently explained by the formation of some amount of a glassy phase, caused by an impurity in the agalmatolite. The results of the bending tests showed that there is a two-fold strength increase after prolonged heat treatment. The determination of specimen shrinkage as a function of temperature showed a small change in linear specimen dimensions up to a temperature of 1380°C. This appears important for obtaining pieces with no deformation.

CONCLUSIONS

1. Cordierite was synthesized using agalmatolite from the Aktashkii deposit. It has a low coefficient of thermal expansion and other interesting properties.

2. The possibility was expressed for use of agalmatolite in obtaining new thermally stable cordierite-base materials.

LITERATURE CITED

1. R. M. Zaionts and G. G. Ul'yanova, Steklo i Keramika, No. 10, p. 25 (1963).
2. G. F. Pankratova, D. N. Poluboyarinov, and R. M. Zaionts, Ogneupory, No. 2, p. 73 (1960).
3. M. B. Gutman, G. A. Kuznetsova, and O. I. Rozhdestvenskii, Élektrotermiya, No. 8, p. 20 (1963).
4. P. W. Lee, Ceramics, New York (1961), p. 50.
5. E. Gugel and H. Vogel, Ber. Dtsch. Keram. Ges., Vol. 41, No. 3, p. 197 (1964).
6. H. H. Reh, Sprechsaal Keramik, Glas, Email, Silik., Vol. 97, No. 7, p. 145 (1964).
7. A. S. Berezhnoi and L. I. Karyakin, Dokl. Akad. Nauk SSSR, Vol. 75, No. 3, p. 423 (1950).
8. V. G. Avetikov, Dokl. Akad. Nauk SSSR, Vol. 58, No. 8, p. 1719 (1947).
9. P. P. Budnikov, V. G. Avetikov, and A. A. Zvyagil'skii, Dokl. Akad. Nauk SSSR, Vol. 81, No. 5, p. 883 (1951).
10. C. A. Sorrel, J. Am. Ceram. Soc., Vol. 43, p. 337 (1960).
11. R. S. Lamar and M. F. Warner, J. Am. Ceram. Soc., Vol. 37, No. 12, p. 602 (1954).
12. M. P. Davis and W. S. Hackler, J. Am. Ceram. Soc., Vol. 40, No. 6, p. 362 (1961).
13. I. Garcia-Vicente and I. Robredo, Science of Ceramics, Vol. 1, p. 277 (1962).
14. I. H. Koenig, Industr. and Engr. Chem., Vol. 40, p. 1782 (1948).
15. J. Mazanec, Sklář a Keramic, Vol. 12, No. 10, p. 309 (1962).

STRUCTURAL STUDY OF
CALCIUM OXIDE CRYSTALS

V. D. Barbanyagre and I. G. Luginina

The present work is a continuation of studies performed by the authors on the structure and properties of certain alkaline-earth oxides [1-3]. This study is devoted to a description of the structure of calcium oxide, CaO, in order to determine the changes occurring in the structure of CaO when it is subjected to different heat treatments.

The CaO was obtained by annealing relatively pure (<0.5% impurity) calcite single crystals from the Kazy-Kurtskii deposit (Southern Caucasus). The annealing was performed in a Silit laboratory furnace at 900-1400°C in 100° intervals by two regimes, gradual and abrupt. In the gradual anneal, the original material was heated at a rate of 7°/min. The abrupt anneal consisted of a 30-min firing at a given furnace temperature with subsequent rapid air cooling. The material was annealed in the form of pressed pellets, 70 mm in diameter and 3-4 mm thick. This specimen shape was dictated by the technique selected for the electrical measurements.

The low-temperature electrical conductivity and the dielectric permeability (DP) of CaO were measured in order to investigate the structure of this material. Precise lattice parameter determinations and direct observation of the CaO surface using an electron microscope were performed.

Calcium oxide nominally belongs to the class of dielectrics. It is known, however, that true crystalline dielectrics are not ideal insulators. On application of a field they display even at room temperature some electrical conductivity which is the result of the continuous diffusion of weakly bound ions at defect sites in the crystal lattice [4]. Measurement of the low-temperature electrical conductivity, whose magnitude is determined chiefly by the lattice defect concentration, can give definite information on the material structure.

The electrical conductivity of the annealed specimens (CaO pellets) was measured at room temperature using a direct current-direct deviation method. The electrodes were applied by silver volatilization in vacuum. The technique used for measuring the electrical conductivity of CaO is described in [3].

Numerous measurements revealed that the dependence of the CaO electrical conductivity on temperature and annealing regime (gradual or abrupt) has the form shown in Fig. 1a. The specific resistivity by volume of very porous, low-annealed (900-1000°C) CaO specimens is not considered, since the latter depends on the surface of contact of the grains, the value of which is undetermined. With an increase in annealing temperature from 1100 to 1400°C, the specific

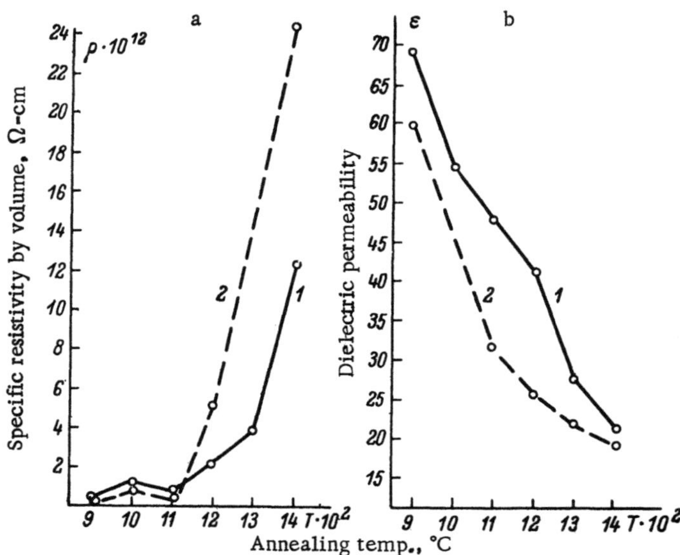

Fig. 1. Effect of temperature and annealing regime on (a) specific
resistivity by volume and (b) dielectric permeability of CaO.
1) Gradual anneal; 2) abrupt anneal.

resistivity by volume of CaO increases an overall two orders of magnitude, and remains this
large for both the abruptly annealed and gradually annealed oxide material.

Additional studies (porosity, crystal size, etc.) established that the decrease in CaO electrical conductivity with increase in annealing temperature corresponds to the net effect of decreasing the internal surface and lattice defect concentration in the oxide. Determination of the
electrical conductivity of CaO using the alternating current and the condenser method (described
by Fleksig [5]) revealed that the specific resistivities by volume obtained using direct and alternating currents agree for CaO within an order of magnitude.

The dielectric permeability was measured in the same specimens in order to determine
the structural changes in the CaO lattice caused by the different heat treatments. The dielectric permeability (DP), which characterizes the polarization of a material per unit of volume,
can give additional information on the material density and the strength or degree of polarization of the ionic bonds in the oxide lattice. Furthermore, the DP of crystals having a strongly
distorted lattice may be considerably higher due to ionic thermal polarization by weakly bound
ions at defect sites in the crystal lattice [6].

The DP was measured by the condenser method. The capacitance of the CaO-pellet specimens (of the order of 8-20 pF) was measured on a "Pimel" apparatus at a frequency of 465
kHz. To reduce stray capacitance in the system, the electrodes were connected to the apparatus by a coaxial cable. In calculating the DP, a formula derived for a plate condenser having
circular electrodes was used [7]. In view of a significant correlation with the density of the
CaO specimens studied (30-80%), the mathematical treatment of the DP values obtained for the
specimens was performed with consideration of the effect of the air in their pores, using the
following logarithmic equation (the DP values of air are added to the DP of the material itself
[8]):

$$\log \varepsilon = x \log \varepsilon' + (1 - x) \log \varepsilon'',$$

where ε is the DP of the CaO specimen, ε' is the DP of the first component (CaO), ε'' is the DP of the second component (air), x is the volume concentration of the first component in unit portions, and $1 - x$ is the volume concentration of the second component.

The DP values obtained by such mathematical treatment are shown in Fig. 1b, where it is evident that the DP of CaO constantly decreases with an increase in annealing temperature. However, it is lower for the abruptly annealed oxide than for the gradually annealed material at all annealing temperatures. The experimental DP values for CaO are considerably greater than those in the literature ($\varepsilon_{CaO} = 12$) [9]. The higher DP values of CaO may be due to thermal ionic polarization, which is observed in defect-ionic crystals. A complex composed of a positive ion impurity and a negative vacancy or of a pair of vacancies of opposite sign has a tendency toward relaxation in the CaO lattice (analogous to what occurs in the alkali-halide and silver-halide crystal lattices). This complex forms a dipole as a result of its charge interaction. The DP of an ionic crystal with relaxing complexes (without considering their mutual interaction), is expressed by the formula [6]

$$\varepsilon - n^2 = 4\pi \left[N_0 \frac{(ze)^2}{\omega_0^2 M^*} + N \frac{(ze)^2 b^2}{12kT} \right], \tag{1}$$

where ε is the DP of CaO, η is the index of refraction, ze is the ionic charge, ω_0 is the intrinsic frequency of optical lattice vibrations, $M^* = M_1 M_2 / (M_1 + M_2)$ is the applied mass, M_1 and M_2 are the masses of the positively and negatively charged ions, b is the distance between two equilibrium positions of a weakly bound ion, k is Boltzmann's constant, T is the temperature (°K), N_0 is the number of ion pairs per cubic centimeter, and N is the concentration of the relaxing complexes in the CaO lattice.

The first term in the right part of Eq. (1) gives the elastic ionic polarization contribution to the DP; the second gives the thermal ionic polarization contribution. However, the first term, according to the improved formula of Born–Sigetti [10] and Fröhlich [9], can be written as follows:

$$\varepsilon_s - n^2 = 4\pi \left(\frac{n^2 + 2}{3} \right)^2 \frac{e^{*2} N_0}{M^* \omega_t^2}, \tag{2}$$

where ε_s is the DP of elastic ionic displacement, e^* is the effective electronic charge, $\omega_t = 2\pi c / \lambda_t$ is the intrinsic frequency of transverse vibrations, λ_t is the wavelength of transverse vibrations, and c is the speed of light.

The values of ε and N_0 in formula (1) are experimentally determined for each annealing temperature; the rest are assumed the same for all annealing temperatures. For CaO: $e^* = 2 \cdot 0.76 = 1.52$, $\lambda_t = 27.4 \cdot 10^{-4}$ cm, as taken from monograph [9]; $b = 2\sqrt{2}a$ (where a is the lattice parameter of CaO), according to calculations for the NaCl-type lattice [11].

The relaxing-complex concentrations in the CaO lattice were computed from Eqs. (1) and (2), treating them as impurities (vacancies and vacancy pairs with opposite sign). The results of the computations (Fig. 2a) show that the complex concentration continuously decreases with increase in temperature, and is lower for all annealing temperatures in the abruptly annealed as opposed to the gradually annealed CaO. The continuous decrease of defect concentration in the CaO lattice with increase of annealing temperature observed in the electrical measurements indicates a continuously increasing ordering of the oxide structure. Ordering occurs in the CaO lattice more vigorously with the abrupt as compared to the gradual specimen anneal.

Changes in the CaO lattice were investigated by precise lattice-parameter measurements performed in the back-reflection angle region ($2\theta = 128\text{–}148°$), using the ionization method. The dependence of the CaO lattice parameter on temperature and annealing regime is shown

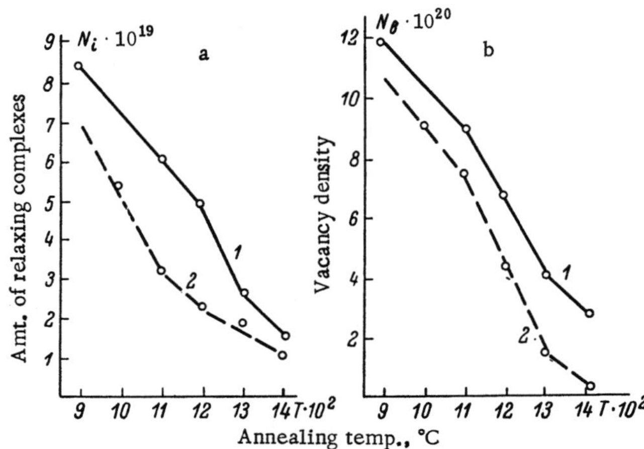

Fig. 2. Dependence of relaxing complex concentration (a)
and vacancy density (b) in the CaO lattice on the temperature
and regime of anneal. 1) Gradual anneal; 2) abrupt anneal.

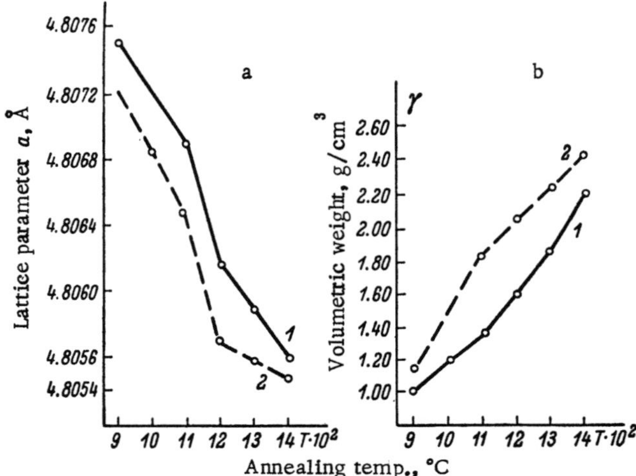

Fig. 3. Effect of temperature and carbonate annealing regime on lattice parameter (a) and volumetric weight (b) of
CaO. 1) Gradual anneal; 2) abrupt anneal.

in Fig. 3a. A continuous decrease is observed for the oxide lattice parameter with an increase
in the annealing temperature. Although this increase is small (0.0004 Å/100°C), the difference
over the temperature limits of anneal (900 and 1400°C) is nonetheless the quite reasonable value
of 0.002 Å. It is usual for the lattice parameter of the abruptly annealed oxide to be smaller
than that of gradually annealed CaO.

Specific gravity measurements on CaO by the pycnometer method taken simultaneous to
the x-ray data enabled determination of the vacancy density in the CaO lattice [12]:

$$\chi_\vartheta = \left(1 - \frac{\rho a^3}{n \bar{A} m_H}\right) 100,$$

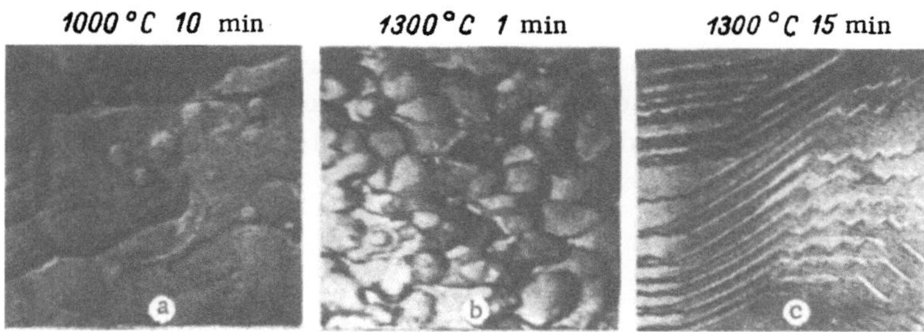

Fig. 4. Electron photomicrograph of the surface of CaO. 15,600 × (preshaded carbon replica). a) 1000°, 10 min; b) 1300°, 1 min; c) 1300°, 15 min.

where χ_ϑ is the vacancy density, %; ρ is the pycnometric density of CaO; a is the lattice parameter of CaO; n is the number of ion pairs per unit cell (n = 4), $\overline{A}$ is the average molecular weight of CaO; and m_H is the atomic mass of hydrogen (m_H = 1.66 · 10^{-24} g).

The vacancy density is readily converted to a vacancy-pair content* in 1 cm^3 (N_b):

$$N_b = 0.25 \cdot \chi_\vartheta N,$$

where N = $1/a^3$ is the number of unit cells in 1 cm^3 and a is the lattice parameter of CaO.

The results of the vacancy density determinations for the CaO lattice (Fig. 2b), show that the general dependence obtained for the vacancy concentration on temperature and annealing regime agrees with the electrical measurements. However, the electrical values are lower by an order of magnitude. The reason for such inconsistency may be the submicroscopic pores which somewhat reduced the pycnometric density of CaO. As a result, a high vacancy concentration is computed on this basis.

The decrease in interatomic spacing on increase of annealing temperature observed by x-ray methods indicates an increase in the energy of the CaO lattice. This process is accompanied by the disappearance (but not by the displacement, as in recrystallization) of subgrain boundaries, and has a dislocation nature [15]. The observations revealed that for a brief (1–5 min) anneal, sintering can run significantly ahead of recrystallization (Fig. 4b) even at relatively high temperatures (1200–1300°C).

The sintering of CaO occurs much more intensely with the abrupt as opposed to the gradual anneal. The defect concentrations computed for the oxide lattice confirm the results of the electrical measurements which suggest a continuous structural ordering of CaO on heating, and the more vigorous CaO ordering processes with the abrupt anneal.

A number of interesting structural characteristics were observed in the crystal morphology of CaO on examination of the surface of a CaO sample with an electron microscope (using the replica method) [13]. It was found that the grain growth in the period preceding recrystallization (Fig. 4a) occurs as a result of coalescence: separate subgrains (0.01–0.1 μ in size) combine into more massive blocks, since this is an extremely nonequilibrium system and self-diffusion processes form the basis for sintering to occur under more favorable conditions. However, on increase in the duration of the high-temperature, abrupt anneal to 15 min, the recrystallization process noticeably accelerates. The intense growth of CaO crystals thereby occurring is accompanied by an ordering of their structure: the total number of dislocations decreases due to annihilation; also single dislocations coalesce into larger ones, forming steps

---

*The number of anionic and cationic vacancies are taken as equal.

(several lattice parameters in altitude) on the crystal surface [14]. After 30 min at temperature, the CaO lattice (as shown above) appears more ordered and denser than with the gradual anneal. The above phenomenon is observed not only in macro- but also in microvolumes: the volumetric weight of abruptly annealed CaO specimens (Fig. 3b) is 15-25% higher than for similar specimens subjected to a gradual anneal.

The data obtained make possible recommendation of abrupt raw material firing when components of increased density are to be produced.

By changing the duration of the anneal-soak period, CaO can be obtained which is very unrecrystallized, more ordered, and denser in micro- and macrovolumes.

CONCLUSIONS

1. Changes occur in CaO with an increase in annealing temperature which lead to stabilization of its structure: its electrical conductivity and dielectric permeability continuously decrease.

2. A gradual decrease is observed in the CaO lattice parameter with an increase in the annealing temperature, attesting to densification and energy increase in the CaO lattice.

3. The defect nature of the CaO lattice was determined by x-ray techniques (as vacancy density) and by measurement of the dielectric permeability, under the assumption that vacancy pairs form relaxing complexes (dipoles) in the CaO lattice. The continuous decrease in the number of point defects in the CaO lattice on increase of annealing temperature, and the decrease in the dislocation density (according to electron microscope data), attest to an intense ordering in the CaO structure.

4. The effect of both temperature and annealing duration on the CaO structure was explained. It was found that the abrupt anneal best changes the CaO structure in both micro- and macrovolumes.

LITERATURE CITED

1. I. G. Luginina, A. I. Saratova, and E. I. Rostovtsev, Izv. Vuzov SSSR, Khimiya i Khim. Tekh., No. 5 (1962).
2. N. A. Toropov and I. G. Luginina, Tsement, No. 2 (1963).
3. I. G. Luginina and V. D. Barbanyagre, Izv. Vuzov SSSR, Khimiya i Khim. Tekh., No. 6 (1963).
4. G. I. Skanavi, Physics of Dielectrics (Weak Field Region) [in Russian], Gostekhizdat, Moscow (1949).
5. V. Fleksig, Electrical Conductivity of Nonmetallic Crystals [in Russian], GONTI, Leningrad (1936).
6. Encyclopaedic Dictionary of Physics, Vol. 4, Moscow (1965), pp. 142-146.
7. Handbook of Electrotechnical Materials, Vol. 1, Part 2 [in Russian], Gosénergoizdat, Moscow (1959), p. 25.
8. A. I. Avgustinik and A. V. Kozlovskii, Zh. Prikl. Khim., Vol. 25, No. 3 (1952).
9. H. Fröhlich, Theory of Dielectrics, Oxford University Press (1958).
10. C. Kittel, Introduction to Solid State Physics, 2nd ed., John Wiley and Sons, New York (1956).
11. A. Lid'yard, Ionic Conductivity of Crystals [Russian translation], IL, Moscow (1962).
12. M. N. Treskina and E. K. Zavodovskaya, Izv. Vuzov SSSR, Fizika, No. 2 (1961).
13. I. G. Luginina and V. D. Barbanyagre, Tsement, No. 2 (1965).
14. Van Bueren, Imperfections in Crystals, 2nd ed., Interscience, New York (1961).
15. J. B. Newkirk and J. H. Wernick, Direct Observations of Imperfections in Crystals, John Wiley and Sons, New York (1962).

PART IV

SILICATE SYSTEMS WITH VOLATILE COMPONENTS

HYDROTHERMAL SYNTHESIS OF
FIBROUS AMPHIBOLES

T. A. Makarova and N. I. Nesterchuk

The production of synthetic fibrous silicates is of scientific and practical interest as applied to the problem of producing and applying inorganic polymeric materials having a number of valuable physiochemical properties. The present work is part of our studies of the synthesis of fibrous silicates, particularly the amphiboles, under hydrothermal conditions.

The structure of the amphiboles, based on infinite double chains of silicon—oxygen tetrahedra, predetermines the fibrous structure of these materials. The chemical composition of the amphiboles is varied, which explains the wide isomorphism characteristic of this group of minerals. The general chemical formula of the amphiboles is $X_{2-3}Y_5Z_8O_{22}(OH)_2$. By complete or partial isomorphous substitution of some ions by others, amphiboles of different compositions which are analogous to natural ones can be produced, and even some that are not. Thus, fibrous synthetic silicates with preassigned properties can be obtained.

The authors succeeded in completely substituting sodium and potassium for group X and magnesium and cobalt for group Y. As a result, three fibrous silicates of amphibole structure were obtained: Na—Mg amphibole, Na—Co amphibole, and K—Mg amphibole. Converting the chemical analyses of the synthetic amphiboles to the crystallochemical formulas, we determined the following composition for the Na—Mg amphibole: $Na_{2.5}Mg_{5.5}Si_8O_{22}(OH)_2$. It crystallizes as white fibers 10^{-5} mm thick and up to 3 mm long. If the syntheses are performed directly in the autoclave without a liner, then the autoclave iron partially substitutes for the magnesium, and a fibrous yellow amphibole of composition $Na_{2.9}Mg_{4.5}Fe^{3+}_{0.6}Si_8O_{22}(OH)_2$ is obtained, while the Na—Co amphibole of composition $Na_{2.5}Co_{5.5}Si_8O_{22}(OH)_2$ crystallizes as rose fibers up to 4 mm long by 10^{-4} mm thick. By substituting potassium for sodium in the Na—Mg amphibole, it was possible to obtain a fibrous K—Mg amphibole.

The Na—Mg and Na—Co amphiboles were obtained in the pure form, i.e., the synthesis product contains up to 98% fiber. The K—Mg amphibole has not yet been obtained in the pure form. We do not carry out a chemical analysis of this amphibole, since the synthesis product also contains, besides potassium amphibole fibers, traces of magnesium hydrosilicate.

The synthesis products of all amphiboles obtained under hydrothermal conditions are an elastic mass of entangled fibers. It is impossible to separate an individual fiber from such a matte, and it is therefore very difficult to determine the length of the fibers. The data quoted for the length of the synthetic fibers were obtained by measuring their length on the specimen surface, where the fibers crystallize in the form of a fuzz.

The amphiboles obtained were studied by crystallo-optical, x-ray, differential thermal, and electron-microscopic methods. The x-ray analysis was performed using comparison of

Table 1. Properties of Synthetic Amphiboles

Properties	Characteristics	$Na_{2.5}Mg_{5.5}Si_8O_{22}(OH)_2$	$Na_{2.5}Co_{5.5}Si_8O_{22}(OH)_2$	K—Mg amphibole
Optical	Ng'	1.591–1.630	1.660–1.682	1.611
	Np'	1.573–1.621	1.649–1.666	1.589
	Ng − Np	0.018–0.009	0.011–0.016	0.022
	Elongation...........	+	+	+
Physical	Fiber length, mm	up to 3	up to 4	up to 4
	Fiber thickness, mm.....	10^{-4}–10^{-5}	10^{-4}	10^{-4}
	Color	White	Rose	White
Thermal	Dehydration temperature,°C	770–800	750	840
	Melting temperature, °C..	1180	1075	1100

powder patterns of the synthetic fibers with the x-ray patterns of natural amphiboles. The indices of refraction of the synthetic amphiboles were determined by the immersion method using bundles of fibers. The melting temperature of the fibers was determined visually on a hot-stage microscope. The table gives the results of these studied.

CONCLUSIONS

1. Fibrous amphiboles of the composition $Na_{2.5}Mg_{5.5}Si_8O_{22}(OH)_2$ and $Na_{2.5}Co_{5.5}Si_8O_{22}(OH)_2$ and a K—Mg amphibole are obtained under hydrothermal conditions at temperatures to 500° and an autoclave fill coefficient of 0.5 to 0.8.

2. It is shown that it is possible to substitute potassium for all of the X group and cobalt for all of the Y group in the amphibole structure.

3. Some of the synthetic amphiboles do not have analogs in nature.

PRODUCTION OF FIBROUS AMPHIBOLES
BY RECRYSTALLIZATION OF SERPENTINE
UNDER HYDROTHERMAL CONDITIONS*

É. N. Korytkova and A. D. Fedoseev

The production of new inorganic materials by the recrystallization of rocks and minerals, and, in particular, the problem of synthesizing fibrous materials through hydrothermal treatment of several natural magnesium silicates (of olivine-serpentine-, dunite-, and peridot-type, etc.) are of scientific and practical interest. Exploratory research of this nature was begun at the Institute of Silicate Chemistry.

Table 1. Chemical Composition of Fibrous Amphibole

Oxides	Content, %
SiO_2	56.24
Al_2O_3	0.85
Fe_2O_3	2.09
CaO	0.25
MgO	26.34
Na_2O	9.22
H_2O	4.59
Total	99.58
Loss at 105°C.	1.13

The mineral chosen as a raw material for producing the fibrous material was one widely distributed in nature — serpentine, or more accurately its varieties of different crystal structure and morphology: antigorite, lizardite, chrysotile, and serophophite.

Samples of these minerals were placed in steel liners after careful grinding. Hydrothermal treatment of the specimens was accomplished in 65-cm^3 stainless-steel autoclaves. The autoclaves were inserted in a vertical muffle furnace with base or wall heaters. The autoclave pressure was found by computational method from the coefficient of water fill in the autoclave.

In the first stage of the work, we tried to reproduce the experiments of Balduzzi et al. [1], who had obtained chrysotile-asbestos by serpentine recrystallization under hydrothermal conditions. The latter researchers treated serpentine with SiO_2 solution from quartz. We repeated these experiments exactly, but did not observe any recrystallization of serpentine under the conditions indicated. Variations of temperature and pressure, an increase in treatment time, and the substitution of silica gel for quartz did not produce the results desired. However, when we began to perform the experiments in an alkaline medium, we were able to obtain some very interesting results.

On hydrothermal treatment of serpentine with sodium silicate, we obtained a fibrous mineral, which upon examination was identified as Na−Mg amphibole. The data of its chemical

---

*Preliminary report.

Table 2. Results of the X-Ray Study of Amphibole

Na−Mg amphibole		Richterite (natural) XRDC (1958)		Na−Mg amphibole		Richterite (natural) XRDC (1958)	
I	d	I	d	I	d	I	d
8	4.53	40	4.52	7	2.17	60	2.17
3	4.07	10	3.96	1	2.03	40	2.03
3	3.87	20	3.85	1	1.95	10	1.95
2	3.54	—	—	—	—	20	1.914
6	3.41	70	3.38	1	1.85	10	1.850
5	3.30	60	3.27	1	1.79	10	1.792
5	3.16	70	3.15	3	1.680	20	1.677
6	3.01	10	3.03	3	1.660	50	1.656
1	2.93	60	2.94	1	1.640	10	1.636
4	2.81	40	2.82	2	1.609	40	1.610
10	2.71	80	2.71	4	1.577	40	1.571
5	2.58	40	2.58	—	—	20	1.546
9	2.53	60	2.53	1	1.532	—	—
4	2.33	50	2.34	6	1.509	60	1.507
6	2.28	50	2.28	2	1.502	—	—

analysis (Table 1) were converted (as oxides) to the crystallochemical formula. The formula obtained is $Na_{2.48}Ca_{0.04}Mg_{5.49}Fe^{3+}_{0.22} \cdot [Al_{0.14}Si_{7.82}O_{22}](OH)_{2.08}$. It is very close to the formula of richterite, $Na_2CaMg_5Si_8O_{22}(OH)_2$. Its optical constants ($Ng = 1.621$, $Np = 1.615$, $Ng - Np = 0.006$, $cNg = 17-19°$) and the x-ray analysis of the resulting mineral (Table 2) confirm it as belonging to that group of asbestos amphiboles. We determined that fibrous amphibole forms from serpentine over a broad range of temperatures (350-650°C) and pressures (400-1500 atm). Thus treatment of serphophite with sodium silicate solution at 350°C and 400 atm also results in formation of very fine fibrous amphibole clusters.

We should remark on a certain aspect of the crystallization of the fibrous product. Clusters always form in the upper part of the liner as longer (∼0.5 mm), thin fibers, while shorter fibers (∼0.1 mm), thicker and less elastic, grow in the lower part. The indices of refraction of these are somewhat lower.

It was impossible to obtain a 100% yield of fibrous material in the experiments with serphophite. Besides amphibole, granules of the initial mineral are always present (up to 10%), together with a partly recrystallized mass, the grains of which display differing degrees of fibrosity. A 100% fibrous product yield was obtained only in the experiments on chrysotile-asbestos, where the conditions indicated resulted in the crystallization of fibrous amphibole (with fiber lengths to 3 mm). The recrystallization of lizardite and that of antigorite occur with more difficulty. A somewhat lesser amount of the fibrous amphibole forms during the hydrothermal treatment of these minerals. Thus the yield of fibrous material in the case of antigorite is only 20%. A considerably larger percentage of fibrous product (∼50%) is obtained with a more prolonged treatment of antigorite at higher temperatures and pressures. It is probable that the stronger, lamellar structure of antigorite is more difficultly rearranged than are the structures of chrysotile and gelatinous serpophite.

In addition to treatment of the above varieties of serpentine with sodium silicate solution, both serpophite and antigorite were treated hydrothermally with Li_2SiO_3 solution. This resulted in a fine, fibrous material, the x-ray and optical examination of which classified it as belonging to the amphibole group of minerals. The fibrous material yield in these experiments attained 50% of the resulting product. The rest consisted of the starting material (digested to varying extents), crystalline lithium metasilicate, and pyroxene. We were still unable to eliminate the latter impurities.

Thus the preliminary explorations we tried concerning the ways and methods of producing fibrous inorganic materials by hydrothermal treatment of several minerals and rocks proved entirely rewarding, and the studies indicate that these materials are due for further development on a broader scale.

LITERATURE CITED

1. F. Balduzzi, W. Epprecht, and P. Niggli, Schweiz. Mineral. und Petr. Mitt., Vol. 31, No. 1, p. 293 (1951).

BEHAVIOR OF
SYNTHETIC FIBROUS FLUORAMPHIBOLES
ON HEATING

O. G. Chigareva and L. F. Grigor'eva

The thermal stability of the amphibole asbestoses is an important property of these minerals. The literature contains much information concerning the thermal behavior of the natural amphiboles [1-5]. At the same time, data on the thermal stability of the synthetic fluoramphiboles is very limited [6, 7].

We studied the thermal properties of fibrous fluoramphiboles of alkaline composition containing cations of differing valence: Cr, Mn, Fe, Ni, Co, and Cu. The natural rezhikite asbestos was studied parallel to these.

The thermal behavior of the amphiboles was studied by means of differential heating curves, weight-loss determinations, and by visual observation under a microscope with a high-temperature camera. The products of anneal were analyzed using microscopic and x-ray methods.

On the basis of the performed studies, we determined that the fluoramphibole thermal decomposition processes (associated with evolution of fluorine) occur slowly. Comparative thermal analyses revealed that the fluoramphibole decomposition rate is 5-8 times below the dehydration rate of their natural analogs. The process of fluorine evolution is accompanied by absorption of heat. However, the endothermic peak on the differential heating curves is not clearly defined and is not always reproducible.

Table 1. Results of a Thermoanalytical Study of Synthetic Fluoramphiboles and a Few Natural Asbestoses

Name	Chemical formula	Temp. range, °C	
		decomp.	melting
Crocidolite (So. Africa)	$Na_2(Fe^{2+}, Mg)_3Fe_2^{3+}Si_8O_{22}(OH)_2$	800—930	No data
Rezhikite asbestos (Urals)	$Na_3(Mg,Fe^{2+})_4Fe^{3+}Si_8O_{22}(OH)_2$	840—910	1010—1040
Ni-fluorichterite	$Na_2Ni^{2+}Mg_5Si_8O_{22}F_2$	960—1000	1070—1110
Co-fluorichterite	$Na_2Co^{2+}Mg_5Si_8O_{22}F_2$	940—1000	1050—1100
Mn-fluorichterite	$Na_2Mn^{2+}Mg_5Si_8O_{22}F_2$	990—1050	1130—1160
Cu-fluoramphibole	$Na_2Cu_{0.5}^{2+}Mg_{5.5}Si_8O_{22}F_2$	910—970	1090—1125
Fe-fluoramphibole	$Na_2(Mg,Fe^{2+})_{5.25}Fe_{0.5}^{3+}Si_8O_{22}F_2$	930—970	1100—1135
Cr-fluoramphibole	$Na_{2.5}Mg_5Cr_{0.5}^{3+}Si_8O_{22}F_2$	1010—1070	1160—1190
Mn-fluorarfvedsonite	$Na_3Mg_4Fe^{3+}Si_8O_{22}F_2$	940—1000	1000—1060

Note. The data for crocidolite are given according to [5]; the remaining data are the authors'.

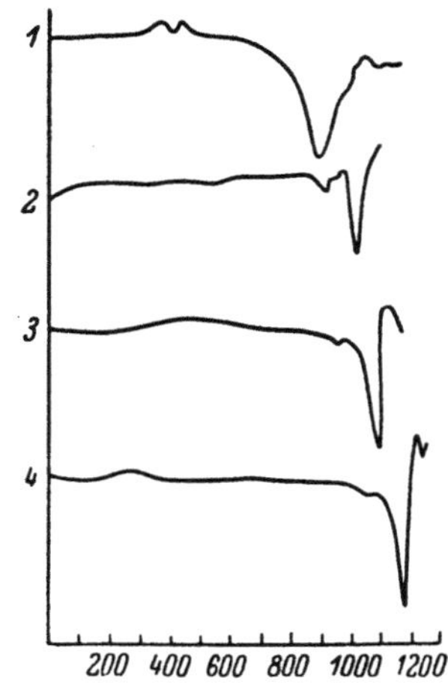

Fig. 1. Differential heating curves of natural and synthetic amphiboles. 1) Crocidolite (So. Africa); 2) rezhikite-asbestos (Urals); 3) Ni-fluorichterite; 4) Cr-fluoramphibole.

The figure shows the heating curves of several of the fluoramphiboles studied, of natural rezhikite-asbestos and of crocidolite. Two endothermic effects are seen on the thermograms: the first results from the evolution of fluorine and decomposition of the mineral; the second is associated with the melting of the decomposition products. Results of the thermogravimetric analyses of the fluoramphiboles showed that their weight loss over the temperature range corresponding to the first effect is about 3%. This amounts to 80% of the total fluorine content of amphibole.

The table gives the temperature ranges for decomposition and melting of all the fibrous fluoramphiboles studied. For comparison, it gives the analogous characteristics for a few natural hydroxyl-containing complexes. As seen from the table, the decomposition temperatures of the synthetic fluoramphiboles are 100-150°C above the dehydration temperatures of the natural amphiboles. On comparison of the decomposition temperatures of the fluoramphiboles, it is not difficult to see that they are positioned in the following ascending series, relative to their degree of thermal stability: Cu-fluoramphibole, Fe^{2+}-fluoramphibole, Ni-fluorichterite, Co-fluorichterite, Mg-fluorarfvedsonite, Mn-fluorichterite, and Cr-fluoramphibole.

The decomposition products of the fluoramphiboles are a mixture of amorphous and crystalline formations. For the case of the Ni-, Co-, and Mn-fluorichterites and the Fe^{2+} fluoramphibole, the crystalline phase is the orthosilicate; for the case of the Mg-fluorarfvedsonite, and the Cr- and Cu-fluoramphiboles, it is the pyroxenes.

Melting of the decomposition products occurs in the 1100-1200°C temperature range, in which the Cr-fluoramphibole proved the most refractory. The cooled melt consists of glass and fine crystallites of forsterite and quartz.

CONCLUSIONS

A thermoanalytical study of synthetic fluoramphiboles revealed that their change with temperature proceeds according to the same scheme as for natural amphiboles. At first fluorine is evolved and the mineral structure disintegrates; then melting of the decay products occurs.

The synthetic fibrous fluoramphiboles have a higher thermal stability than natural hydroxyl-containing amphiboles of similar composition.

LITERATURE CITED

1. A. F. Korzhinskii, Dokl. Akad. Nauk SSSR, Vol. 111, No. 2, p. 445 (1956).
2. A. F. Korzhinskii, Trudy Vost.-Sib. Fil. Akad. Nauk SSSR, Ser. Geol., No. 16, p. 245 (1961).
3. L. I. Ovchinnikov, A. S. Shur, and N. T. El'kina, Proceedings of the First Congress of Thermography [in Russian], Izd. Akad. Nauk SSSR, Moscow (1955), p. 250.
4. N.I. Toker, L.K. Abend, and É.P. Kraineva, Trudy VNII Asbest., No. 3, p. 51 (1962).
5. F.H.S. Vermaas, Trans. and Proc. Geol. Soc. So. Africa, Vol. 55, p. 199 (1952).
6. A.D. Fedoseev, L.F. Grigor'eva, and Z.V. Krupenikova, in: Silicates and Oxides in High-Temperature Chemistry [in Russian], Izd. Vses. Khim. Obshch. im. D. I. Mendeleeva, Moscow (1963), p. 180.
7. A.D. Fedoseev, L.F. Grigor'eva, O.G. Chigareva, Z.V. Krupenikova, and G.A. Rozhnova, Izv. Akad. Nauk SSSR, Neorg.Mat., Vol. 1, No. 11, p. 2031 (1965).

SYNTHESIS AND STUDY OF
CHROMITIC HYDROGARNETS

É. G. Klimenko and V. A. Tikhonov

In 1941, Flint, McMurdie, and Wells [1] attempted to synthesize chromitic hydrogarnets at atmospheric pressure. However, their attempt was unsuccessful and neither were they able to synthesize manganitic hydrogarnets. After successfully synthesizing manganitic hydrogarnets, we set out to prepare the chromitic ones.

This problem is proposed on the following basis. Aluminum hydrogarnet ($3CaO \cdot Al_2O_3 \cdot 6H_2O$), which does not contain SiO_2, is unstable in a sulfate environment. The introduction of SiO_2 with part of the chemically combined water, and especially the introduction of Fe_2O_3 with part of the Al_2O_3, sharply increase the stability of the hydrogarnets in presence of sulfates, and also visibly increase their stability in carbonate environments. On the other hand, the partial or complete substitution of the Al_2O_3 in the hydrogarnets by Fe_2O_3 somewhat lowers their thermal stability. It was accordingly of interest to synthesize hydrogarnets having a R_2O_3 structural element which would be more stable than Fe_2O_3.

Such an element is Cr_2O_3, a highly refractory, dark green material, which is insoluble in water and acids. For this reason, Cr_2O_3 is stable under atmospheric conditions.

A greyish blue precipitate of $Cr(OH)_3$ (difficultly soluble in water) is obtained by the action of ammonia on aqueous solutions of chrome alum. The latter has a very apparent amphoteric character. With acids it gives chromium-oxide salts, while with strong bases it gives salts of chromous acid ($HCrO_2$), known as chromites. In synthesis of the chromitic hydrogarnets, we are interested in a second type of reaction: $Cr(OH)_3 + KOH = KCrO_2 + 2H_2O$.

The hydrate of chromium oxide (chromium hydroxide) can also be produced by the action of strong reducing agents on any Cr^{6+} ion in a neutral or weakly alkaline environment. Thus, on heating, there results the reaction:

$$2K_2Cr_2O_7 + 3(NH_4)_2S + 8H_2O = 2Cr(OH)_3 + 3S + 4KOH + 6NH_4OH.$$

However, $Cr(OH)_3$ is difficultly soluble. Formation of the chromitic calcium hydrogarnets thus occurs very slowly. As starting material for the chromitic hydrogarnet synthesis it is convenient to use chromium chloride as the crystalline hydrate $CrCl_3 \cdot 6H_2O$, which we can dissolve in water. The anhydrous $CrCl_3$ salt is unheard of.

Technique of Synthesis. The syntheses were performed according to the same technique used for the aluminum, ferritic, and ferritic alum hydrogarnets (at 100°C under atmospheric pressure [1-3]). The syntheses were also performed under pressures of 2.5 to 8.0

Table 1. Dependence of Index of Refraction of the Hydrogarnets $3CaO \cdot Cr_2O_3 \cdot mSiO_2 \cdot (6 - 2m)H_2O$ on SiO_2 Content

Formula	Index of refraction	Comment
$3CaO \cdot Cr_2O_3 \cdot 6H_2O$	1.706	Synthesized by us
$3CaO \cdot Cr_2O_3 \cdot 0.22SiO_2 \cdot 5.56H_2O$	1.712	The same
$3CaO \cdot Cr_2O_3 \cdot 0.4SiO_2 \cdot 5.2H_2O$	1.722	» »
$3CaO \cdot Cr_2O_3 \cdot 0.61SiO_2 \cdot 4.78H_2O$	1.733	» »
$3CaO \cdot Cr_2O_3 \cdot 0.8SiO_2 \cdot 4.4H_2O$	1.742	» »
$3CaO \cdot Cr_2O_3 \cdot 0.91SiO_2 \cdot 4.18H_2O$	1.751	» »
$3CaO \cdot Cr_2O_3 \cdot 3SiO_2$	1.870	From [4].

atm. At the end of a synthesis, the solid phase was separated by filtration and twice washed in absolute alcohol. It was then dried under freshly heated calcium chloride. The preparation thus obtained was subjected to microscopic, electron microscopic, and thermographic investigations.

STUDY OF CHROMITIC HYDROGARNETS

A microscopic study of hydrogarnets produced at atmospheric pressure revealed that the chromitic hydrogarnets form in a highly dispersed fashion at 100°C. It is impossible to identify the crystal habit under the microscope. The crystals are distinctly visible with the electron microscope. They are penetration twins of two cubic tetrahedra.

It is possible to obtain comparably more massive chromitic hydrogarnet crystals (to 5 μ) at 175°C and 8 atm pressure. These are also penetration twins of two cubic tetrahedra (Fig. 1).

The chromitic hydrogarnet crystals resemble in shape the ferritic [3] and manganitic hydrogarnets; however, they differ from the latter by being more dispersed.

We note that the aluminum hydrogarnet crystals are the most massive, followed (with decreasing size) by the manganitic, ferritic, and chromitic hydrogarnets.

As for chemical composition, the chromitic hydrogarnets synthesized at 8 atm have the formula $3CaO \cdot Cr_2O_3 \cdot 0.4SiO_2 \cdot 5.2H_2O$. Their index of refraction is approximately 1.722. Besides the hydrogarnet of the above composition, we also synthesized at 8 atm hydrogarnets with various SiO_2 contents and determined their indices of refraction (Table 1).

The index of refraction of a hydrogarnet increases with increase in SiO_2 content (Fig. 2), changing from 1.706 for $3CaO \cdot Cr_2O_3 \cdot 6H_2O$ to 1.870 for the natural mineral uvarovite ($3CaO \cdot Cr_2O_3 \cdot 3SiO_2$). The solid line in Fig. 2 shows the dependence of the index of refraction on the amount of silica in the hydrogarnet molecule. The dashed line shows the dependence proposed for the composition range from $3CaO \cdot Cr_2O_3 \cdot 0.9SiO_2 \cdot 4.2H_2O$ to $3CaO \cdot Cr_2O_3 \cdot 3SiO_2$.

The differential thermal analysis of the chromitic hydrogarnets was performed on a Kurnakov

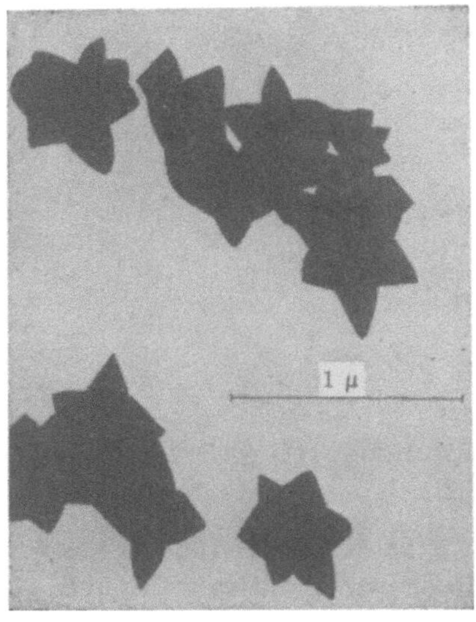

Fig. 1. Electron photomicrograph of the chromitic hydrogarnet $3CaO \cdot Cr_2O_3 \cdot 0.4SiO_2 \cdot 5.2H_2O$.

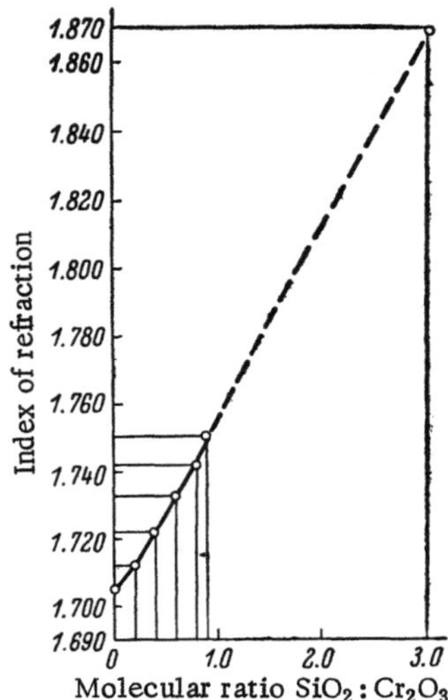

Fig. 2. Dependence of index of refraction on SiO_2 content in the chromitic hydrogarnets $3CaO \cdot Cr_2O_3 \cdot mSiO_2 \cdot (6-2m)H_2O$.

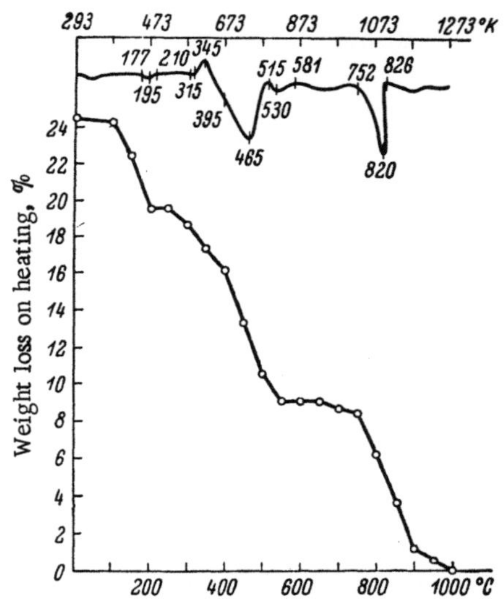

Fig. 3. Thermogram and weight loss on heating for a chromitic hydrogarnet of composition $3CaO \cdot Cr_2O_3 \cdot 0.4SiO_2 \cdot 5.2H_2O$.

Table 2. Weight Losses on Heating a Chromitic Hydrogarnet
of Composition $3CaO \cdot Cr_2O_3 \cdot 0.4SiO_2 \cdot 5.2H_2O$

Temp., °C	Volatile matter, %	Comments
100	24.2	Overall amount of volatile matter is
150	22.4	24.2% Separation of zeolitic water,
200	19.5	in the amount of 24.2 − 19.5 = 4.7%
250	19.5	
300	18.6	
350	17.3	Separation of constituent water (at
400	16.1	300–500°C), in the amount of
450	13.3	19.5 − 8.9 = 10.6%
500	10.5	
550	8.9	
600	8.9	
650	8.9	
700	8.6	
750	8.3	Separation of combined CO_2, in the
800	6.2	amount of 8.9%.
850	3.7	
900	1.2	
950	0.6	
1000	0.0	

pyrometer. A weight-loss-on-heating curve was determined simultaneously with the thermograms.

The thermogram of the chromitic hydrogarnets of composition $3CaO \cdot Cr_2O_3 \cdot 0.4SiO_2 \cdot 5.2H_2O$ is shown in Fig. 3. The weight-loss-on-heating curve is shown in the same figure.

Five endothermic effects are observed in the thermogram (at 177–210°, 305–315°, 350–515°, 515–581°, and 752–826°C), and one exothermic effect (at 330–350°C). The endothermic effect at 177–210°C corresponds to decomposition of the complex compound $3CaO \cdot Cr_2O_3 \cdot mSiO_2 \cdot nH_2O$, resulting in formation of $Ca(OH)_2$ and $Cr(OH)_3$. These evolve H_2O with an endothermic effect at 305–315°C and transform to Cr_2O_3, the crystallization of which gives the exothermic effect at 330–350°C.

The endothermic effect at 350–515°C corresponds to evolution of H_2O from the chromitic hydrogarnets, with their decomposition to Cr_2O_3, calcium silicate, and $Ca(OH)_2$. The dehydration gives the endothermic effect at 515–581°C. Consequently, the decomposition of the chromitic hydrogarnets occurs at a higher temperature than that of the aluminum, ferritic, and manganitic hydrogarnets.

A small amount of $CaCO_3$ forms during the hydrogarnet synthesis due to the interaction of atmospheric CO_2 and CaO. The endothermic effect at 752–826°C corresponds to the decomposition of $CaCO_3$.

The thermograms of the other chromitic hydrogarnets of different SiO_2 content are similar and are therefore not recorded here.

The endothermic effect occurring in the temperature range 350–515°C is thus typical.

The bends in the weight-loss-on-heating curves (Fig. 3) correspond to the endothermic effects observed on the thermogram. It is true that each weight loss occurs at a somewhat lower temperature than the corresponding endothermic effect on the thermogram. The overall weight loss on heating is 24.2%, of which 4.7 and 10.6% are the values respectively for the zeolitic and chemically combined water. About 8.9% is for the CO_2 (Table 2).

Consequently, the chromitic (as well as the aluminum and ferritic) hydrogarnets have partially zeolitic properties.

CONCLUSIONS

The following principal conclusions can be drawn from the studies performed:

1. Chromitic hydrogarnets were for the first time synthesized as $3CaO \cdot Cr_2O_3 \cdot 6H_2O$—$3CaO \cdot Cr_2O_3 \cdot 3SiO_2$ solid solutions.

2. A hydrogarnet of composition $3CaO \cdot Cr_2O_3 \cdot 0.4SiO_2 \cdot 5.2H_2O$ was studied in the above series of solid solutions. The crystal shape and index of refraction were determined, and a thermogram and weight-loss-on-heating curve were obtained.

3. On substitution of SiO_2 for $2H_2O$ in the chromitic hydrogarnet molecule, the index of refraction increases from 1.706 for the $3CaO \cdot Cr_2O_3 \cdot 6H_2O$ hydrogarnet to 1.870 for the $3CaO \cdot Cr_2O_3 \cdot 3SiO_2$ hydrogarnet.

4. Substitution of Cr_2O_3 for the Al_2O_3, Fe_2O_3, or Mn_2O_3 in a hydrogarnet increases its thermal stability.

LITERATURE CITED

1. E. P. Flint, H. F. McMurdie, and L. S. Wells, J. Res. Natl. Bur. Stnd., Vol. 26, No. 1, p. 13 (1941).
2. N. A. Toropov, Chemistry of Cements [in Russian], Promstroiizdat, Moscow (1956).
3. Z. K. Klimenko and V. A. Tikhonov, Synthesis and Study of the Aluminum, Ferritic, and Aluminum Hydrogarnets, First Scientific Conference on Chemistry and Chemical Technology [in Russian], Izd. L'vovsk. Gos. Univ. (1958).
4. I. A. Preobrazhenskii and S. G. Sarkisyan, Minerals of Sedimentary Rocks [in Russian], Gostoptekhizdat, Moscow (1954).

INTERACTION OF
TITANIUM AND GERMANIUM TETRACHLORIDES
WITH HYDRATED SILICA

A. M. Shevyakov, G. N. Kuznetsova,
and V. B. Aleskovskii

The recently widely used chemical modification of the surface of silica and numerous silicates, which generally produces materials of valuable practical properties, is brought about through hydroxyl groups which form under certain conditions of acid or base functions [1]. The general scheme for reactions between hydrated silica and chlorides is

$$xn\,[\equiv Si-OH] + n MeCl_4 \rightarrow xn HCl + n\,[(\equiv Si-O)_x MeCl_{4-x}],$$

where

$$x = 1,\ 2,\ 3;\quad Me = Ge,\ Ti,\ Si.$$

We used the technique described in [2] for studying the reactions between silica gel and $TiCl_4$ or $GeCl_4$. The silica gel was exposed to the chloride vapors in a current of dry air. The optimum exposure had been previously determined for each reagent. The reactions were conducted at 180°C, without physical adsorption of water on the SiO_2 surface. The HCl forming in the reactions and any surplus chlorides was carried away by the dry air. The chlorine groups forming on the surface were purged with water vapor until cessation of HCl evolution. On completion of hydrolysis, the silica gel was dried at 180°C and the exposure—purging—hydrolysis—drying cycle was again repeated. Silica gel containing 1.48 ± 0.01 millimole/g of combined water was used for the experiments with $TiCl_4$ ($t_{melting} = -23°$, $t_{boiling} = 136°C$). Silica gel having 1.62 ± 0.01 millimole/g was used for the experiments with $GeCl_4$ ($t_{melting} = 26°$, $t_{boiling} = 186°C$). The reaction products were analyzed for Cl^- (by the Volgard method), as well as for titanium (experiments with $TiCl_4$ [3]) or germanium (experiments with $GeCl_4$ [4]). The results obtained are cited in Tables 1 and 2.

The interaction between silica gel and the gaseous Cl, Ti, or Ge species is a type reaction involving surface chemical modification by functional groups. Since the reactions considered were carried out under saturation of reaction-susceptible surface groups, the above coatings result only because hydroxyls are introduced from the chloride hydrolysis, i.e., for each condensation-reaction cycle, the silica gel surface is covered by a single layer of $(-O)_x MeCl_{4-x}$ groups (where x is the number of reacting chlorine atoms). Thus the reactions under study may be considered as reactions of monomolecular overlap [2].

Table 1. Content of Ti (mA/g*) in the Reaction Products of Silica Gel and $TiCl_4$

Reaction No.	No. of treatments									
	1	2	3	4	5	6	7	8	9	10
I	0.99	1.50	2.22	2.79	3.34	3.52	3.91	4.09	4.28	4.40
II	1.01	1.52	2.28	2.70	3.30	3.57	3.92	4.07	4.30	4.45
III	1.00	1.64	2.30	2.78	3.18	3.67	3.95	4.12	4.35	4.42
Average	1.00	1.55	2.27	2.76	3.27	3.59	3.93	4.09	4.31	4.42
Deviation from average { I	+0.01	+0.05	+0.05	−0.03	−0.07	+0.07	+0.02	0.00	+0.03	+0.02
II	−0.01	+0.03	−0.01	+0.06	−0.03	+0.02	+0.01	+0.02	+0.01	−0.03
III	0.00	−0.09	−0.03	−0.02	+0.09	−0.08	−0.02	−0.03	−0.04	0.00
TiO_2, mg/g SiO_2	80.0	131.2	184.0	222.1	254.0	294.0	316.0	330.0	344.0	354.0

Table 2. Content of Ge (mA/g) in the Reaction Products of Silica Gel and $GeCl_4$

Reaction No.	No. of treatments						
	1	2	3	4	5	6	7 **
I	0.60	1.06	1.25	1.50	1.69	1.88	—
II	0.58	0.97	1.22	1.54	1.79	1.94	2.22
III	0.56	0.99	1.25	1.50	—	—	—
IV	0.53	1.10	1.28	1.63	—	—	—
Average	0.57	1.03	1.25	1.57	1.74	1.91	2.22
Deviation from average { I	−0.03	−0.03	0.00	+0.07	+0.05	+0.03	—
II	−0.01	+0.06	+0.03	+0.03	−0.05	−0.03	—
III	+0.01	+0.04	0.00	+0.07	—	—	—
IV	+0.04	−0.07	−0.03	−0.06	—	—	—
ΔGe	0.57	0.46	0.22	0.32	0.17	0.17	0.31
GeO_2, mg/g	59.6	107.7	130.8	164.1	182.0	199.6	232.0

*mA = milligram-atom.
** The seventh treatment was performed after a 250°C hydrolysis of the product of the sixth treatment.

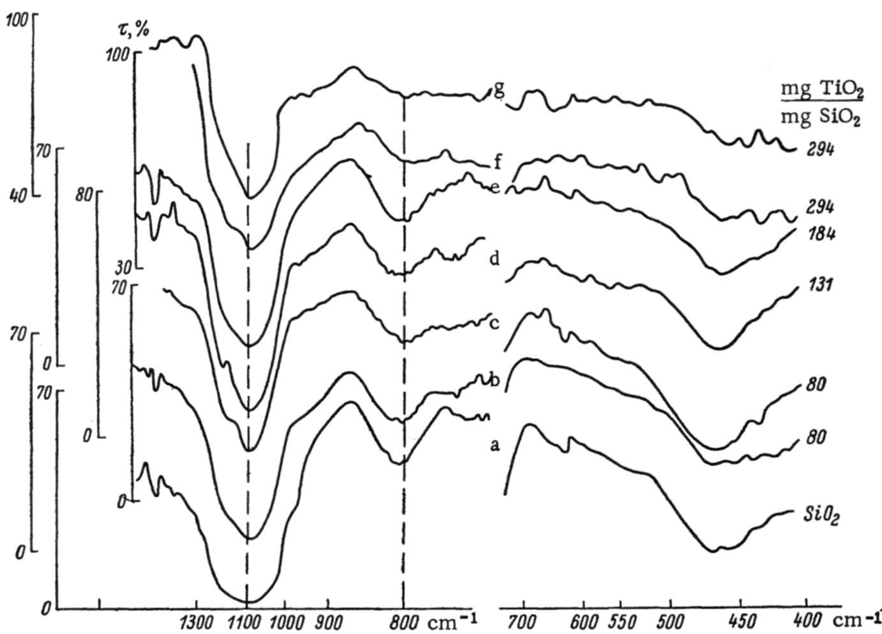

Fig. 1. Infrared transmission spectra of: a) hydrated SiO_2; b) product of first $TiCl_4$ treatment; c) same after hydrolysis; d) product of second treatment after hydrolysis; e) product of third treatment after hydrolysis; f) product of sixth treatment; g) same after hydrolysis.

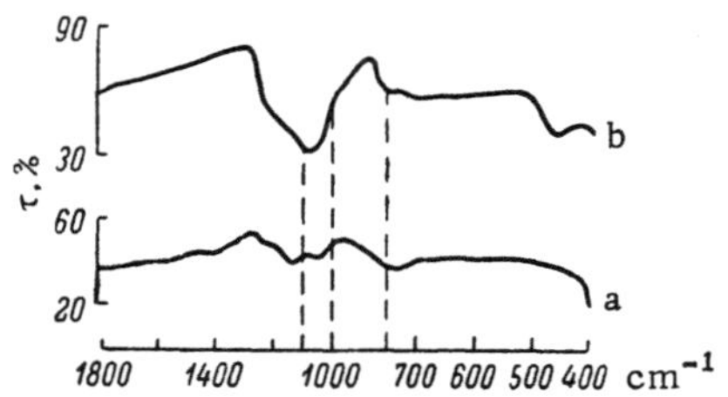

Fig. 2. Infrared transmission spectra of: a) synthetic anatase, produced by hydrolysis of $TiCl_4$; b) the product of tenth treatment of silica gel with $TiCl_4$ containing 354 mg TiO_2/ g SiO_2.

Data from the physicochemical study also attest the validity of the latter assumption on the mechanism of interaction.

The infrared spectra of the samples synthesized are shown in Figs. 1-3.

Figures 1 and 2 show the spectra of products with different amounts of titanium, as well as the spectrum of synthetic anatase prepared by hydrolysis of $TiCl_4$ (technique of [5]). According to the literature data [6], the Ti—O—Ti valence vibrations cover the wavelength range 820-720 cm⁻¹. The band at 919-925 cm⁻¹, in the opinion of Zeitler and Brown [7], is a result of the valence vibrations of Si—O—Ti. The position of the principal absorption maximum obtained in the spectra of our specimens (1080-1090 cm⁻¹) does not change on increase of titanium content. The intense absorption after 800 cm⁻¹ in the spectrum of a sample with a SiO_2 : TiO_2 ratio of 2.8 : 1 (Fig. 2, curve b) is evidently associated with the Ti—O vibrations.

The x-ray patterns of the interaction products between silica gel and $TiCl_4$, contain the lines of anatase (Table 3) starting at the product of the fourth treatment.

The sensitivity of the x-ray method is adequate for showing the presence of TiO_2 even after the first treatment (cf. data of Table 1). However, the anatase lines appear only in the x-ray pattern of a sample having a TiO_2 content twice that of the given mixture. The infrared spectra of the Ge-containing specimens synthesized are given in Fig. 3. The same figure also

Table 3. Interplanar Spacings, d

| Anatase I | | Anatase II | | Rutile I | | Rutile II | | Mixture of 100 mg TiO₂ per g SiO₂ | | Composition of products of treatment of silica gel (mg TiO₂/g SiO₂) | | | | | | | | | |
| | | | | | | | | | | 224 | | 282 | | 328 | | 352 | | 352 * | |
I	d_α/n	I	d_α/n	I	d_α/n	I	d_α/n	I	d_α/n	I	d_α/n	I	d_α/n	I	d_α/n	I	d_α/n	I	d_α/n
3	3.928	10	3.542	4	4.264	2	3.584	10	3.556	10	3.542	10	3.556	10	3.570	10	3.570	2	3.928
10	3.556	5	1.892	5	3.717	2	3.261	3	2.394	2	2.394	3	2.394	3	2.394	2	2.388	6	3.514
2	2.388	5	1.678	10	3.346	2	2.742	6	1.899	2	1.670	3	1.899	5	1.907	3	1.899	10	3.346°
5	1.896			4	2.464	7	2.491	3	1.704			3	1.488	2	1.701	2	1.675	2	2.663
2	1.701			3	2.283	2	2.323	3	1.672					2	1.675	2	1.486	2	2.451°
2	1.672			3	2.133	3	2.213	3	1.488					2	1.494			4	1.889
3	1.484			10	1.821	10	1.954	3	1.267									2	1.698
2	1.264			4	1.672	6	1.689											2	1.667
1	0.896			8	1.541	2	1.486											3	1.481
				10	1.376													2	1.402
				3	1.199													2	1.337
				3	1.181														

Note. Anatase I = commercial preparation; rutile I = natural mineral (Leningrad State University Museum); anatase II, rutile II = TiO₂ specimens synthesized by [5]; asterisk = composition after heating at 1000°C.

Table 4. Interplanar Spacings, d

| Mixture of 100 mg/g GeO₂ * | | Mixture of 50 mg/g GeO₂ † | | GeO₂, mg/g | | | |
| | | | | 199.6 | | 199.6 ‡ | |
I	d_α/n	I	d_α/n	I	d_α/n	I	d_α/n
2	7.127	10	3.447	10	3.474	10	3.501
2	4.389	2	2.618	3	2.177	7	2.8620
2	3.860	3	2.364	3	1.539	7	2.3640
10	3.474	2	2.272			5	2.1674
3	2.376	3	2.157			7	1.759
2	2.294	3	1.678			7	1.593
3	2.177	2	1.725				
2	1.885	3	1.578				
2	1.725	3	1.417				
3	1.573						
2	1.417						

*Commercial preparation.
†Germanium dioxide obtained by hydrolysis of GeCl₄.
‡After annealing at 800°C.

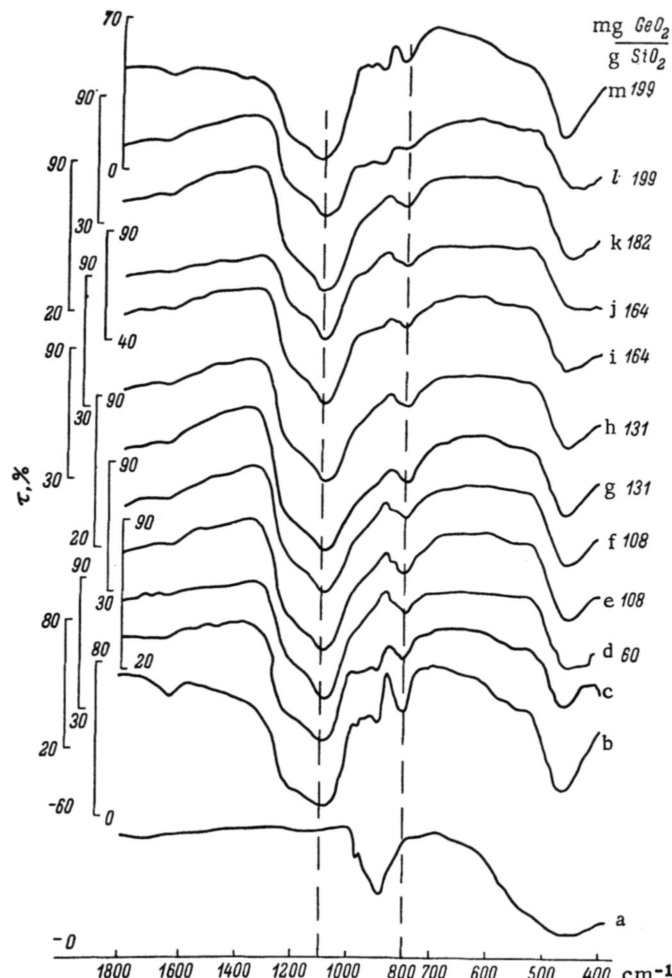

Fig. 3. Infrared transmission spectra of: a) GeO_2; b) 100 mg GeO_2/g SiO_2 (mixture); c) 50 mg GeO_2/g SiO_2 (mixture); d) product of first treatment; e) product of second treatment; f) same after hydrolysis; g) product of third treatment; h) same after hydrolysis; i) product of fourth treatment; j) same after hydrolysis; k) product of fifth treatment; l) product of sixth treatment; m) same after hydrolysis.

shows the spectra of GeO_2 obtained by hydrolysis of $GeCl_4$, as well as the spectra of mixtures of SiO_2 and GeO_2. An absorption from the Ge—O valence vibrations is observed at 880 cm^{-1}, agreeing with literature data [8]. Such absorption in this region is even noticeable in the spectrum of a mixture with 50 mg GeO_2/g SiO_2. This band is particularly intense in the spectrum of a mixture with 100 mg GeO_2/g SiO_2.

In the spectra of the samples, only the product with one-fourth more GeO_2 (sixth treatment) has approximately the same absorption.

In the x-ray patterns, the GeO_2 lines appear only in a sample containing ~190 mg GeO_2/g SiO_2. Lines characteristic of Ge are even present in the x-ray pattern of a mixture with 5% GeO_2 (50 mg/g SiO_2) (Table 4).

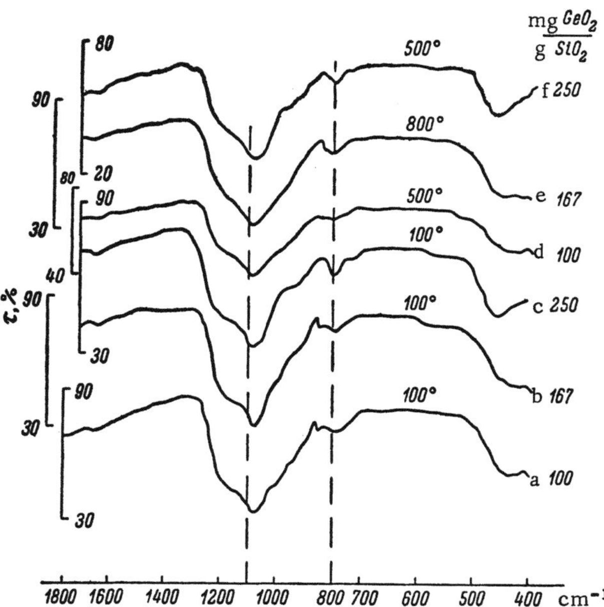

Fig. 4. Infrared transmission spectra of: a) sample containing 100 mg GeO_2/g SiO_2, treated at 100°C; b) sample containing 167 mg GeO_2/g SiO_2, treated at 100°C; c) sample containing 250 mg GeO_2/g SiO_2, treated at 100°C; d) sample containing 100 mg GeO_2/g SiO_2, treated at 500°C; e) sample containing 167 mg GeO_2/g SiO_2, treated at 800°C; f) sample containing 250 mg GeO_2/g SiO_2, treated at 500°C.

The fact that infrared spectroscopy as well as x-ray studies show a decreased (relative to chemical analysis data) content of TiO_2 and GeO_2 indicates a definite effect of the silica-gel lattice on the structure of the coating being formed. The effect of the lattice becomes less with an increase in the number of treatments. At a specific stage Ti and Ge begin to form characteristic structures.

The effect of the silicon—oxygen lattice on coordination of the metal—oxygen bonds is also clearly revealed in the spectra of Ge-containing samples produced by coprecipitation of a mixture of hydrolyzed tetratosisiline and $GeCl_4$ (method of Borisenko [9]). The introduction of 250 mg GeO_2/g SiO_2 does not give the distinct Ge—O spectral absorption band at 880 cm⁻¹ (Fig. 4).

The position of the principal ≡ Si—O band at 1100 cm⁻¹ also does not change, attesting to the absence of excitation at the Si—O bond.

CONCLUSIONS

1. The monomolecular oxide layer on the surface of hydrated silica forms by the action of the hydrogen ions in hydroxyl groups and of the chloride ions in the original tetrachlorides.

2. Further development of the oxide structure results on generation of hydrogen ions in hydroxyl groups forming during the hydrolysis of additional chlorine groups (chloride ions of the tetrachlorides).

3. The effect of the silicon—oxygen anion on the development of the oxide structure was shown, attesting to its definite role in the structural formation of the compounds examined.

LITERATURE CITED

1. M. G. Voronkov, E. A. Lasskaya, and A. A. Pashchenko, Zh. Prikl. Khim., Vol. 38, No. 7, p. 1483 (1965).
2. S. I. Kol'tsov, Abstract of Dissertation, 1962; Theses of the Report to the Scientific-Technical Conference of the Lensoveta Leningrad Technical Institute [in Russian] (1963), p. 27.
3. A. K. Babko and A. T. Pilipenko, Colorimetric Analysis [in Russian], Goskhimizdat, Moscow (1951).
4. V. A. Nazarenko, N. V. Lebedeva, and R. V. Ravitskaya, Zavodsk. Lab., Vol. 1, p. 9 (1958).
5. Funaki, Esitomoén, Saéki, and Yudzo, Kogyo Kagaku Zasshi, Vol. 59, No. 11, p. 1291 (1965).
6. S. A. Giddings, Inorg. Chem., Vol. 3, No. 5, p. 684 (1964).
7. W. A. Zeitler and C. A. Brown, J. Phys. Chem., Vol. 61, No. 6, p. 1174 (1957).
8. E. R. Lippincott, A. V. Valkenburg, C. E. Weir, and E. W. Bunting, J. Res. Natl. Bur. Stnd., Vol. 61, No. 1 (1958).
9. A. I. Borisenko, Thin-Layered Coatings, Gosstroiizdat, Moscow (1958).

PROCESSES OCCURRING ON THE INTERACTION OF POLYORGANOSILOXANES AND LAMELLAR SILICATES WHILE HEATING FROM 400 TO 1600°C

A. I. Boikova, N. A. Toropov, and E. S. Sher

Besides the typical inorganic coatings of complex phase and chemical composition, the organosilicate materials developed at the Institute of Silicate Chemistry are now being used widely in various branches of technology and construction.

They are produced on the reaction of such seemingly different materials as the polymeric organosilicon compounds (the polyorganosiloxanes [1, 2]) and natural inorganic minerals (finely dispersed hydrosilicates such as muscovite, asbestos, or talc). The reaction is activated through an appropriate heat treatment. Refractory oxide powders sometimes enter the composition of the organosilicate materials.

The organosilicate products combine very successfully the valuable properties of the polyorganosiloxanes (imperviousness to moisture, good electrical insulation, elasticity when formed into films) with the high thermal and chemical stability typical of the silicates and several oxides. Many thermoelectric water-resistant materials already have broad practical application. They are further modified in many variations, known to industry under various names [3]. They are used in the manufacture of radiological components, wire resistors, condensers, etc., since, in many respects, they are superior to the normal glass enamel insulation.

Thin-layered organosilicate coatings are used for protecting thermocouples at 800-900°C. Many materials have lately been developed to insulate electrical leads and cables intended for operation at elevated temperatures [4]. New types of glass and metal sheet have been produced by combining glass fibers and metal wire, or glass and metal mesh. The possibility is being studied for producing new materials by compacting organosilicate powders.

The organosilicates are very resistant toward many types of corrosion. They are very resistant to radiation. They are hydrophobic and stable against biological corrosion from fungus mold and other types of biological decay (particularly important when the organosilicate insulation is used under tropical conditions). These materials are now being even more widely used as high-temperature adhesives for tensimetric equipment. Long experiment has shown the reliability of their use for high-temperature tensimetry at temperatures up to 500-600°C.

The physical and chemical properties of organosilicate materials depend on both the compositional and structural characteristics of the starting components and on the thermal characteristics of the polyorganosiloxanes and the lamellar silicates. Their properties can widely vary according to their quantitative compositional ratios and the conditions of previous heat treatment.

The most important structural element of the polyorganosiloxanes, which plays a large role in high-temperature processes, is a silicon—oxygen chain of —Si—O—Si— molecules, with the free silicon valences filled by organic radicals. Polydimethylphenylsiloxane acts as the polymer for the organosilicate materials studied. The combination of the siloxane bonds and the aromatic and aliphatic radicals is a source of many valuable properties of the polyorganosiloxanes. The —Si—O— groups have large binding energies, which ensures that compounds containing them have high thermal stability compared to others. However, the thermal stability of the siloxane bond in the polyorganosiloxanes is considerably less than that of the different forms of silica and the silicates. It depends on the nature of the silicon—organic radicals coupled to the silicon atoms.

To produce high-quality materials, the organosilicates are subjected to hot drying (to 300°C), in which a final "polycondensation" occurs, forming a net-like, three-dimensional structure. The net-like three-dimensional polymers are thermally more stable, since breaking one of their links involves breaking a chain at two or three points. The formation of the branched —Si—O—Si— complexes is not only essential for increasing the thermal stability of the polyorganosiloxanes, but also (as will be stated below) it acts to produce the valuable engineering properties of the organosiloxane materials.

It follows to consider the important fact that the siloxane radical structure becomes more complex on heating the polyorganosiloxanes, due to destructive thermal oxidation as, for example, occurs on its transition from a chain-like to a net-like, or even trimeric configuration.

Oxygen replaces the organic radicals of the polyorganosiloxanes liberated on the destructive thermal oxidation. At the same time, new siloxane bonds characterized by high thermal stability are created. As a result, the polymer is not disrupted at specific links (or their combinations), and acquires the more branched siloxane structure [5]. Another very essential aspect of the destructive thermal oxidation process is the formation of OH radicals.. These react with a silicon atom in the polymeric molecule according to the scheme

$$\begin{array}{ccc} & & \text{OH} \\ & & | \\ -\text{O}-\overset{|}{\underset{|}{\text{Si}}}-\text{O}- + \text{HO}^{\cdot} \rightarrow & -\text{O}-\overset{|}{\underset{|}{\text{Si}}}-\text{O}-\ldots \\ \text{O} & & \text{O} \\ | & & | \end{array}$$

Thus there results, on the one hand, a possibility for condensation of products forming longer —Si—O—Si— chains and, on the other, there results the possibility for the hydroxyl dissociation products of the polyorganosiloxanes to actively combine with the hydroxyls of the lamellar silicates constituting the stable complexes.

Great attention is given to observations which show that the molecular chains in polydimethylphenylsiloxane have a spiral structure, with three to six silicon atoms in a spiral loop, and that at elevated temperatures it is possible to produce cyclical structure elements as six- or three-member rings. This circumstance is of particularly great significance relative to the detailed study of the interaction between the polyorganosiloxanes and the thermal dissociation products of the complex silicates present in the formation of organosilicate materials.

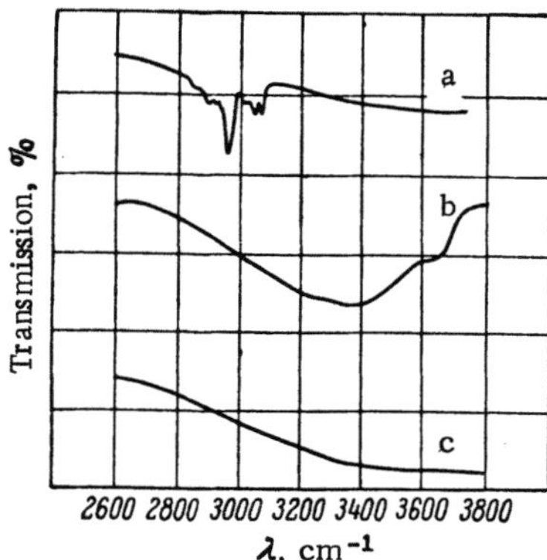

Fig. 1. Infrared absorption spectra of polydimethylphenylsiloxane (K-44 varnish) and the products of its anneal. a) Original product; b) annealed at 600°C; c) annealed at 700°C.

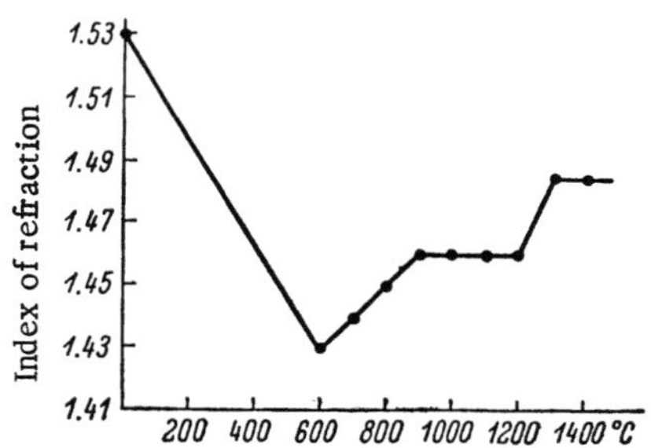

Fig. 2. Temperature-dependence curve for the index of refraction of polymer annealing-products.

The structure of the lamellar silicates is composed of layers of silicon—oxygen tetrahedra and intermediate octahedral layers of hydroxyl—oxygen octahedra, populated by the cations Al^{3+}, Mg^{2+}, Fe^{2+}, Cr^{3+}, and others of corresponding cationic radii. The crystallochemical correspondence between the six-member rings formed by the thermal dissociation products of the complex hydrosilicates (on the one hand) and the polyorganosiloxanes (on the other) should result in formation of reasonably stable secondary structures.

The formation of these stable structures is also facilitated by one of the structural characteristics of the silicon—oxygen elements — their pliability, which enables their adaptation to the basic structural motif [6, 7].

EXPERIMENTAL

The material at our disposition was B-58, the principal component of which was activated muscovite previously subjected to thermomechanical treatment. It contained a considerably lesser amount of polydimethylphenylsiloxane (K-44 varnish), and also a small amount of chromium oxide. The material can be prepared as a suspension, paste, or powder, depending on its purpose.

We used both material which had been previously heated to 270°C and pulverized, and material as a pliable powder which had not been subjected to heat treatment.* The specimen under study was heated slowly to 300°C and then by 50-100°C stages to 1600°C. It is well to emphasize that the heating regime to a large extent determines the state of the material, and particularly its carbon content. A sample was taken at each temperature. An annealing period of 3-4 h was adequate to ensure equilibrium for a given temperature. The annealing was performed in a platinum—rhodium resistance furnace.

An analogous heat treatment was performed separately on muscovite and polydimethylphenylsiloxane in a parallel fashion to permit comparison of their thermal decomposition products with the interaction products for the muscovite in the B-58 material. It is known that the decomposition of polydimethylphenylsiloxane involves primarily the disruption of the —Si—C— bonds, beginning with disruption of the methyl, and then of the phenyl radicals. The decomposition product of polyorganosiloxane which has completely lost its organic radicals is amorphous silica. On subsequent heating to temperatures above 1200°C, the amorphous SiO_2 transforms to crystobalite.

Chief attention was devoted in our work to studying the intermediate products forming with increased annealing temperatures, since such products result from the interaction with muscovite.

---

*The authors thank V. A. Krotikov for the preparation of the specimens.

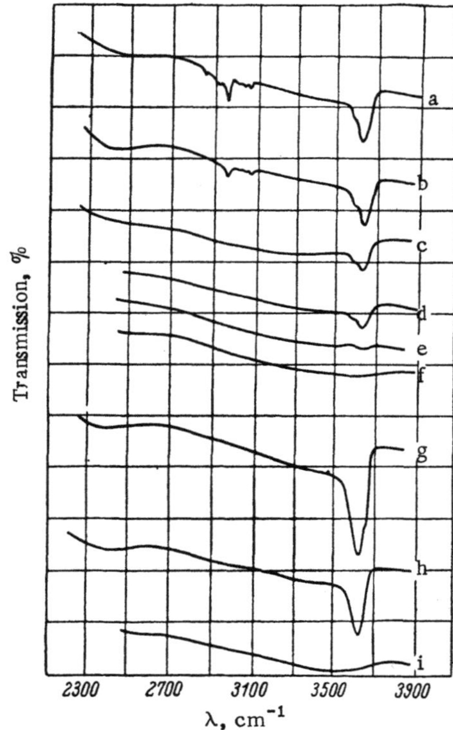

Fig. 3. Infrared absorption spectra of B-58 (spectra a-f), of muscovite (spectra g-i), and of the products of their heat treatment. a) Original B-58; b) annealed at 400°; c) at 500°; d) at 600°; e) at 650°; f) at 700°C; g) original muscovite; h) annealed at 650°; i) at 750°C.

According to DTA, infrared spectroscopic analysis, and chemical analysis, the breaking away of the organic radicals is most intensive at 350-500°C. At 500-600°C there are no organic radicals in the product. Indeed, the absorption bands in the 2100-2900 cm^{-1} wavelength region (corresponding to vibrations of the Si—CH$_3$ and Si—C$_6$H$_5$ bonds), are not observed in this sample (Fig. 1b).

Amorphous silica (contaminated by hydroxyl groups) forms in the 500- to 600°C temperature range in one of the decomposition stages of polydimethylphenylsiloxane. The infrared absorption spectrum of this product displays the absorption bands characteristic of the Si—OH bond vibrations (Fig. 1b). The hydroxyls separate at higher temperatures, and the spectrum of amorphous silica annealed at 700°C does not have the OH absorption bands (Fig. 1b).

The curves showing the index of refraction of the polymer-heat-treatment products as a function of temperature (Fig. 2) are typical. At room temperature, polydimethylphenylsiloxane appears under the microscope as isotropic plates with an index of refraction of 1.53 ± 0.003. The index of refraction decreases to 1.43 ± 0.003 for the product obtained in the 500- to 600°C range during hydroxyl enrichment, as a result of temperature increase and separation of organic radicals from the polymeric structure. The annealing products of polydimethylphenylsiloxane appear under the microscope as isotropic plates. Above 600°C, the hydroxyl groups break away and the silica structure becomes more dense, with the index of refraction increasing to 1.46 ± 0.003. Recrystallization of the amorphous silica structure occurs in the 900-1200°C temperature range, with no apparent increase in density. At 1150-1200°C, the high-temperature crystobalite lattice forms. The latter transforms to low-temperature crystobalite on cooling.

The hydroxyl-enriched decomposition product produced at 500-600°C was studied in greater detail. Its index of refraction is between 1.42 and 1.43, its pycnometer density 1.98 g/cm^3, and its silica content 94.6% (approximating the data for the cubic modification of SiO$_2$ — metanopalogite [8]). Determination of the benzene-vapor sorption isotherms showed that this product does not absorb at all.

While determining the specific surface for nitrogen adsorption at 195°C under a pressure of 10^{-2} mm Hg, it was found that the specific surface of this sample is not large — approximately 50 m^2/g. The specific surface becomes even less with increase of sample annealing temperature. We earlier published a study on the heat-treatment products of muscovite [9].

Having data on the thermal transformation of muscovite and the decomposition of polydimethylphenylsiloxane, we shall examine the thermal transformation process of B-58 material in the 400-1600°C temperature range. Figure 3a-f compares the infrared absorption spectra of the B-58 material with those of muscovite (Fig. 3g-i).

The absorption bands at wavelengths between 3500 and 3700 cm^{-1} correspond to the Si—OH bond vibrations in muscovite. The bands with maxima between 2800 and 3100 cm^{-1} belong to the polymeric radical. Based on the spectroscopic analysis, it can be said that not only does an intensive liberation of the methyl and phenyl groups from the polymer occur at 400-500°C (nearly totally complete by 500°C), but also that the muscovite hydroxyl groups are involved in the reaction. For a sample annealed at 500° (Fig. 3c), not only is the organic radical absorption band absent, but also the intensity of the muscovite absorption bands decreases. Separation of the hydroxyls is nearly complete by 650°C (Fig. 3e), when the dehydroxylation process is just beginning in the original muscovite (Fig. 3h).

The carbon content in a sample annealed at 500°C is 2.37 and 2.19% according to chemical analysis (two parallel determinations). In a sample annealed at 700°C it is 0.09%, and at 900°C it is 0.04%. This is for slow heating of the material. Rapid heating results in some carbon remaining in the material, evidently as carbides.

Microscopic observation revealed that corrosion of the muscovite crystals in the B-58 material already starts at 400°C. With further temperature increase the corrosion becomes more intense. Corrosion of pure mica crystals is observed only above 700°C.

In our opinion, the following occurs in the 400-650°C temperature range. The hydroxyl radicals forming on the oxidizing thermal decomposition of the polymer partly participate in the condensation reactions and partly interact with the hydroxyl groups of the mica. This explains the fact that the dehydroxylation of muscovite in the B-58 material starts at temperatures approximately 200°C lower than for the original mica. This corresponds to the temperature range where free hydroxyl radicals form in the decomposing polymer. Chemical bonding develops between the muscovite and the polymeric-decomposition product (thin layers of amorphous silica). Interaction between the mica hydroxyls and the polymeric-decomposition product hydrolysis is reduced because the surface of the mica plates is highly imperfect, owing to previous mechanical treatment. Since polydimethylphenylsiloxane has a three-dimensional structure, many points of contact develop, and interaction between the polydimethylphenylsiloxane decomposition products and muscovite occurs over the entire mutually contiguous surface of these materials.

The surface layers of the mica plates are involved chiefly with interaction with the hydroxyls of the decomposing polymer. Such a change in the mica structure would produce a reflection in the x-ray patterns. The results of the x-ray ionization study (Fig. 4), reveal that the ionization intensity curves of the B-58 material retain the basic character of the muscovite curves (curves II, III) to a temperature of 900°C. Until this temperature the ionization curves display no additional maxima attributable to polymeric decomposition products, since these products are amorphous. Neither are such products observed under the microscope as the isotropic plates we observed for the pure polymer. The amorphous decomposition products are near the surface of the mica plates as a thin reaction layer. A compact structure occurs both as a result of the chemical bond between the polymeric decomposition products and mica and, as noted above, as a result of the crystallochemical correspondence between the six-member silicon—oxygen rings of the hydrosilicate and those of the polyorganosiloxane decomposition products (epitaxial growth). The closer the structural similarity of the particle surfaces, the more numerous the contacts between them. The mica plates are interspersed and cemented by thin layers of amorphous silica, resulting in a strong, stable structure. The silicon—oxygen complexes, which are elastic, strong microarmatures, play a large role in strengthening the resulting structure.

This structure disintegrates above 900°C at the start of the structural rebuilding process in muscovite: amorphization, new phase formation, and recrystallization of several of the phases. The following new phases are observed to form in the material at 1000°C: γ-Al_2O_3,

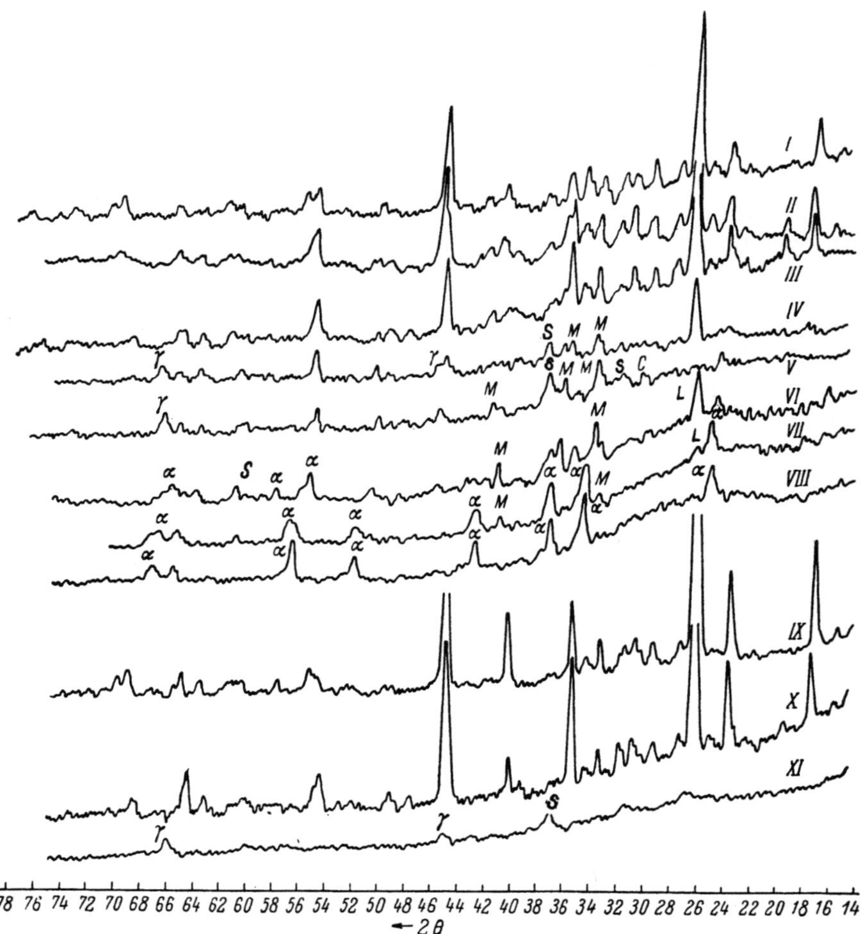

Fig. 4. Ionization intensity curves of B-58 material (curves I-VIII), muscovite (curves IX-XI), and the products of their heat treatment. I) Original B-58; II) annealed at 600°; III) at 900°; IV) at 1000°; V) at 1100°; VI) at 1200°; VII) at 1300°; VIII) at 1600°C; IX) original muscovite; X) annealed at 900°; XI) at 1000°C; $\gamma = \gamma\text{-}Al_2O_3$; $\alpha = \alpha\text{-}Al_2O_3$; S) magnesium aluminum spinel $(MgO \cdot Al_2O_3)$; M = mullite $(3Al_2O_3 \cdot 2SiO_2)$; C = sanidine $(K_2O \cdot Al_2O_3 \cdot 2SiO_2)$; L = leucite $(K_2O \cdot Al_2O_3 \cdot 4SiO_2)$.

mullite $(3Al_2O_3 \cdot SiO_2)$, and spinel $(MgO \cdot Al_2O_3)$ Fig. 4, IV, V). The amount of each increases with rise in temperature (with the exception of $\gamma\text{-}Al_2O_3$, which transforms to the α-form) (Fig. 4, VI-VIII).

The next transition stage is melting at 1200-1300°C. The fused material is a completely porous, opaque glassy mass of greenish-brown color, interspersed by fine crystallites, chiefly $\alpha\text{-}Al_2O_3$ (Fig. 4, VI, VIII).

The present work does not consider the role of chromium oxide in the interaction between muscovite and the polymer. The solution to this problem can be the goal of an independent study. It can be presumed that chromium is a component which ensures a strong, detailed coating and which also plays a large role as catalyst in the hardening of the coating.

Neither did we consider the stage of enamel hardening occurring below 300°C. The important coating engineering-properties developing in this stage (when the polymer is not destroyed), probably conform to laws other than those for higher temperatures.

CONCLUSIONS

The interaction at 350-400°C between the components of organosilicate materials, containing lamellar silicates, polyorganosiloxanes, and several oxides, and which involves the liberation of organic radicals, is based chiefly on the nature of the polyorganosiloxane decomposition process, the formation of OH radicals at points of disruption of the —Si—C— bonds. These radicals guarantee both the condensation of silicon—oxygen links (formation of a siloxonated structure) and their vigorous reaction with the hydroxyls of the lamellar silicate.

The lamellar silicate is involved in the interaction with the OH-enriched polymeric decomposition products at the decomposition stage when OH groups are present in the decomposing polymer. Although structural analysis indicates that the lamellar silicates (muscovite, talc, asbestos) do not have hydroxyl groups on their surface, nevertheless previous mechanical treatment increases the defect character of the surface layers of the mineral, at the same time making the hydroxyls of these minerals accessible for reaction.

The interaction between the hydroxyls of the polymeric decomposition products and the hydroxyls of the lamellar silicate (for the case of muscovite), indeed explains the fact that the dehydroxylation of the muscovite in the material begins 200°C lower than for the pure mineral. The dehydroxylation begins simultaneous to the appearance of OH radicals in the decomposing polymer. The three-dimensional character of the polymer guarantees an interaction between the silicate and the polymeric-decomposition product over the entire surface of the silicate.

Strong growth occurs not only as a result of the interaction of the silicate hydroxyls and the polymeric-decomposition products, but also on account of the epitaxial growth of crystallochemically similar silicon—oxygen fragments having the structure of the silicate mineral or its decomposition products. With close similarity of the surface of the intergrowing components, many contacts occur. The silicon—oxygen radicals are elastic and stable microarmatures, easily tending to a basic structural composition. As a result of the above factors, a strong, stable reaction-layer forms in the given temperature interval, which cements the crystals of the lamellar silicate. The picture resembles lime—silicate materials subjected to autoclave treatment. The chief bulk of these materials consists of SiO_2 crystals firmly cemented by a reaction layer of the interaction products of calcium hydroxide, with the surface layers of silica (calcium hydrosilicates). In some cases the microscopic pictures are very similar, while the x-ray patterns are as follows: in the case of the organosilicate material the x-ray pattern is of the silicate material; in the case of the lime—silicate material, the x-ray pattern is of SiO_2. The transformation products have a structural similarity — silicon—oxygen structural elements — both for the organosilicate materials and for the lime-silicate transformation products.

The next stage in the structural changes of the B-58 material is associated with the structural rebuilding of the lamellar silicate, amorphization, and finally formation of new phases. At this point the material properties should change; in particular, the strengths should decrease. With increase in temperature, several of the initially forming phases can undergo polymorphic transformations, others decompose, etc.

Finally, the melting stage is accompanied by formation of a glassy material, interspersed with crystallites of different phases whose composition depends on the original silicate used.

In our opinion, for all organosilicate materials of the type studied, the interaction processes of lamellar silicates (talc, asbestos, muscovite) with the polyorganosiloxanes occur chiefly according to the scheme proposed above. Several deviations for the temperature regions of the different stages of transformation of the phase composition of the crystalline products forming may be associated with the properties of the original lamellar silicate, polyorganosiloxane, and oxide.

LITERATURE CITED

1. K. A. Andrianov, Organosilicon Compounds [in Russian], Goskhimizdat, Moscow (1955).
2. K. A. Andrianov, I. Khaiduk, and L. M. Khananashvili, Uspekhi Khim., Vol. 34, No. 1, p. 27 (1965).
3. V. B. Tikhomirov, Polymerized Coatings in Atomic Technology [in Russian], Atomizdat, Moscow (1965).
4. G. V. Belinskaya, I. B. Peshkov, and N. P. Kharitonov, Heat-Resistant Insulation of Conductor Wrappings [in Russian], Izd. Nauka, Moscow (1965).
5. K. A. Andrianov, Polymers with Inorganic Principal Chains of Molecules, Izd. Akad. Nauk SSSR, Moscow (1962).
6. N. V. Belov, Crystal Chemistry of Large-Cation Silicates, Consultants Bureau, New York (1963).
7. N. V. Belov, Mineralog. Sb. L'vovsk. Geol. Obshch., No. 13, p. 23 (1959).
8. B. J. Skinner and D. E. Appleman, Am. Mineral., Vol. 48, No. 7-8, p. 854 (1963).
9. N. A. Toropov, E. S. Sher, and A. I. Boikova, Izv. Akad. Nauk SSSR, Neorg. Mat., Vol. 2, No. 8, p. 1487 (1966).

PART V

SINTERING AND DEFORMATION IN OXIDE SYSTEMS

KINETICS AND DEFORMATION MECHANISM OF
HIGHLY REFRACTORY OXIDES
AND THEIR COMPOUNDS
DURING COMPONENT FORMATION

I. G. Orlova

One of the basic physicochemical processes in formation of ceramic and refractory components, sintering, is always accompanied by deformation resulting from diverse forces. It is a phenomenon which has almost never been studied for the oxides, from either the applied or theoretical standpoint, but which is well known in powder metallurgy. We studied experimentally the general kinetic laws and deformation mechanism in the sintering of the highly refractory oxides Al_2O_3, MgO, and ZrO_2 and the compound $MgO \cdot Al_2O_3$. The practical significance of this study is indicated in this paper.

1. If stresses occur in a solid, porous body at high temperatures, it self-consistently deforms in a diffusionally viscous manner [6-11] through volume diffusion of vacancies within the boundaries of its element of structure (grain, block). This in part depends on the type of defect acting as a source or sink for vacancies [12-15].

The rate $\dot{\varepsilon}$ of diffusional viscous flow is determined by the coefficient of viscosity η_D, and is proportional to $KTL^2/D\Omega$, and to the applied stress σ:

$$\dot{\varepsilon} \simeq \frac{\sigma}{\eta_D} \simeq \frac{D\Omega}{KT} \cdot \frac{\sigma}{L^2} \quad [8, 9],$$

where D is the coefficient of self-diffusion, Ω is the volume of the unit cell, K is the Boltzmann constant, and L is the linear dimension typical of the structure element.

In the study of the deformation kinetics of unfired ceramics, the stress range was established for occurrence of a proportional deformation rate with the first degree of stress.

These and all the other studies were performed on dross-cast specimens compacted at a pressure of 1000 kg/cm^2 using a technique developed by I. Kainarskii and I. Orlova for materials with a constituent oxide content in excess of 99.8%.* The stresses varied from less than 20 to 230 g/mm^2, depending on the temperature and object studied.

---

*Except ZrO_2, which contained $\sim$1.5% impurities.

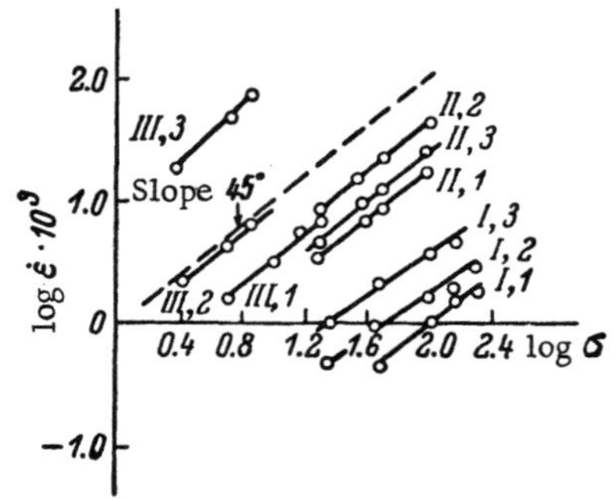

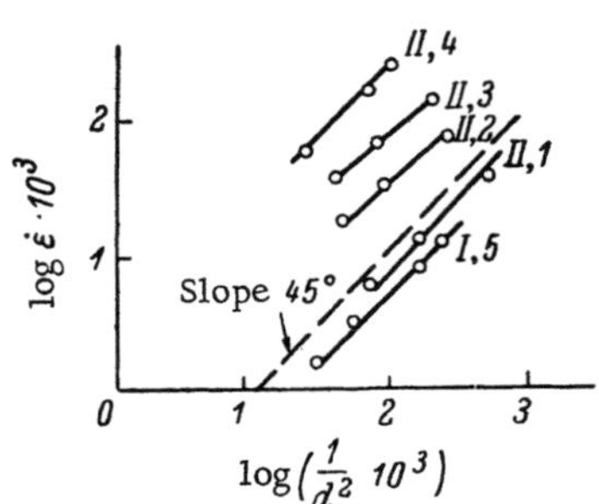

Fig. 1. Dependence of logarithm of deformation rate on logarithm of stress. I) Corundum (Al_2O_3); II) MgO; III) ZrO_2; 1) 1300°; 2) 1400°; 3) 1500°C.

Fig. 2. Dependence of logarithm of deformation rate on logarithm of the reciprocal square grain size. I) Corundum; II) Mgo; 1) 800°C; 2) 1000°C; 3) 1100°C; 4) 1300°C; 5) 1400°C.

Table 1. Comparison with Published Data of the Coefficients of Self-Diffusion D and Activation Energies E Computed from Deformation Data in the 1300–1500°C Temperature Range

Oxide	D, cm^2/sec		E, kcal/mole	
	present data	published data	present data	published data
Al_2O_3 . . .	$2 \cdot 10^{-10}$—$5 \cdot 10^{-12}$	$9 \cdot 10^{-11}$—$2 \cdot 10^{-12}$ [18]; 10^{-12}—10^{-13} [19]	45—98	46—118 [20, 21]; $\sim$ 130 [17, 22]
MgO . . .	$1.6 \cdot 10^{-11}$—$1.4 \cdot 10^{-12}$ (1300—1400°)	$(2-6) \cdot 10^{-14}$ [23, 24]; 10^{-11}—$3 \cdot 10^{-12}$ [25]	54.5	58—71 [23—25]; 76—162 [26]; 60—88 [27—30]
ZrO_2 . . .	$2.5 \cdot 10^{-10}$—$5.1 \cdot 10^{-12}$	$5 \cdot 10^{-10}$ [31] for 700—1000°	45	28 [31] for 700—1000°

An $\dot{\varepsilon} \sim \sigma^n$ dependence, where n = 1 (Fig. 1), is maintained by corundum samples for stresses decreasing from 200-230 to 35-100 g/mm^2 in the 1300–1500°C temperature range. For MgO samples this dependence holds at stresses up to 100 g/cm^2* in the same temperature range. For ZrO_2, it only holds to 10 g/mm^2 at 1400°C.

For stresses exceeding the above, the $\dot{\varepsilon} \sim \sigma^n$ dependence is observed with n changing from ~3 to 4.2, i.e., it increases in value as deformation occurs, by the mechanism of dislocation creep and by restricted diffusion [16]. Another mechanism is also possible, for example, by segregation and slip along grain boundaries [17].

The practicality of the conditions for the diffusional viscous flow theory ($\dot{\varepsilon} \sim 1/L^2$) is shown for corundum and MgO, where the grain size maintains a characteristic linear dimension L (Fig. 2). The proportionality $\dot{\varepsilon} \sim \sigma/L^2$ permits use of the diffusional-viscous-flow rate equation for calculation of the coefficient of self-diffusion D and for its comparison with published data. It also makes possible calculation of the activation energy of the process from the temperature dependence of $\dot{\varepsilon} \sim e^{-E/RT}$ (Table 1).

---

*Larger stresses were not studied.

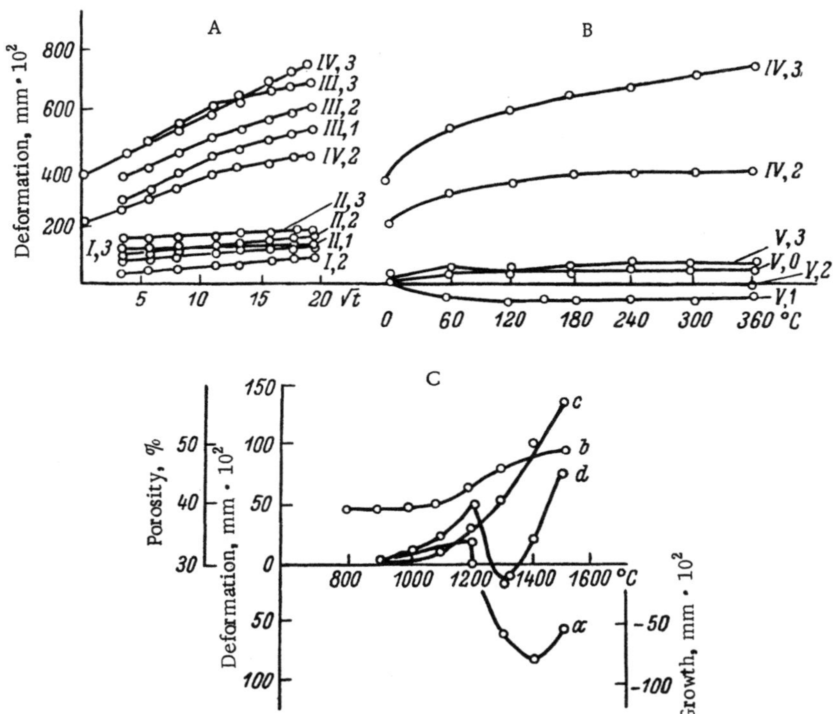

Fig. 3. Kinetics of deformation with coordinates. A) Deformation = √t; B) deformation = t. I) Corundum; II) periclase; III) ZrO_2 stabilized by CaO; IV) synthesized spinel; V) stoichiometric mixture of α-Al_2O_3 and periclase. 1) 1300°; 2) 1400°; 3) 1500°C. C) Change of observed deformation (d), porosity (b), growth (a), and intrinsic deformation (c) as a function of temperature of isothermal anneal after 2 h.

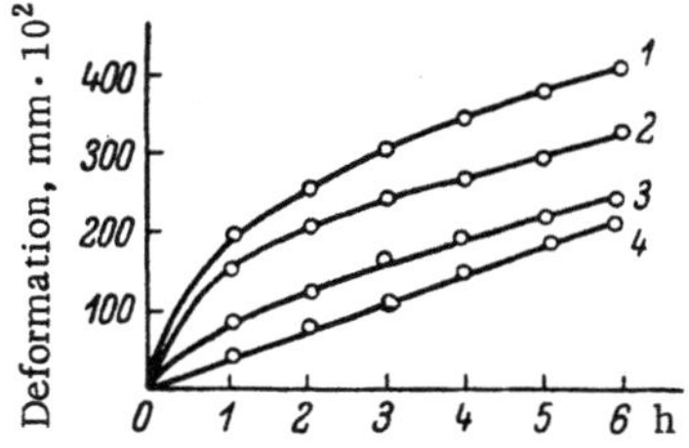

Fig. 4. Effect of time of previous anneal of corundum samples at 1400°C on deformation. Kinetics at a stress of 100 g/mm².

The values obtained for the coefficient of self-diffusion are entirely reasonable, and are characteristic, as follows from Table 1, of the self-diffusion coefficients for temperatures of intense sintering [32].

The two-order increase in the self-diffusion coefficient may be associated with the considerably larger particle surface, as well as with the presence of other defects. These factors cause an essential change in the path of vacancies (and consequently in the normal linear dimensions [6, 12-14]) relative to those used in the computation. A one-order total decrease in this dimension results in a two-order increase in the diffusion coefficient, i.e., in some cases it even gives values above the equilibrium values.

2. The deformation kinetics of refractory oxide and compound samples on sintering is described by the proportionality ~√t (where t is the duration of isothermal soaking), i.e., the deformation kinetics are similar to the shrinkage kinetics on sintering refractory and metallo-ceramic pellets [20, 21, 28, 33, 34] (Fig. 3A). The break in the lines is linked to the heating rate, and is eliminated by slow, gradual heating, according to [21] and [32].

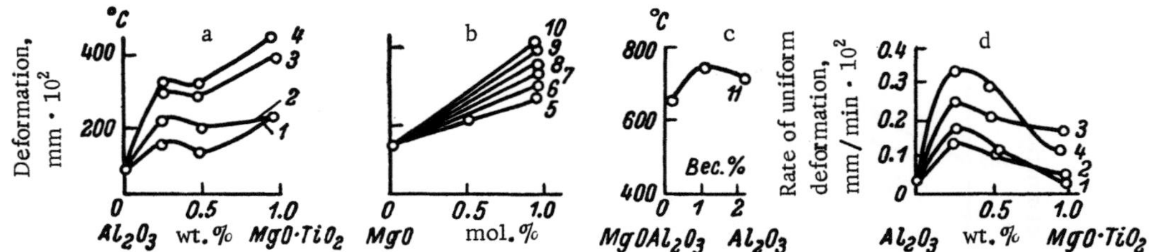

Fig. 5. Effect of small amounts of additives on the deformation and uniform deformation rate of a,d) α-Al_2O_3; b) deformation of MgO; c) of $Al_2O_3 \cdot MgO$, at 1500°C. 1) Spinel additive; 2) forsterite; 3) talc; 4) magnesium titanate; 5) ZrO_2; 6) ZnO; 7) SiO_2; 8) FeO; 9) Al_2O_3; 10) TiO_2; 11) Al_2O_3.

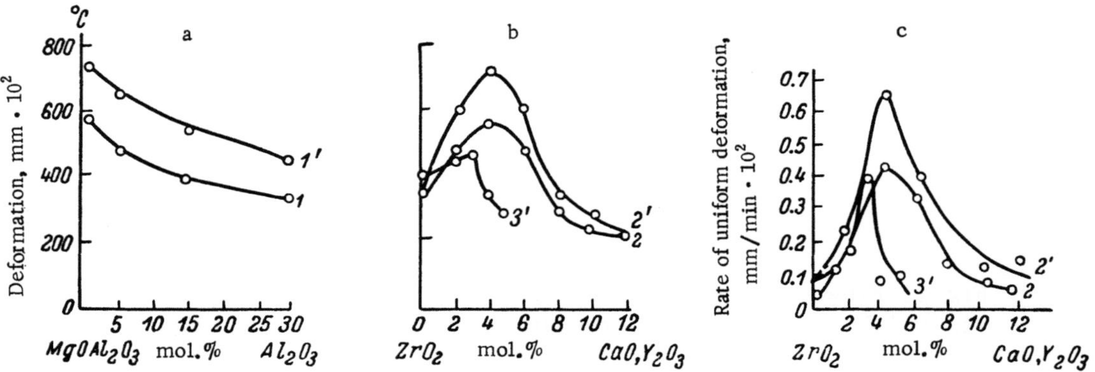

Fig. 6. Effect of large amounts of additives forming solid solutions with the base component on the deformation and resulting deformation rate of: a) aluminum spinel; b, c) zirconium dioxide. 1, 1') Al_2O_3 additive at 1400 and 1500°C; 2, 2') CaO additive at 1400 and 1500°C; 3') Y_2O_3 additive at 1500°C.

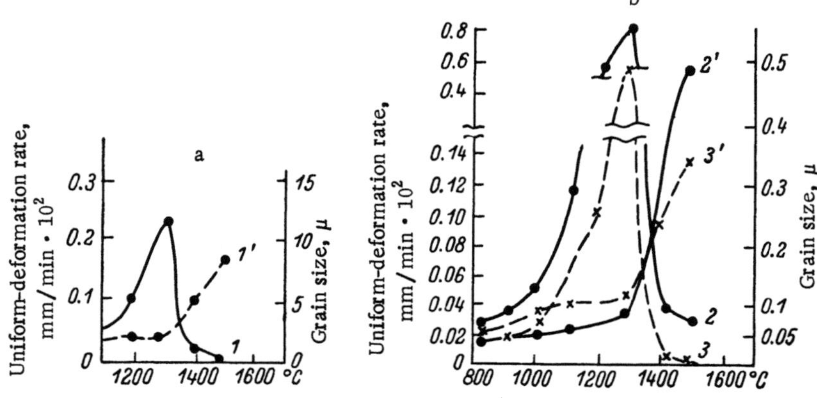

Fig. 7. Effect of recrystallization on uniform deformation rate. a) Alumina with $Al_2O_3 \cdot TiO_2$ additive; b) phosphor-purity MgO (2,2'); MgO from hydrocarbonate base (3, 3'). The numerals without apostrophe are the uniform deformation rates; the numerals with apostrophe are crystal sizes.

The introduction of a second component as small amounts of mineral additives of the same oxide having a very defective crystal lattice, or of additives having considerable solubility in the principal oxide (stabilizing additives in ZrO_2, Al_2O_3 in spinel) does not change the $\sqrt{t}$ proportionality.

A change in the deformation kinetics of refractory oxide mixtures can result in simultaneous occurrence of reactive heterodiffusion. Surplus vacancies arise as a result of the difference in the partial coefficients of heterodiffusion. These coagulate and result in a diffusional porosity — the Frenkel effect [35-38]. As a result, the volume of individual grains increases, and at relatively low pellet porosity, boundary displacement occurs for the entire body.

For this reason, we observed a change in deformation kinetics while studying the deformation of stoichiometric mixtures of corundum and periclase. This can be of interest in the production of components which involves a simultaneous synthesis of spinel (Fig. 3B).

The kinetic deformation curves (Fig. 3B) of this mixture (V) change their usual shape (IV, 2 and 3) as a result of the development of diffusional porosity, stipulated by the progress of the deformation process. The curves in Fig. 3 show values for 2-h anneals: the growth effect at 1200-1400°C (curve a) is associated with development (in the same temperature range) of diffusional porosity (curve b). Curve c is the intrinsic deformation, while curve d is the total (observed) effect.

It should be emphasized that with high initial pellet porosity and use of powders of defect structure for one or both of the synthesis components (e.g., a heat-treated, very lightweight magnesium oxide, γ-Al_2O_3), the appearance of diffusional porosity is not accompanied by a growth which leads to change in the deformation kinetics. This is also observed in [35].

A change in deformation kinetics is observed, with sagging of a specimen to the side opposing the action of the drawing forces. Macroinhomogeneities appear as a result of particle stratification due to difference in size or specimen gravity. We observed such an effect while studying the deformation kinetics of mixtures of α-Al_2O_3 with a small admixture of γ-Al_2O_3 [39]; and of corundum with spinel (the more additive the higher the degree of inhomogeneity).

3. The actually observed kinetic deformation curves do not follow the theories predicted for diffusional-viscous flow. The latter have it that deformation should be proportional to time, to the first degree. Such a discrepancy is associated with the lack of system equilibrium existing in unannealed porous materials. As a result, there is increased diffusional activity in the initial stages of isothermal deformation.

A preliminary thermal treatment of the material (or specimen) substantially lowers the irregular stage of deformation, and can lead to its elimination (Fig. 4). This was first observed in [40-43].

A substantial reduction in deformation, both as a result of a considerable decrease in its initial (irregular) stage, and as a result of a lowering of the deformation rate constant, was also observed for MgO specimens.

These studies made it possible to expose and rule out a nonlinear interdependence between deformation and sintering. As a result it was possible effectively to reduce deformation for the same degree of sintering in corundum [4, 5].

4. The introduction of small (up to 1%) amounts of additives to the base oxide (compound) can reduce the threshold deformation temperature, and generally increases both the deformation in the irregular stage and increases the rate of uniform deformation (Fig. 5). More detailed studies revealed that deformation does not increase proportionally to additive concentration. It increases quite sharply on introduction of an additive in the minimum amount studied

Table 2. Effect of Magnesium Oxide Additive, Produced by Thermal
Decomposition, on the Deformation and Uniform Deformation Rate of
Phosphor-Purity Magnesium Oxide

"Active" MgO content, %	Deformation at $t°C \cdot 10^2$ mm			Uniform deformation rate at $t°C \cdot 10^2$/mm/min		
	1300	1400	1500	1300	1400	1500
—	100	120	140	0.04	0	0
20	120	180	230	0.062	0.10	0.13
30	200	270	320	0.139	0.20	0.25
50	220	310	400	0.195	0.33	0.40

[as for example in corundum (Fig. 5a)]. Moreover, the maximum deformation rate in the uniform stage corresponds to this additive concentration (Fig. 5d). Thus, small amounts of additives activate the diffusional deformation process, especially in its irregular stage. However, the details of the mechanism of its effect is still not clear.

The introduction of large (>1-2 mol.%) amounts of additives forming high-concentration solid solutions (e.g., of CaO, MgO, Y_2O_3, CeO_2 additives to ZrO_2 [44-48], and the addition of corundum to aluminum spinel [49]) can reduce deformation either starting at the introduction of more than 1-2% additive, or after some period of activation of the self-diffusion process. This leads to the spread of deformation and to an increase of its rate in the uniform stage.

For the spinel—corundum mixtures, a decrease in the deformation of spinel is observed already at 2 mol.% Al_2O_3. This actively continues up to 30% Al_2O_3 in the mixture (Fig. 6a). This corresponds to a predominance of heterodiffusional processes (to which belongs the process of solid-solution formation) with high values of the heterodiffusion coefficients as compared with the self-diffusion coefficients [38]. Therefore, these processes impede the spread of deformation.

The deformation and uniform deformation rate increase during formation of CaO and Y_2O_3 solid solutions in ZrO_2 right up to, respectively, 4 and 3 mol.%; only then does their decrease begin (Fig. 6b, c).

The reason for the different effect for the solid solutions evidently is that solid-solution formation in the $MgO \cdot Al_2O_3—Al_2O_3$ mixtures studied leads to an increase in the total level of vacancy concentrations. In mixtures of ZrO_2 and the stabilizing additives examined, it leads to an increase in the total vacancy concentration gradient, causing an increase in the self-diffusion processes. The latter essentially can occur only up to certain additive concentrations.

5. Addition to the base component of an oxide of the same composition produces a significant increase in the diffusional processes and, consequently, in the deformation itself. However, with a loose and deformed crystal lattice this is a very fine-grained additive (Table 2). The increase in the diffusional activity is the result of a considerable growth in intergranular (interblock) particles, which act as vacancy sinks [15, 50]. As a result, the characteristic size L determining the diffusional viscosity of the system decreases.

It can be shown that the observed increase in the uniform deformation rate (e.g., by 1.25, 2.8, and 3.9 times at 1300°C) can occur if the characteristic dimension L of the system (for the given case we take L as d, the grain size) amounts to 0.22, 0.15, and 0.13 μ [since $\dot{\varepsilon}_1/\dot{\varepsilon}_2 \approx (d_2/d_1)^2$]. Moreover, the average grain sizes computed for the mixture are 0.23, 0.21, and 0.18 μ, i.e., very close, considering their means of computation.

Table 3. Effect of Reagent Heterodiffusion on Deformation

Composition	Deformation during 6-h isothermal soaking at °C ($\times 10^2$ mm)		
	1300	1400	1500
Magnesium spinel synthesized at 1750°C.	100	400	480
The same (free MgO ~ 5.5%).	25	80	100
Stoichiometric mixture of MgO and α-Al$_2$O$_3$.	80	140	250

Thus, the concepts regarding the relative causes of the effect of finely-dispersed-defect powders appear to be established.

6. Reactive heterodiffusion, as shown above, leads to a decrease in the self-diffusion processes, as a result of which the deformation decreases (Table 3). Even the relatively small increase in reactive heterodiffusion, which occurred during final specimen deformation of spinels synthesized at a low temperature, led to a substantial specimen shrinkage compared to the high-temperature spinel over the temperature range studied.

7. The temperature range for intense deformation corresponds to the shrinkage on sintering, during which intergranular porosity is eliminated and recrystallization begins.* The latter reduces the uniform deformation rate to practically negligible values (Fig. 7). We observed this effect in the temperature range studied for (a) corundum specimens with Al$_2$O$_3 \cdot$ TiO$_2$ and, (b) different types of MgO whose recrystallization begins above 1300°C. The reduction in the uniform deformation rate as a result of the progress of recrystallization can be explained by the significant increase in the diffusional viscosity of the system. This occurs in connection with the increase of the characteristic size L as a result of a generally decreased degree of development of particles, grains, blocks, and other defects [12-15].

Therefore, it does not follow to expect considerable growth in deformation over a period of intense recrystallization in a material.

The study conducted on the kinetics and the laws of deformation of refractory oxides and their compounds during sintering allows the conclusion that the deformation of oxide materials in the absence of a liquid phase (within threshold stress limits) obeys the laws of diffusional-viscous flow and is dependent on the volume diffusion of vacancies.

It also opens scientifically founded possibilities for controlling the deformation in pieces made of oxides and their compounds during high-temperature anneal.

LITERATURE CITED

1. I. G. Orlova and I. S. Kainarskii, Dokl. Akad. Nauk SSSR, Vol. 157, No. 2, p. 331 (1964).
2. I. G. Orlova, I. S. Kainarskii, and R. E. Mirkina, Ogneupory, No. 1, p. 40 (1965).
3. I. S. Kainarskii, I. G. Orlova, and R. E. Mirkina, Theoretical and Technological Studies in the Area of Refractories (Scientific Transactions of the Ukrainian Scientific Research Institute for Refractories, No. 8), Izd. Khar'kovsk. Gos. Univ. (1965), p. 3.
4. I. S. Kainarskii, I. G. Orlova, and É. V. Degtyareva, Poroshk. Metallurg., Vol. 5, p. 82 (1965).
5. I. S. Kainarskii, I. G. Orlova, and É. V. Degtyareva, Dokl. Akad. Nauk SSSR, Vol. 164, No. 6, p. 1283 (1965).
6. B. Ya. Pines, Fiz. Metal. i Metalloved., Vol. 16, p. 557 (1963).
7. B. Ya. Pines, E. E. Badiyan, and V. P. Khizhkovyi, Fiz. Tverd. Tela, Vol. 5, p. 2859 (1963).
8. F. R. N. Navarro, Ref. Conf. on Strength of Solids, Phys. Soc. London (1947), p. 47.

---

*Cf. É. V. Degtyareva, this volume, p. 188.

9. C. Herring, J. Appl. Phys., Vol. 21, p. 37 (1950).
10. Ya. E. Geguzin and I. M. Lifshits, Fiz. Tverd. Tela, Vol. 4, p. 1326 (1962).
11. I. M. Lifshits and V. B. Shikin, Fiz. Tverd. Tela, Vol. 6, p. 1735 (1964).
12. V. L. Idenbom and A. N. Orlov, Usp. Fiz. Nauk, Vol. 76, p. 557 (1962).
13. Ya. E. Geguzin and V. L. Rabets, Phys. Solid State, Vol. 9, p. 893 (1965).
14. V. V. Skorokhod, Poroshk. Metallurg., Vol. 1, p. 12 (1964).
15. B. Ya. Pines, Fiz. Metal. i Metalloved., Vol. 20, p. 84 (1965).
16. J. Wertman, J. Appl. Phys., Vol. 26, p. 210 (1956).
17. S. I. Warshaw and F. H. Norton, J. Am. Ceram. Soc., Vol. 45, p. 10 (1962).
18. R. L. Coble and J. E. Burke, Proc. 4th JSRS (1960).
19. A. E. Paladino and W. D. Kingery, J. Chem. Phys., Vol. 37, p. 5 (1962).
20. I. S. Kainarskii, É. V. Degtyareva, and S. B. Totsenko, Ogneupory, No. 9, p. 40 (1964).
21. É. V. Degtyareva and I. S. Kainarskii, Dokl. Akad. Nauk SSSR, Vol. 156, p. 4 (1964).
22. R. C. Folweiler, J. Appl. Phys., Vol. 32, p. 5 (1961).
23. I. Oishi and W. D. Kingery, J. Chem. Phys., Vol. 33, p. 905 (1960).
24. I. Oishi and W. D. Kingery, J. Chem. Phys., Vol. 34, p. 688 (1961).
25. R. Lindner and G. D. Parfitt, J. Chem. Phys., Vol. 26, p. 182 (1957).
26. R. L. Cummurow, J. Appl. Phys., Vol. 34, p. 6 (1963).
27. P. P. Budnikov, M. A. Matveev, and V. K. Yanovskii, Dokl. Akad. Nauk SSSR, Vol. 159, p. 872 (1964).
28. P. P. Budnikov, M. A. Matveev, and V. K. Yanovskii, Ogneupory, No. 4, p. 1965.
29. P. P. Budnikov, M. A. Matveev, V. K. Yanovskii, and F. Ya. Kharitonov, Izv. Akad. Nauk SSSR, Neorg. Mat., Vol. 1, No. 8, p. 1349 (1965).
30. A. U. Daniels, R. C. Lowrie, R. L. Gibby, and J. B. Cutler, J. Am. Ceram. Soc., Vol. 45, p. 6 (1962).
31. W. D. Kingery, J. Pappis, M. E. Doty, and D. C. Hill, J. Am. Ceram. Soc., Vol. 42, p. 8 (1959).
32. B. Ya. Pines, Usp. Fiz. Nauk, Vol. 52, p. 4 (1954).
33. I. S. Kainarskii, É. V. Degtyareva, and S. B. Totsenko, Theoretical and Technological Studies in the Area of Refractories (Scientific Transations of the Ukrainian Scientific Research Institute for Refractories, No. 8), Izd. Khar'kovsk. Gos. Univ. (1965), p. 9.
34. B. Ya. Pines and Ya. E. Geguzin, Zh. Tekh. Fiz., Vol. 23, p. 11 (1953).
35. B. Ya. Pines and A. F. Sirenko, Zh. Tekh. Fiz., Vol. 29, p. 5 (1959).
36. B. Ya. Pines and A. F. Sirenko, Zh. Tekh. Fiz., Vol. 28, p. 8 (1958).
37. Ya. E. Geguzin, Uchebn. Zap. Khar'kovsk. Gos. Univ., Fiz. Otdel, Vol. 7, p. 267 (1958).
38. Ya. E. Geguzin, Fiz. Metal. i Metalloved., Vol. 2, p. 3 (1956).
39. I. S. Kainarskii, É. V. Degtyareva, I. G. Orlova, A. G. Karaulov, and G. E. Gnatyuk, Ogneupory, No. 11, p. 27 (1965).
40. B. Ya. Pines and A. F. Sirenko, in: Papers on the 80th Anniversary of Academician N. N. Davidenkov [in Russian], Izd. Akad. Nauk SSSR, Moscow (1955), p. 87.
41. B. Ya. Pines and A. F. Sirenko, Fiz. Metal. i Metalloved., Vol. 7, No. 5, p. 766 (1959).
42. B. Ya. Pines and A. F. Sirenko, Izv. Vuzov SSSR, Chern. Metallurg., No. 5, p. 121 (1960).
43. B. Ya. Pines and A. F. Sirenko, Izv. Vuzov SSSR, Chern. Metallurg., No. 2, p. 81 (1960).
44. I. I. Vishnevskii, A. M. Gavrish, and B. Ya. Sukharevskii, X-Ray Study of Mineral Raw Materials, No. 2 [in Russian], Gosgeolizdat, Moscow (1962), p. 5.
45. B. Ya. Sukharevskii, I. I. Vishnevskii, and A. M. Gavrish, Dokl. Akad. Nauk SSSR, Vol. 140, No. 4, p. 884 (1961).
46. Fan Fu-k'ang, A. K. Kuznetsov, and É. K. Keler, Izv. Akad. Nauk SSSR, Otdel. Khim. Nauk, Vol. 4, p. 601 (1963).
47. S. F. Pal'guev, S. I. Alyamovskii, and Z. V. Volchenkova, Zh. Prikl. Khim., Vol. 4, No. 11, p. 2571 (1959).

48.   D. Viechnicky and V. S. Stubicăn, J. Am. Ceram. Soc., Vol. 48, p. 6 (1965).
49.   D. M. Roy, R. Roy, and E. F. Osborn, Am. J. Sci., Vol. 251, p. 5 (1953).
50.   Ya. E. Geguzin, Fiz. Metal. i Metalloved., Vol. 9, p. 842 (1960).

KINETICS AND MECHANISM OF
SINTERING AND RECRYSTALLIZATION
IN REFRACTORY OXIDES AND THEIR COMPOUNDS

É. V. Degtyareva

Sintering is a basic physicochemical process in the annealing of refractory oxides. It has now been established that sintering of crystalline powders (in the absence of a liquid phase) occurs by volume diffusion of vacancies [1-9] and the corresponding equation for shrinkage kinetics during sintering is now under discussion. According to [1-3], the basic proportionality describing shrinkage on sintering of metallic powders is $\Delta l / l \sim t^{1/2}$ [5, 6] for a volume diffusion mechanism. According to oxide sintering data, the shrinkage $\Delta l / l$ is proportional to about $t^{2/5}$, where t is the time of isothermal soaking.

A sintering theory was recently developed [10-12] for the particular case of diffusional decomposition of a supersaturated solid-solution of vacancies, according to which the kinetics of pore volume contraction is written as $V \sim t^{1/3}$.

In a systematic study of the sintering kinetics of corundum samples [13-15], the authors determined that the kinetics of total shrinkage [occurring on removal of open (intergranular) porosity as well as by deformation], is described by the proportionality $\Delta l / l \sim t^{1/2}$. This was later confirmed for high purity MgO [19-21].

The present work investigates the shrinkage kinetics on sintering the important refractory oxides, Al_2O_3, MgO, and ZrO_2, and the compound $MgO-Al_2O_3$. The study was performed at temperatures up to 1500°C (where sintering occurs by removal of open porosity) and at temperatures of 1600-1750°C (where sintering occurs by removal of closed porosity). The effect of pressure on the sintering of the above oxides was also studied. The interrelation between sintering and recrystallization was examined through the kinetics of sintering by heterodiffusion (for solid-solution formation) and by reactive diffusion (for compound formation).

The specimens were prepared by pressing the pure oxides at a pressure of 1000 kg/cm^2. The content of the base component was not less than 99.8% for MgO and Al_2O_3, and over 98.5% for ZrO_2. The magnitude of the shrinkage was measured using the dilatometer described in [14, 15]; a microscopic study was carried out using slides, polished specimens, and immersion preparations.

The study revealed that during gradual specimen heating according to [2], the kinetics of shrinkage due to removal of open porosity is determined (for all oxides studied and magnesium

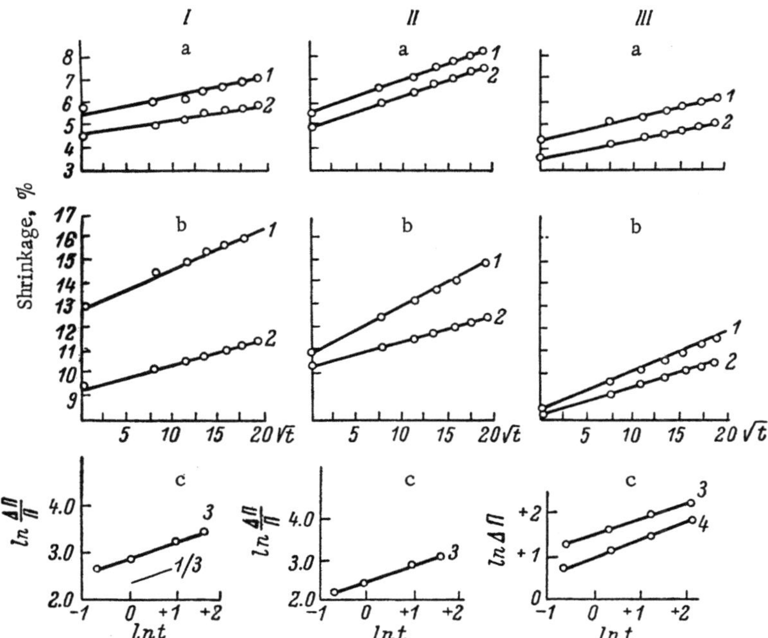

Fig. 1. Kinetics of shrinkage at temperatures of: 1) 1500°C; 2) 1400°C, for specimens of: I) MgO; II) ZrO_2; III) Al_2O_3. a) Annealed at 1750°C; b) unannealed; c) the densification kinetics of the same oxides at a temperature of 1700°C (3), and for corundum with 0.1% MgO additive (4). According to the shrinkage data of (b), the following were calculated: $D_{MgO} = 1.15 \cdot 10^{-13}$ cm^2/sec, 1400°C; $D_{MgO} = 4.15 \cdot 10^{-13}$ cm^2/sec, 1500°C; $D_{ZrO_2} = 1.30 \cdot 10^{-13}$ cm^2/sec, 1400°C; $D_{ZrO_2} = 7.80 \cdot 10^{-13}$ cm^2/sec, 1500°C; $D_{Al_2O_3} = 3.70 \cdot 10^{-14}$ cm^2/sec, 1400°C; $D_{Al_2O_3} = 1.82 \cdot 10^{-13}$ cm^2/sec, 1500°C; $Q_{MgO} = 78$ kcal/mole; $Q_{ZrO_2} = 75$ kcal/mole; $Q_{Al_2O_3} = 114$ kcal/mole.

spinel) by the proportionality $\Delta l / l \sim t^{1/2}$ (Fig. 1), and is described by the equation

$$\frac{\Delta l}{l} = m + n \sqrt{t}.$$

Here, $\Delta l / l$ is the overall shrinkage, m is a coefficient representing the amount of shrinkage up to the moment of isothermal soaking, and n is a coefficient characterizing the rate of isothermal shrinkage.

The coefficients in the equation of shrinkage kinetics depend on the composition of the material, its previous history, and a number of technological factors. Corundum has the lowest isothermal shrinkage rate. The coefficients of the isothermal shrinkage equation for MgO are intermediate in value, and those for ZrO_2 have a maximum value. This attests to the fact that the rate of isothermal shrinkage increases in the series $Al_2O_3 \rightarrow MgO \rightarrow ZrO_2$ at temperatures up to 1500°C. Correspondingly, the overall specimen shrinkage also changes (Fig. 1a, b).

The diffusion coefficients and sintering activation energies computed from data on the shrinkage of finely dispersed powders have entirely reasonable values (Fig. 1). They agree with data obtained in other studies both by direct measurement of the volume diffusion coefficients,

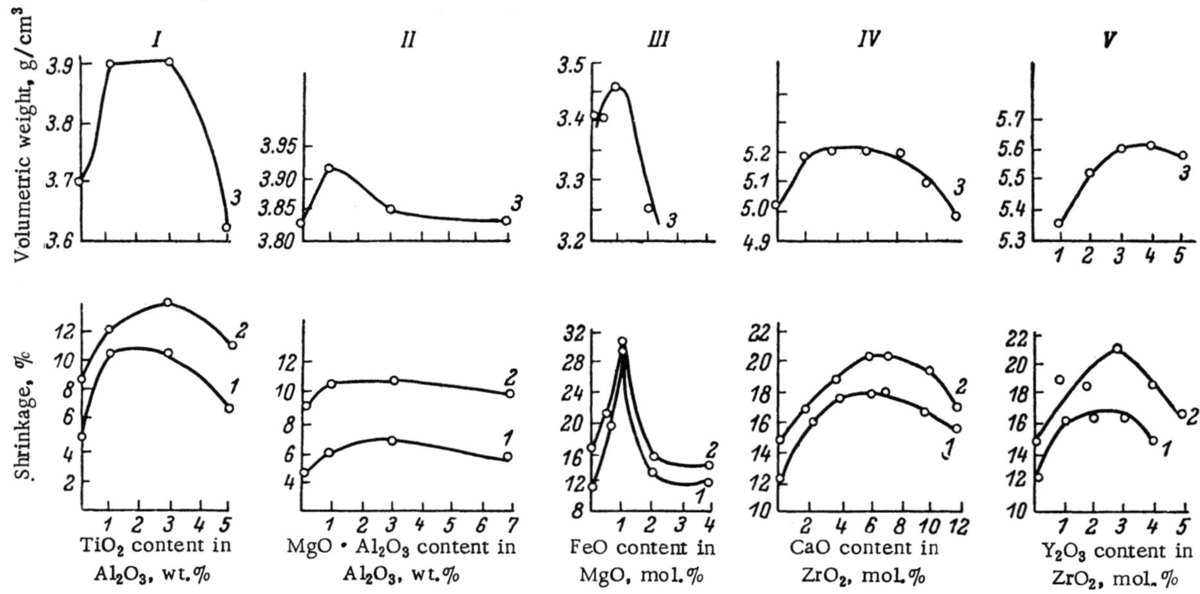

Fig. 2. Dependence of shrinkage at 1400°C (1) and 1500°C (2); 3) volumetric weight during 1500°C sintering of Al_2O_3, as a function of the amount of TiO_2 (I), $MgO \cdot Al_2O_3$ (II) additives; for sintering of MgO as a function of the amount of iron (III); and for sintering of ZrO_2 as a function of CaO (IV) and Y_2O_3 (V) additives.

Table 1. Effect of Introduction of Powders Produced by Decomposition of Basic Magnesium Carbonate and Zirconium Nitrate on the Coefficients of the Equation of Shrinkage Kinetics for Sintering, Respectively, of MgO and ZrO_2 at 1500°C

Amt. of activated powder introduced	Coefficients of the equation of shrinkage			
	m		n	
	MgO	ZrO_2	MgO	ZrO_2
0	13.0	11	0.188	0.257
20	28.5	27	0.272	0.305
30	32.5	31	0.322	0.333
50	36.2	38	0.322	0.361

as well as by sintering data [6, 9, 19-25]. This allows confirmation that the sintering of refractory oxides and their compounds is accomplished by volume diffusion of vacancies in the temperature range studied.

The coefficients of the kinetic equations of shrinkage and the overall shrinkage (Fig. 1b) sharply increase with use of "activated" powders — particularly on introduction of activated powders produced by low-temperature decomposition of salts (Table 1). However, the character of the kinetics does not change during this. This is associated with the presence of an excess of structural defects in the "activated" powders, which play a role in the stages of sintering. These can be intergranular (interblock) boundaries [26, 27], whose expanse rises with an increase in the content of finely dispersed particles. The original defects are removed

Table 2. Effect of Reactive Heterodiffusion during Formation
of Spinels on Coefficients of the Shrinkage Equation

Spinel	Coefficients of shrinkage equation			
	m		n	
	1400°	1500°	1400°	1500°
Previously synthesized at 1750°C.	5.4	10.0	0.128	0.144
Forming during synthesis	5.1	8.6	0.100	0.106

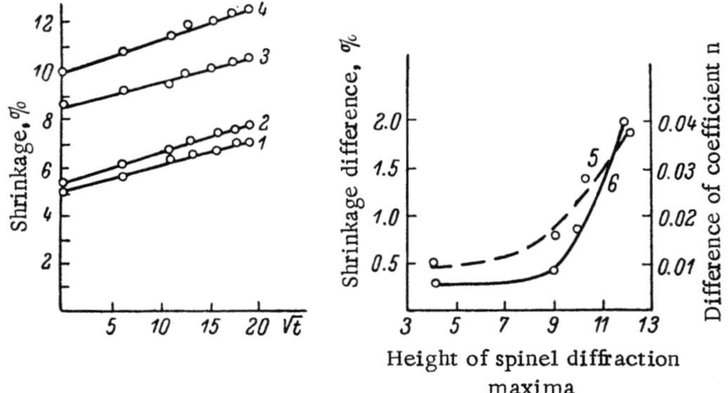

Fig. 3. Shrinkage kinetics at: 1,2) 1400°C; 3,4) 1500° for
spinels previously synthesized at 1750° (1, 3) and in the
process of synthesis (2,4), as a function of the difference
in overall shrinkage (6) and the coefficient n (5) of the
equation of shrinkage kinetics of a previously synthesized
spinel and in the process of synthesis from an amount of
the resulting spinels, as expressed by the amplitude of
diffraction lines.

extremely quickly [4]; however, an excess of defects can occur during the final anneal as a result of creep of dislocations and scattering of dislocation boundaries [28].

A study of sintering at higher temperatures (1650-1750°C), where the open porosity is already largely eliminated and sintering occurs through removal of closed porosity, revealed that
the fundamental proportionality of the densification kinetics is $\Delta\pi/\pi \sim t^{1/3}$ (Fig. 1c), where π
is the porosity. This was theoretically predicted in [10, 11] and experimentally confirmed for
NaCl in [29] and for corundum by ourselves [30]. Our study of the kinetics of change of the width
free of pores in the crust attests to a $\xi \sim t^{1/3}$ dependence (where ξ is the crust width), i.e.,
the kinetics for removal of closed pores is determined by the same proportionality [30].

The formation of solid solutions in refractory oxides does not change their shrinkage
kinetics of $\Delta l/l \sim t^{1/2}$ for open (intergranular) and $\Delta\pi/\pi \sim t^{1/3}$ for closed (intragranular) porosity.

Furthermore, the formation of small amounts of solid solutions, which increase the
amount of defects in the crystal lattice, facilitates an increased degree of sintering in the basic
oxide (Fig. 2). However, further increase in the amount of additive leads to an intensive growth
of solid-solution formation by the process of heterodiffusion (as indeed occurs in the

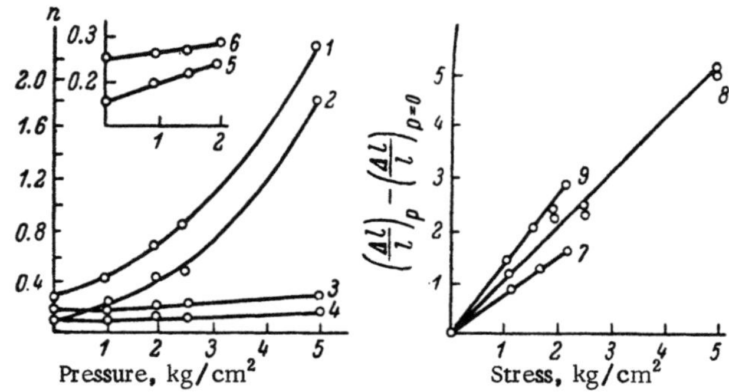

Fig. 4. Dependence of the coefficient n of the equation for
shrinkage kinetics: of corundum under pressure at 1400°C
(2, 4) and 1500°C (1, 3), of MgO (5) and ZrO_2 (6) at 1500°C
for direct (1, 2) and gradual (3-6) specimen anneal, and dependence of the additional shrinkage effect on sintering of
corundum (8), MgO (7), and ZrO_2 (9) under pressure on the
amount of such pressure.

Al_2O_3—$MgO \cdot Al_2O_3$, MgO—FeO, ZrO_2—CaO, and ZrO_2—Y_2O_3 systems [44-46]). It is known that
the coefficients of heterodiffusion and reactive diffusion are larger than the coefficient of self-
diffusion [31]. It is therefore possible for sintering to be retarded by the intensive growth of
the processes of heterodiffusion. It is usual for a decrease in the degree of sintering of ZrO_2
to be observed on introduction of different amounts of CaO and Y_2O_3 additives, and for the same
amount of cubic solid solutions to result.

Reactive diffusion during the formation of aluminum titanate in the Al_2O_3—TiO_2 system
(the presence of which was confirmed by x-ray analysis) also reduces the degree of sintering
of corundum (Fig. 2I).

A reduction in the degree of sintering by reactive diffusion is demonstrated in the study
of sintering during formation of magnesium spinels from stoichiometric mixtures of MgO and
corundum. The overall shrinkage (Fig. 3, curves 1-4) and the coefficients of the kinetic equation of shrinkage (Table 2) are considerably lower for the sintering of mixtures of the original
components than for sintering the previously synthesized spinels. This appears to be
the result of reactive diffusion during the spinel synthesis which outstrips the sintering process.
In this process, the more intense the spinel formation (gauged by the heights of the spinel diffraction maxima), the larger the difference in the value of the overall shrinkage and the isothermal shrinkage rate of specimens of previously synthesized spinels or their stoichiometric
oxide-mixtures (Fig. 3, curves 5 and 6).

Thus, reactive diffusion, which does not alter the kinetics, sharply reduces the degree of
sintering in refractory oxides and their compounds.

The application of small external loads with hydrostatic pressure sharply increases the
degree of refractory oxide sintering, while not changing the character of the shrinkage kinetics
[32]. The coefficient n, characterizing the rate of isothermal shrinkage of corundum for the
usual (not gradual) sintering of specimens, increases exponentially with increase in load (Fig. 4,
curves 1 and 2). The absence of the linear proportionality (typical of diffusional sintering), between the coefficient n and the load appears to be the result of a nonequilibrium state in the
system (pellets of finely pulverized powder).

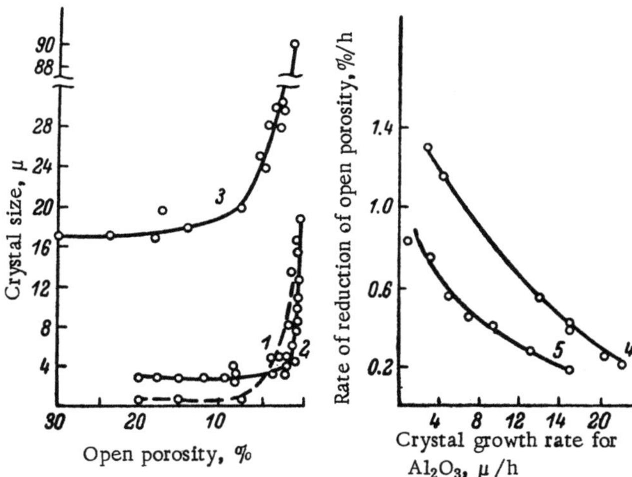

Fig. 5. Dependence of crystal size during the sintering
of specimens of MgO (1), corundum (2), and spinel (3)
on the values of their open porosity, and the dependence
of the rate of reduction of closed porosity of corundum
at 1600°C (4) and 1700°C (5) on the rate of its crystal
growth.

A prior anneal of the corundum samples (using gradual heating), which brings the system closer to equilibrium, sharply reduces the isothermal shrinkage rate. During this the dependence of the coefficient n on the applied stress remains linear. This attests to the fact that the sintering process of corundum on application of small loads (up to 5 kg/cm^2) occurs in the same fashion as without loading, i.e., by means of vacancy diffusion. The establishment of a proportionality in shrinkage during the gradual anneal, and the deformation of corundum samples also attest to this.

Furthermore, a linear dependence is observed for corundum between the additional shrinkage effect under action of applied pressure, and the value of such pressure (Fig. 4, curve 8), such as occurs for metalloceramic compacts [34]. The corresponding dependences are obtained for MgO and ZrO$_2$ at loads up to 2 kg/cm^2 (Fig. 4, curves 5, 6, 7, 9).

The sintering process of the refractory oxides and their compounds predominates over the process of their recrystallization, such that (according to [35]), the approximate average grain size (D_f), the average diameter of inclusions (d) as pores, and their specific fraction (f) in the specimen are interrelated by the equation $D_f = d/f$. For metalloceramic compacts, the retarding effect of macroscopic pores on the grain boundary motion and the overall recrystallization process was conclusively shown [36]. The pore size is usually 0.1 times the average grain diameter. Therefore, according to [37, 38], the grain growth can begin with a volumetric fraction of pores equalling $f = d/D = 0.1$, which we confirmed for corundum [39]. Our data on the recrystallization of corundum, MgO, and spinel (Fig. 5) also attest to this. The recrystallization of MgO begins at a somewhat higher porosity (~15%) than does recrystallization of corundum, while the recrystallization of spinel occurs at even a higher porosity (~18%). Since the force actuating grain growth is proportional to the surface energy [36, 40], it is possible that the large surface energy of MgO and spinel indeed demands their recrystallization with a large open porosity.

For the last stage of sintering in corundum, the growth rate of its crystals determines their internal porosity. On the one hand, a small residual amount of fine pores partially results in their inclusion within the growing crystal with intensive crystal growth [4, 38, 41-43]. On considerable supersaturation of the crystal lattice with vacancies the occurrence of diffusional porosity is also possible [4, 10]. On the other hand, a crust devoid of pores comes into being as a result of the diffusion of pores situated near the crystal boundaries [4]. The pores distributed in the central portion of the crystals remain, and their coalescence can occur. Thus, the removal of closed pores and the formation of a poreless crust on sintering should occur faster when the pores are nearer to the grain boundary, i.e., the slower the crystals grow. An increase in the crystal growth rate of corundum actually markedly reduces the decrease in closed porosity (Fig. 5, curves 4 and 5).

Therefore, in presence of a second phase surrounding the crystals of the primary oxide, the possibility exists for preparing a poreless body with theoretical density when the crystal growth rate is very small.

LITERATURE CITED

1. B. Ya. Pines, Zh. Tekh. Fiz., Vol. 16, No. 6, p. 738 (1946).
2. B. Ya. Pines, Usp. Fiz. Nauk, Vol. 52, No. 4, p. 501 (1954).
3. B. Ya. Pines and Ya. E. Geguzin, Zh. Tekh. Fiz., Vol. 23, No. 11, p. 2078 (1953).
4. Ya. E. Geguzin, Macroscopic Defects in Metals [in Russian], Metallurgizdat, Moscow (1962).
5. G. C. Kuczynski, J. Metals, Vol. 2, p. 169 (1949).
6. G. C. Kuczynski, L. Abernathy, and J. Allan, Kinetics of High-Temperature Processes, New York (1959).
7. W. D. Kingery and M. Berg, J. Appl. Phys., Vol. 26, No. 10, p. 1205 (1955).
8. E. A. Aitken, J. Am. Ceram. Soc., Vol. 43, No. 12, p. 627 (1960).
9. R. L. Coble, J. Appl. Phys., Vol. 32, No. 5, p. 787 (1961).
10. I. M. Lifshits and V. V. Slezov, Zh. Éksp. i Teor. Fiz., Vol. 35, No. 2(8), p. 473 (1958).
11. I. M. Lifshits and V. V. Slezov, Fiz. Metal. i Metalloved., Vol. 13, No. 6, p. 937 (1962).
12. R. I. Garber, V. S. Kogan, and I. M. Polyakov, Zh. Éksp. i Teor. Fiz., Vol. 35, No. 6(12), p. 1364 (1958).
13. É. V. Degtyareva and I. S. Kainarskii, Dokl. Akad. Nauk SSSR, Vol. 156, No. 4, p. 937 (1964).
14. É. V. Degtyareva, I. S. Kainarskii, and S. B. Totsenko, Ogneupory, No. 9, p. 400 (1964).
15. I. S. Kainarskii, É. V. Degtyareva, and S. B. Totsenko, Theoretical and Technological Studies in the Field of Refractories (Scientific Transations of the Ukrainian Scientific Research Institute for Refractories, No. 9), Izd. Khar'kovsk. Gos. Univ. (1965), p. 8.
16. I. G. Orlova and I. S. Kainarskii, Dokl. Akad. Nauk SSSR, Vol. 157, No. 2, p. 331 (1964).
17. I. G. Orlova, I. S. Kainarskii, and R. E. Mirkina, Ogneupory, No. 1, p. 40 (1965).
18. I. S. Kainarskii, I. G. Orlova, and R. E. Mirkina, Theoretical and Technological Studies in the Area of Refractories (Scientific Transactions of the Ukrainian Scientific Research Institute for Refractories, No. 8), Izd. Khar'kovsk Gos. Univ. (1965), p. 3.
19. P. P. Budnikov, M. A. Matveev, and V. K. Yanovskii, Dokl. Akad. Nauk SSSR, Vol. 159, No. 4, p. 872 (1964).
20. P. P. Budnikov, M. A. Matveev, and V. K. Yanovskii, Ogneupory, No. 4, p. 32 (1965).
21. P. P. Budnikov, M. A. Matveev, V. K. Yanovskii, and F. Ya. Kharitonov, Izv. Akad. Nauk SSSR, Neorg. Mat., Vol. 1, No. 8, p. 1349 (1965).
22. Y. Oishi and W. D. Kingery, J. Chem. Phys., Vol. 33, No. 2, p. 480 (1960).
23. Y. Oishi and W. D. Kingery, J. Chem. Phys., Vol. 33, No. 3, p. 905 (1960).
24. W. H. Rhodes and R. E. Carter, Am. Ceram. Soc. Bull., Vol. 41, No. 4, p. 283 (1962).

25. R. Lindner and G. D. Parffitt, J. Chem. Phys., Vol. 26, No. 1, p. 182 (1957).

26. B. Ya. Pines, Fiz. Metal. i Metalloved., Vol. 20, No. 1, p. 84 (1965).

27. Ya. E. Geguzin, Fiz. Metal. i Metalloved., Vol. 9, No. 6, p. 842 (1960).

28. Ya. E. Geguzin and V. L. Rabets, Phys. Solid State, Vol. 9, p. 893 (1965).

29. R. I. Garber, V. S. Kogan, and I. M. Polyakov, Zh. Eksp. i Teor. Fiz., Vol. 35, No. 6(12), p. 1364 (1958).

30. É. V. Degtyareva, Izv. Akad. Nauk SSSR, Neorg. Mat., Vol. 1, No. 2, p. 281 (1965).

31. B. Ya. Pines, Outlines on Metal Physics [in Russian], Izd. Khar'kovsk. Gos. Univ. (1961).

32. É. V. Degtyareva and I. S. Kainarskii, Izv. Akad. Nauk SSSR, Neorg. Mat., Vol. 2, No. 2, p. 239 (1966).

33. I. S. Kainarskii, I. G. Orlova, and É. V. Degtyareva, Dokl. Akad. Nauk SSSR, Vol. 164, No. 6, p. 1283 (1965).

34. Ya. E. Geguzin and B. Ya. Sukharevskii, Zh. Tekh. Fiz., Vol. 34, No. 9, p. 1613 (1954).

35. C. S. Smith, Trans. AIME, Vol. 175, p. 15 (1948).

36. Ya. E. Geguzin and L. N. Paritskaya, Fiz. Metal. i Metalloved., Vol. 12, No. 6, p. 900 (1961).

37. J. E. Burke, J. Am. Ceram. Soc., Vol. 40, No. 3, p. 80 (1957).

38. J. E. Burke, Processes of the Ceramic Industry [Russian translation], IL, Moscow (1960).

39. É. V. Degtyareva, Dokl. Akad. Nauk SSSR, Vol. 165, No. 2, p. 372 (1965).

40. J. E. Burke and D. Turnbull, in: Progress in Metal Physics [Russian translation], Metallurgizdat, Moscow (1956), p. 368.

41. B. H. Alexander and R. W. Balluffi, Acta Metallurgica, Vol. 5, No. 11, p. 666 (1957).

42. R. J. Bernard, Powder Metallurgy Bull., Vol. 3, p. 86 (1959).

43. S. G. Tresvyatskii, Ogneupory, No. 3, p. 130 (1960).

44. I. I. Vishnevskii, A. M. Gavrish, and B. Ya. Sukharevskii, X-ray Analysis of Mineral Raw Materials, No. 2 [in Russian], Gosgeolizdat, Moscow (1962).

45. B. Ya. Sukharevskii, I. I. Vishnevskii, and A. M. Gavrish, Dokl. Akad. Nauk SSSR, Vol. 140, No. 4, p. 884 (1961).

46. I. I. Vishnevskii and V. M. Skripak, Fiz. Tverd. Tela, Vol. 7, No. 10, p. 2925 (1965).

ACTIVATION OF SINTERING PROCESSES
IN MAGNESIUM OXIDE
AND IN SOME MgO-BASE COMMERCIAL PRODUCTS

V. A. Bron, M. I. Diesperova, L. P. Sudakova,
and I. A. Stepanova

The question of sintering low-fired MgO obtained from the crystalline carbonate or hydrate has considerable practical significance in connection with the problem of preparing fired magnesite powder. The problem has no less significance as applied to special, pure magnesium oxide ceramics.

The ease with which MgO sinters depends on many factors, including its origin, degree of purity, dispersion, and the structural characteristics of its particles [1-10].

A method which intensifies sintering of MgO is low-temperature annealing, which enables activation of subsequent sintering and recrystallization processes. There are different opinions as to the optimum temperature of preliminary anneal, ranging from 700-1300°C [2, 4, 5, 8, 11, 12]. The effect of the activation process rate on subsequent anneal is not yet clear.

As objects of the study we used varieties of MgO produced from the carbonate or hydrate. These differed with respect to their dispersion.

A study of the capacity of these products to sinter revealed that caustic magnesite (dust from a rotating furnace), the hydrate of such magnesite, and the chemically pure MgO all sinter especially poorly (Table 1). Thermal activation (intermediate calcination at 700-1200°C) was used to intensify the sintering processes in MgO and it was found that the most favorable activation temperatures for chemically pure MgO and $MgCO_3$ are 1100-1000°C. For the hydrates, the optimum calcination temperature is 700-1000°C, depending on particle dispersion and the nature of the starting material.

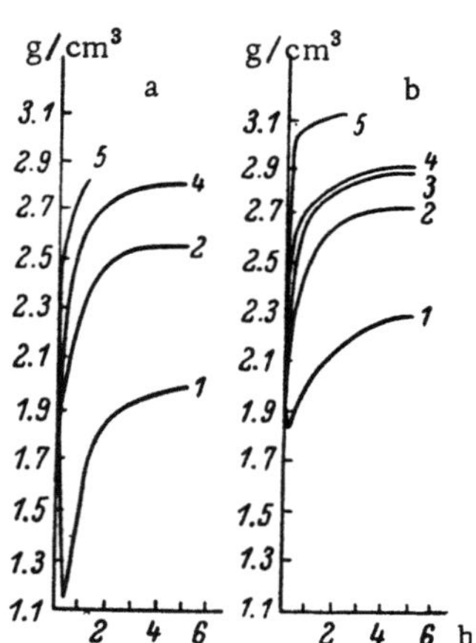

Fig. 1. Kinetics of isothermal sintering of magnesium oxide produced from the crystalline carbonate. a) Nonactivated; b) activated. Sintering temperature: 1) 1300°C; 2) 1400°C; 3) 1500°C; 4) 1600°C; 5) 1750°C.

196

Table 1. Porosity and Dispersion of Fired MgO Samples Obtained from Carbonates and Hydrates

Sample characteristics	Chem. pure MgO	Chem. pure MgCO$_3$	Caustic magnesite (dust waste)	Dispersed caustic magnesite	Hydrate of MgO from natural crystalline carbonate, annealed at 800°	Hydrate of MgO from dispersed carbonate (48-h grinding)	Hydrate of MgO from caustic magnesite
Specific surface of original product, m^2/g. . . .	2.23	1.68	0.12	0.27	0.30	1.51	0.98
Porosity of MgO samples (1600°C anneal), %. . .	48.8	7.8	36.2	5.3	17.2	1.1	32.4
Porosity of MgO samples subjected to intermediate activation at 1000° (1600°C anneal), %	21.8	0.7	14.5	0.3	0.8	–	5.4

Table 2. Effect of Heating Rate during Thermal Activation of MgO on Subsequent Sintering at 1600°C

Starting material	Conditions of heating and soaking	Density of samples after firing at 1600°C	
		porosity, %	volumetric wt., g/cm^3
Chemically pure MgO	Heating rate, deg/h:		
	120	21.1	2.81
	300	18.8	2.87
	600	12.7	3.13
Caustic magnesite (dust waste)	120	38.0	2.21
	300	21.1	2.46
	600	28.2	2.27
Chemically pure MgO	Soaking temperature on heating, °C:		
	400	17.4	2.96
	700	14.0	3.06
	1000	12.2	3.13
Caustic magnesite (dust waste):	Soaking on cooling:		
heating to 600°C	400°	26.3	2.62
	without soaking	18.4	2.86
heating to 800°	400°	21.3	2.78
	without soaking	4.4	3.14
heating to 1000°	400°	5.4	3.19
	without soaking	2.1	3.21

The heating rate of the material during the thermal activation has a substantial effect on sintering. When reduced heating or cooling rates are used in the comparatively low-temperature region (400-1000°C), the sintering becomes worse (Table 2).

An effect which we called thermal aging of MgO was observed in both the chemically pure and in the commercial products. During this it was found that MgO activated at 1000°C did not display a tendency toward poorer sintering, as a result of the slower low-temperature cooling. The aging effect on prolonged heating or cooling of MgO depends on the ordering of structural defects, as confirmed by a decrease in diffuseness of the periclase lines in the x-ray patterns.

Relative to the kinetic curves of low-fired MgO from the crystalline carbonate, there is a characteristic sharp rise in density over a very short time interval, especially on isothermal sintering at 1600°C or more. This permits clear delineation of two stages of the process: a

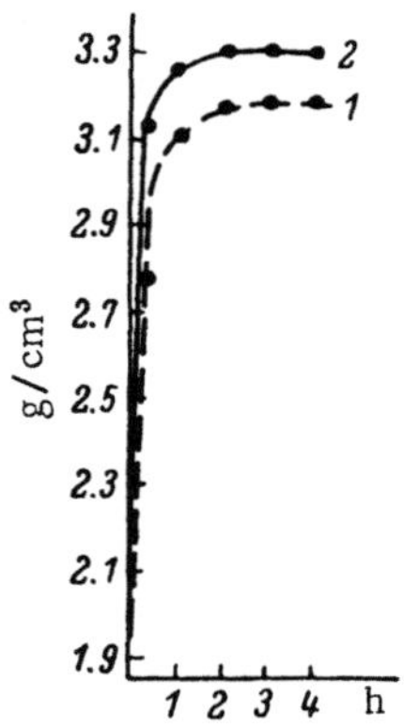

Fig. 2. Kinetics of isothermal sintering (at 1600°C) of dispersed MgO obtained from the crystalline carbonate. 1) Nonactivated; 2) activated.

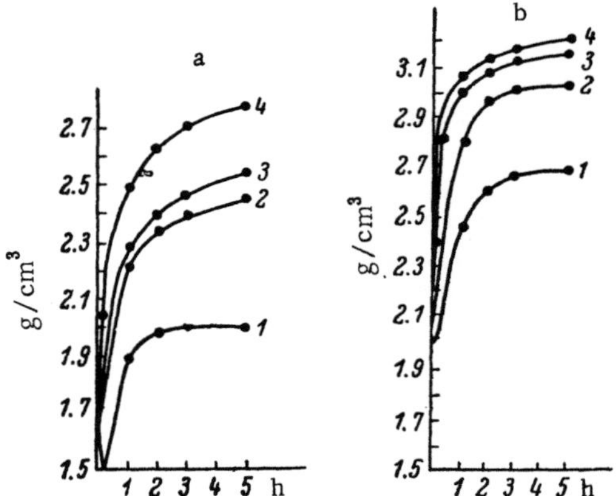

Fig. 3. Kinetics of isothermal sintering of MgO produced from the hydrate. a) Nonactivated; b) activated. Sintering temperature 1) 1300°C; 2) 1400°C; 3) 1500°C; 4) 1600°C.

Table 3. Effect of Heating Temperature of Magnesium Carbonate on the Crystal Lattice Parameter and Specific Gravity of Magnesium Oxide

Annealing temp. of $MgCO_3$, °C	Crystal lattice parameter, A	Specific gravity, g/cm^3	No. of atoms in a unit cell of MgO
600	Not determined	2.967	—
800	4.2135	3.333	7.35
1000	4.2127	3.512	7.90
1200	4.2122	3.525	7.95
1600	4.2105	3.567	8.00
	Normal periclase		

primary stage, with a very high sintering rate, and a secondary stage, with a substantially lower sintering rate (Fig. 1). With activation of MgO, the sintering rate in the primary stage increases. It rises even more with preliminary dispersion of the magnesium oxide (Fig. 2). Very similar kinetic curves are obtained for isothermal sintering of the MgO produced from magnesium oxide hydrate (Fig. 3).

The difference of sintering rates for the primary and secondary stages (Fig. 4) is not so sharply defined for the sintering of chemically pure MgO. This agrees with the literature data. The activation energy for the sintering process of MgO produced from the crystalline carbonate (computed from kinetic curves for the 1400–1600°C temperature range) equals 88.1 kcal/mole for the secondary sintering stage. The activation energy is considerably lower for activation in the secondary heat treatment of MgO at 1000°C, and amounts to 45.7 kcal/mole. The value of the activation energy for processes in the primary sintering stage is considerably lower. It was found to equal 11.5 kcal/mole for nonactivated and 39 kcal/mole for activated MgO. For magnesium oxide hydrate, the activation energies in the primary sintering stage were 26 kcal/mole for the nonactivated and 8.2 kcal/mole for the activated hydrate. Thermal activation and dispersion heighten the sintering efficiency in the primary stage.

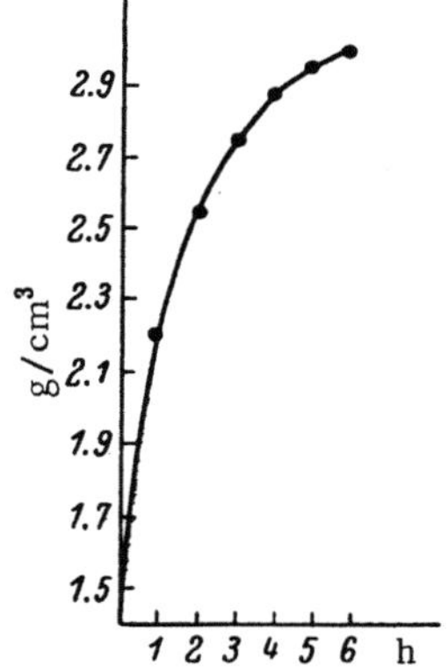

Fig. 4. Kinetics of isothermal sintering (at 1600°C) of chemically pure MgO.

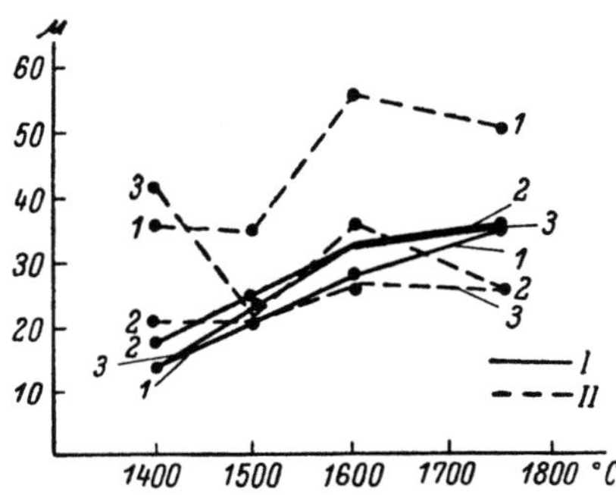

Fig. 5. Kinetics of the size variation of
the crystals and pores of periclase produced by sintering MgO at various temperatures. I) Average crystal size; II)
average pore size. 1) After grinding for
1 h; 2) after grinding for 48 h; 3) of activated MgO after grinding for 1 h.

X-ray determinations revealed that the crystal lattice parameter of MgO obtained at 600–1000°C
from the crystalline carbonate is distinct from the
parameter of periclase (Table 3). The specific
gravity (determined with the pycnometric method
by weighing samples in kerosene) systematically
rises with an increase in annealing temperature.
However, it remains less than that of periclase
even after an anneal at 1200°C.

Thus, MgO annealed at low temperatures has
a distorted crystal lattice with larger parameter
values. Computation of the number of atoms per
unit cell revealed the tendency toward a decrease
of the latter with lower calcination temperatures.

A study of the microstructure * of periclase
made possible delineation of the growth kinetics of
periclase crystals on sintering of MgO (produced
from the crystalline carbonate and hydrate) as a
function of temperature. It was established that
activation favors an increased growth rate (Fig. 5),
while the crystal size is approximately the same
(for the highest temperature in the conditions of
the experiment) independent of the properties of the
starting material. The kinetics of the change in
average pore size was also determined. At first
the latter decreases, but subsequently it begins to
rise with further increase in temperature (this is
caused by a decrease in the number of fine pores
as a result of coalescence).

A further increase in temperature, which
stipulates the migration of vacancies to intercrystal-
lite boundaries, leads to a new decrease in the average pore size due to removal of coalescing pores in
the periclase recrystallization process. The dis-
persion of MgO has an especially favorable effect on the microstructure of periclase – stipulating development of close contact between the crystals, as well as a minimal pore size. This is
well illustrated by the example of fired magnesium oxide hydrate produced from caustic magnesite after 48 h of grinding in a ball mill (Fig. 6).

Study of the intensely ground (48-h grinding) and briefly ground (1-h grinding) hydrates
under an electron microscope revealed a sharp difference in their structures. The briefly
ground hydrate has a crystal structure with well-developed brucite crystals in the form of hexagonal plates. The poor packing of the plates in different directions is consistent with the loose
structure of the hydrate. The brucite crystals are comparatively large (>1 μ).

After 48 h of grinding, the structure of the hydrate changes substantially. The crystal
fragments are held together by colloidal particles of semicircular or irregular shape. Although
it was not feasible to define the structure of the colloid, the formation of the colloidal particles

---

*Conducted by M. V. Medyakova.

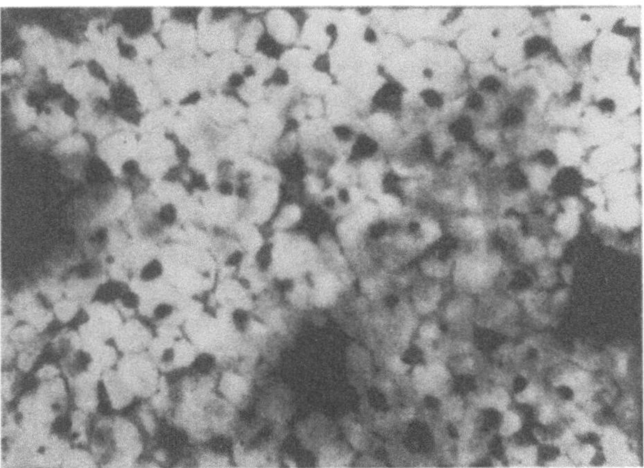

Fig. 6. Microstructure of MgO from the hydrate fired
at 1750°C, subjected to prolonged grinding (48 h). Reflected
light, 170×.

is evidently associated with the disintegration of the hydrate structure and the formation of a
complex containing the H^+ ion. The decomposition of the hydrate structure is confirmed by a
thermographic study. A considerable decrease in the endothermic effect caused by decomposition of brucite on heating is observed after a 48-h grinding.

The possibility for formation of a colloid containing a positive H^+ charge is also confirmed by the fact that the pH of the environment decreases with increase in duration of wet
grinding; consequently, additional positive charges are being generated, although the environment remains alkaline.

Thus the sintering of magnesium oxide hydrate obtained after prolonged wet grinding is
associated not only with increased dispersion, but also with the abrupt change of crystal structure and formation of a colloidal hydrate.

CONCLUSIONS

1. Sintering processes were studied for low-temperature magnesium oxide obtained from
the carbonate or hydrate, and for several commercial products with it as base.

2. Thermal activation, a very effective means of intensifying sintering, is independent of
the structural characteristics and particle size of the magnesium oxide being sintered.

3. Intense sintering is achieved without thermal activation where there is a high degree
of dispersion in the magnesium oxide hydrate.

4. For isothermal sintering, the low-temperature MgO (obtained from the natural crystalline carbonate) sinters intensely in a primary stage, characterized by a short time interval.
Thermal activation or dispersion increases the sintering rate.

5. The sintering characteristics of the low-temperature magnesium oxide are dependent
on a defect structure in the periclase lattice which has increased parameter values.

6. The dependence on sintering temperature of the kinetics of periclase recrystallization
and pore formation was explained.

7. The intense sintering of highly dispersed magnesium oxide hydrate is dependent not
only on the grinding of the brucite crystals, but also on the formation of a colloidal hydrate containing the positive H^+ ion.

LITERATURE CITED

1. A. S. Berezhnoi, Transactions of the Third All-Union Conference on Refractory Materials [in Russian], Izd. Akad. Nauk SSSR, Moscow (1947), p. 161.
2. G. V. Kukolev, I. E. Dudavskii, N. P. Volobueva, and G. Z. Dolgina, in: Proceedings of the Khar'kov Institute of Refractories, No. 46 [in Russian], Metallurgizdat, Khar'kov (1941).
3. W. R. Eubank, J. Am. Ceram. Soc., Vol. 34, No. 8, p. 225 (1951).
4. S. G. Tresvyatskii and P. L. Volodin, in: Scientific Work on the Chemistry and Technology of the Silicates [in Russian], Promstroiizdat, Moscow (1956), p. 248.
5. P. P. Budnikov, Dokl. Akad. Nauk SSSR, No. 5, p. 339 (1950).
6. S. G. Tresvyatskii, Ogneupory, No. 5, p. 229 (1958).
7. I. S. Kainarskii and I. D. Nazarenko, Ogneupory, No. 2, p. 66 (1956).
8. P. W. Clark, I. H. Cannon, and I. White, Trans. Brit. Ceram. Soc., Vol. 52, No. 1, p. 1 (1953).
9. P. W. Clark and I. White, Trans. Brit. Ceram. Soc., Vol. 49, No. 7, p. 305 (1950).
10. P. P. Budnikov, Zh. Prikl. Khim., Vol. 20, No. 4, p. 319 (1947).
11. D. N. Poluboyarinov, P. L. Popil'skii, and Chiang Tung-hua, Ogneupory, No. 2, p. 80 (1961).
12. P. P. Budnikov, P. L. Volodin, and S. G. Tresvyatskii, Ogneupory, No. 2, p. 7 (1960).

PART VI

STUDY OF CEMENT CLINKER

ORDERED AND DISORDERED STRUCTURES IN TRICALCIUM SILICATE AND ITS SOLID SOLUTIONS

A. I. Boikova, N. A. Toropov, V. T. Vavilonova, G. N. Samsonkina, and Yu. G. Sokolov

The solid solutions and polymorphism of the highly basic silicate $3CaO \cdot SiO_2$, which provide the engineering advantage of cement, constitute one of the most complex problems in the chemistry of cement and other inorganic binding materials. The structural transformations in tricalcium silicate solid solutions, known as alites in cement clinker, are of particularly great importance.

The study of the polymorphic modifications of $3CaO \cdot SiO_2$ and its solid solutions is directly associated with the current development of ideas regarding the existence of states of varying degrees of crystal lattice ordering. It is also related to ideas regarding the frequent occurrence of complex silicate structures [1] (feldspar [2], cordierite [3, 4]).

This applies to the structures involved in the transition of the less-ordered to the more-ordered state, and vice versa. This question is of much practical importance, since a change in material properties depends on a redistribution of atoms to form a crystal lattice of a different degree of ordering.

A theory of the ordering and disordering of states has been adequately developed for metals which partially explains the great simplicity of their structures as compared to the silicates. Although the most definite ordered or disordered state can be characterized only after the fine structure of the crystal investigated is known, nevertheless in some cases the disordering of a crystal can be predicted from thermographic studies and from x-ray data. Thus, the transition from order in the low-temperature forms of sanidine ($KAlSi_3O_8$) and high albite ($NaAlSi_3O_8$) to disorder in the high-temperature forms is entirely confirmed by data from a thermographic study, and can be predicted on the basis of these data [2].

Our work characterizes the state of a different degree of tricalcium silicate solid-solution ordering on the basis of the results of ionization x-ray studies and differential thermal analysis (DTA).

The ideal order of particles in a crystal is only one of the possible ways for the spatial distribution of particles relative to each other. However, it is necessary to mention such factors as the concentration (for the case of solid solutions) and the thermal aspects affecting the degree of crystal lattice ordering, since it becomes apparent that a much larger number of configurations correspond to the disordered structure. A unique diffraction pattern will correspond to each distribution of particles in the crystal (to the ordered, disordered, and their intermediates). Such are distinguished from each other by their line intensities, by line splitting or by new superstructural lines. The formation of a superstructure in crystals is predominantly characterized by an increase in the crystal lattice unit-cell parameters. However, the reverse transition (from the more complex structure with a large unit cell to the simpler one) is accompanied by the formation of a substructure related to the original. This has a corresponding multiple decrease in the unit cell parameters.

Thus, the nature of the polymorphic transformations in tricalcium silicate, as based on the development of a superstructure, was explained by the x-ray investigation of the polymorphism in $3CaO \cdot SiO_2$. We simultaneously studied the $3CaO \cdot SiO_2$ polymorphism by a thermal method and under a high-temperature microscope, by observation of the change of the crystal interference colors at the point of transition of one modification to another.

The scheme revealed for the tricalcium silicate transformations by precision, high-temperature x-ray analysis, thermal analysis, and high-temperature microscopy is the following [5, 6, 7]: triclinic form $TI \underset{}{\overset{620°}{\rightleftarrows}}$ triclinic form $TII \underset{}{\overset{920°}{\rightleftarrows}}$ triclinic form $TIII \underset{}{\overset{980°}{\rightleftarrows}}$ monoclinic form $\underset{}{\overset{990°}{\rightleftarrows}}$ orthorhombic form $\underset{}{\overset{1050°}{\rightleftarrows}}$ hexagonal form.

A change of symmetry is observed in the pure tricalcium silicate at 1050°C, but a thermal effect does not occur on the DTA curve. The characteristic (220), (201), and (204) lines undergo a slight splitting, and remain closely crowded together. Another change of lattice symmetry occurs at 990°C, the transition to the monoclinic form. During the 980°C transition from the monoclinic to trigonal form, the (220), (201), and (204) lines form a triplet. A doubling of the a and b parameters is observed, and the superstructure I results. An additional doubling of parameters is possible at 920°C, with the development of superstructure II. This is related to the ordered tricalcium silicate and reflects one of the basic principles characterizing the ordered state of crystals. This is founded on the spatial expanse of the order, repeating itself by a measurable amount in a chosen cell direction [8]. A small expansion of the lattice occurs on heating through 620°C.

The mechanism of superstructure formation during polymorphic transformations is such that regroupings occur, in consequence of which new planes arise which give additional reflections or lines on the x-ray patterns [9, 10].

The transition from the less- to more-ordered state can be accompanied by a change of crystal symmetry.

The process of superstructure formation involves a change in the energetic state of the system, which is revealed by the corresponding thermal effect on the heating or cooling curve.

The following factors have a predominant effect on the change in the degree of ordering during solid-solution formation: previous heat treatment (anneal, quenching and annealing temperature), the chemical nature of the added component, and its concentration. An increase in temperature promotes an increase in disorder and dispersion of the cations. The cooling regime (anneal, quench) has a strong effect on the ordering process of particles in the solid solution lattice. With certain cooling conditions the atoms tend to occupy definite sites in the lattice, i.e., they change from a disordered to ordered arrangement. The ordering process is diffusional (the transformation involves atomic substitution; therefore, slow cooling promotes ordering).

Substitutional disordering is typical for all types of solid solution disordering. It occurs when a number of crystallographically equivalent sites contain atoms of more than one chemi-

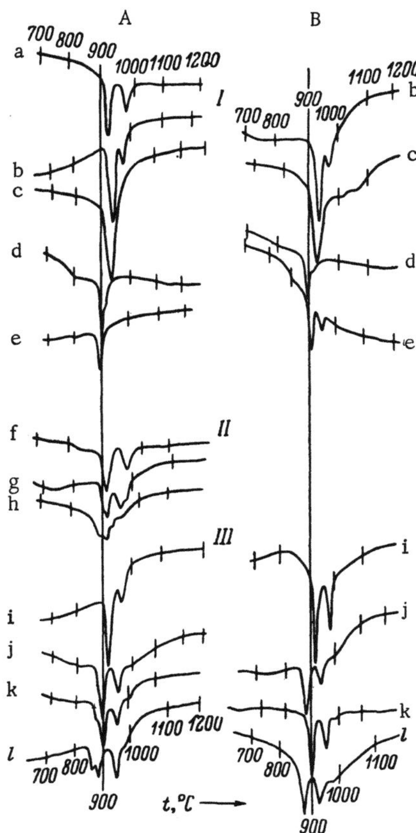

Fig. 1. DTA curves of quenched (A) and annealed (B) solid solutions. a) Pure $3CaO \cdot SiO_2$; b) with 0.5% MgO; c) with 1.0% MgO; d) with 4.0% $Y_2O_3 \cdot SiO_2$; e) with 5.0% $Y_2O_3 \cdot SiO_2$; f) with 1.0% Cr_2O_3; g) with 2.0% Cr_2O_3; h) with 1.0% Al_2O_3; i) with 3.0% $Sc_2O_3 \cdot SiO_2$; j) with 2.0% $La_2O_3 \cdot SiO_2$; k) with 3.0% $La_2O_3 \cdot SiO_2$; l) with 5.0% $La_2O \cdot SiO_2$.

cal species, distributed in a disorderly array over all the sites represented by this number. In the solid solutions, specific atoms of the solvent or matrix structure are replaced by solute atoms of similar chemical properties to form a continuous series of compositions.

The lattice of tricalcium silicate in industrial clinkers always contains some amount of impurities, which form solid solutions. We studied the effect of the oxides of aluminum, chromium, magnesium, strontium, yttrium, and of the rare-earth elements (lanthanum, neodymium, gadolinium, erbium, and scandium), on the structural transformations in the resulting solid solutions.

Although there are no complete data on the direct x-ray determination of the fine structure of tricalcium silicate and its solid solutions, nevertheless the results obtained by combined methods (ionization x-ray exposure using a Geiger—Müller counter and DTA studies) enabled an assessment as to the occurrence of a disordered state in the solid solutions.

The effect of atomic impurities on the structural transformations of tricalcium silicate is of a very diverse nature. We examined the effect of the chemical nature of the added impurity, its concentration, and the regime of heat treatment (anneal, quench, annealing temperature).

We divided the additives studied into three groups (Fig. 1 I, II, III), according to their effect on the thermal transformations in tricalcium silicate solid-solutions.

As seen from the figure, in the region most characteristic of the polymorphic transformations in tricalcium silicate (900–1000°C), additives such as Y, Mg, and also Gd (the DTA curves of the solid solutions with Gd are not presented) not only cause the polymorphic transformation temperatures to decrease, but also lead to lessening of the temperature interval between the two endothermic effects (Fig. 1b, d). The temperatures of the two endothermic transformations run together as a result of this process. This is achieved within the composition limits of solid solutions which contain respectively 2% MgO and 5% $Y_2O_3 \cdot SiO_2$.

If large amounts of the latter components enter the sample, then they appear in the preparation as independent phases. The presence of one endothermic transformation in the solid solutions examined is observed in the samples with magnesium (both quenched and annealed). However, in samples with yttrium at a $Y_2O_3 \cdot SiO_2$ concentration exceeding 4%, the annealed samples again display the two polymorphic transformations. As is known, slow cooling results in an ordering of the particle distribution in the solid-solution lattice. Solid solutions with two endothermic transformations correspond to the more ordered state. In the group of solid solutions

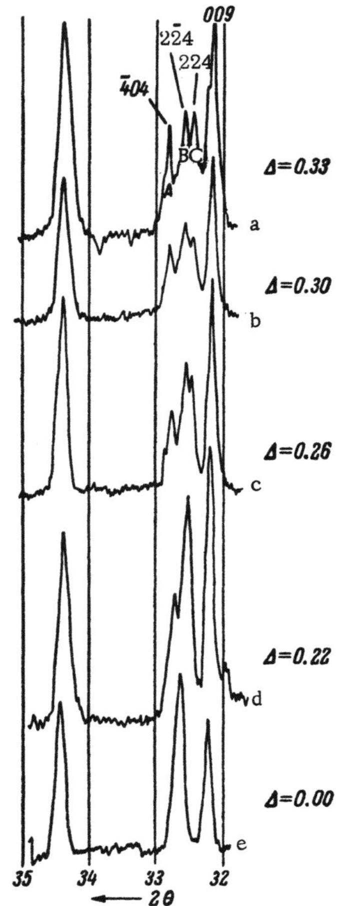

Fig. 2. Ionization intensity curves of quenched solid-solutions with impurities of different chemical nature. a) Pure $3CaO \cdot SiO_2$; b) with .5% MgO; c) with 1% MgO; d) with 4% $Y_2O_3 \cdot SiO_2$; e) with 5% $Y_2O_3 \cdot SiO_2$; f) with 1% Cr_2O_3; g) with 2% Cr_2O_3; h) with 1% Al_2O_3; i) with 3% $Sc_2O_3 \cdot SiO_2$; j) with 2% $La_2O_3 \cdot SiO_2$; k) with 3% $La_2O_3 \cdot SiO_2$; l) with 5% $La_2O_3 \cdot SiO_2$.

examined, the effect of both temperature and a concentration factor is observed on solid-solution ordering.

The second group of impurities, to which we assign chromium and aluminum, causes a sharply different effect on the thermal transformation in the solid solutions (Fig. 1f-h). Two endothermic transformations are observed at definite concentrations of each component (up to 1% Cr_2O_3 and 0.5% Al_2O_3), just as for the pure silicate, but at somewhat lower temperatures. However, an increase of Cr_2O_3 concentration above 1.5% and Al_2O_3 above 1% results in a splitting of the endothermic effects. This splitting is associated with a disruption in the regularity of atomic distribution in the solid-solution lattice. Such a state proved unstable, and on increase of impurity concentration led to decomposition of the solid solutions and formation of new phases. The picture depicted for the transformations characterizes both the quenched and annealed solid solutions.

The third group of impurities — strontium, scandium, and lanthanum — when added to the solid solution as the respective orthosilicates (Sr_3SiO_5, Sc_2SiO_5, and La_2SiO_5) do not basically change the character of the thermal transformations (Fig. 1i-l). Two endothermic transformations are observed over the temperature range examined in the annealed and quenched samples, as for the undoped tricalcium silicate. The solid solutions with lanthanum have transformations of a more complex nature (Fig. 1j-l) than those with scandium. A splitting of one of the endothermic effects occurs in the quenched samples. Slow cooling removes this effect, and the DTA curve again shows the presence of the two polymorphic transformations.

For the solid solutions examined, a complete series of ordered and disordered substitutions evidently occurs, since the unit cells are not altogether identical but differ in the atoms entering their composition. The impurity atoms differ from the atoms substituted in the matrix structure not only in size but also in electrostatic charge (divalent, Mg, Sr; trivalent, Al, Cr, Ln, Sc, Y; tetravalent, Si). The impurity atoms destroy both the anionic and cationic structural frameworks. Thus Cr^{6+} and Al^{3+} apparently can replace Si^{4+}, while Mg^{2+}, Y^{3+}, Sc^{3+}, La^{3+}, and Sr^{2+} can replace Ca^{2+}. Also, there lies the solid-solution composition in the voids of the structure. These factors indeed guarantee the possibility for formation of a complete series of disordered structures.

A change in the degree of ordering in the solid solutions studied was established by x-ray ionization techniques. The x-ray patterns were taken on a URS-50I apparatus, assembled in the "DRON" fashion with a BSV-8 tube. A Geiger—Müller counter was used together with Cu $K\alpha$ radiation and a 0.02-mm thick Ni filter. The counter-rotation rate was 0.1-0.2 deg/min.

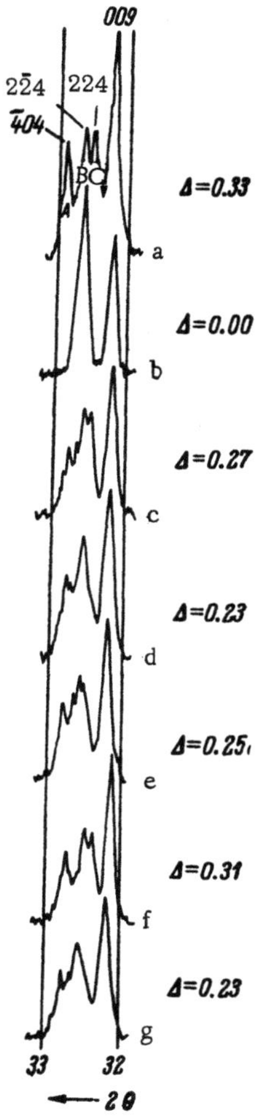

Fig. 3. Ionization intensity curves of solid solutions subjected to a different heat treatment. a) $3CaO \cdot SiO_2$ without impurity, annealed at 1500°C; b) with 5.0% $La_2O_3 \cdot SiO_2$, quenched; c) with 5.0% $La_2O_3 \cdot SiO_2$, annealed; d) with 1.0% Al_2O_3, quenched; e) with 1.0% Al_2O_3, annealed; f) with 1.0% Cr_2O_3 annealed at 1500°C in air atmosphere; g) with 1.0% Cr_2O_3 annealed at 1800°C in argon.

The analysis of the ionization intensity curves showed that within phase homogeneity limits, a splitting of the several maxima, or inversely their union, occurs as a function of the chemical nature and concentration of the added impurities (Fig. 2), and sample heat-treatment procedure. The most distinct occurrence of the splitting or union of maxima is observed in our determinations in the 2θ angular region of 32.5-33.5°, 52-53°, and 62-63°. The figures show the x-ray patterns taken in the angular region of 32-35°. As seen from Figs. 2 and 3, the ($2\bar{2}4$), (224), and ($\bar{4}04$) lines diverge or grow closer and combine depending on the above factors. In the x-ray pattern of a solid solution with $Y_2O_3 \cdot SiO_2$ (Fig. 2b, c), the maxima corresponding to the reflections from the ($\bar{4}04$), ($2\bar{2}4$), and (224) planes grow together more closely than the above $Y_2O_3 \cdot SiO_2$ concentration. The maxima combine on the x-ray pattern of a sample with MgO and Al_2O_3 (Fig. 2d), corresponding to the ($0\bar{4}4$) and ($\bar{4}44$) reflections, while the x-ray pattern of a solid solution with $La_2O_3 \cdot SiO_2$ (Fig. 2e), displays a combination of all three maxima.

The x-ray patterns shown in Fig. 3 show that the positions of the A, B, and C maxima depend on the regime of solid-solution heat-treatment (cooling regime and annealing temperature). The largest changes evidently occur in the structure of the solid solution with lanthanum. If the above maxima in the x-ray pattern of a quenched solid-solution combine (Fig. 3b), then the x-ray pattern of the annealed sample (Fig. 3c) again reveals reflections corresponding to the ($\bar{4}04$), ($2\bar{2}4$), and (224) planes. The structure of solid solutions with Al_2O_3 (Fig. 3d, e) evidently changes much less.

In order to characterize the degree of solid-solution ordering, we used the value Δ, described in this work, as the index of ordering. This was computed from the change of the 2θ angles between the additionally arising lines. The distance between the A, B, and C maxima changes with the degree of structural ordering. These distances are less for larger structural disorder. Thus, these distances or some values describing the relationship between them can be used as an index of the degree of ordering. It was found in [4] that the index of the degree of ordering which is most satisfactory is that computed by the formula

$$\Delta = 2\theta_A - \frac{2\theta_B + 2\theta_C}{2},$$

where $2\theta_A$, $2\theta_B$, and $2\theta_C$ are the angles of slide (in deg) of the maxima A, B, and C. Thus, this index is zero for a quenched solid-solution with lanthanum oxyorthosilicate, while for the annealed sample it is 0.27. A temperature rise promotes an increase in the degree of disorder in the cationic distribution. The index of ordering for solid solutions annealed at 500°C is 0.31; it is 0.23 for those annealed at 1800°C. The index of ordering changes as a function of the chemical nature of the impurity added from a value of $\Delta = 0.31$ in solid solutions with chromium to $\Delta = 0.00$ for solid solutions with lanthanum.

CONCLUSION

In the present work an attempt was made to describe the state of the different degree of ordering in tricalcium silicate solid solutions which form with a different chemical nature and composition of added impurities and with the regime of heat treatment.

Although we are not able to present exact structural data on this problem, the material obtained on the basis of ionization intensity curves and thermographic analysis can nevertheless be used for assessing the structural characteristics of tricalcium silicate, its polymorphism, its solid solutions, and for deciphering its fine structure.

LITERATURE CITED

1. G. Donnay, Y. Wyart, and G. Sabatier, Z. Krist., Vol. 112, p. 161 (1959).
2. H. D. Megaw, Mineral. Mag., Vol. 32, No. 246, p. 226 (1959).
3. A. Miyashiro, T. Jiyama, M. Yamasaki, and T. Miyashiro, Am. J. Sci., Vol. 253, p. 185 (1955).
4. A. Miyashiro, Am. J. Sci., Vol. 255, p. 43 (1957).
5. A. I. Boikova and N. A. Toropov, Dokl. Akad. Nauk SSSR, Vol. 156, No. 6, p. 1428 (1964).
6. N. Yannaguis, M. Regourd, C. Mazières, and A. Guinier, Bull. Soc. Franc. Mineral et Crystallogr., Vol. 85, p. 271 (1962).
7. M. Regourd, Bull. Soc. Franc. Mineral. et Crystallogr., Vol. 87, p. 241 (1964).
8. J. D. Bernal, Z. Krist., Vol. 112, p. 4 (1959).
9. A. Guinier, Heterogeneities in Solid Solutions, New York (1959).
10. A. Guinier, Théorie et Technique de la Radiocrystallographie, Paris (1956).

CRYSTALLIZATION OF CLINKER MELTS
IN THE 1900-1200°C TEMPERATURE RANGE

Yu. M. Butt, V. V. Timashev
and N. S. Panina

The present work is devoted to an investigation of a comparatively little-studied stage of the process of manufacture of crystallized Portland cement clinker by the method of melting. It also considers the crystallization process of fused clinkers on cooling. Its purpose is to improve the methods of effecting crystallization and mineral formation processes.

Portland cement raw materials, the characteristics of which are given in Table 1, were selected for the study. Mixture No. I was calculated to give a high-alite clinker, while mixtures Nos. II and III were designed to yield a medium- and low-alite clinker. The Portland cement clinkers were synthesized from chemically pure Al_2O_3, SiO_2, Fe_2O_3, and $CaCO_3$.

The initial mixtures were prepared by the wet homogenization method. They were dried and decarbonated 4 h at 900°C until constant weight. Mixtures prepared by this method were fired in a TVV-4 tungsten-resistance furnace in a helium atmosphere. The material was held in a molybdenum crucible. The maximum firing-temperature, at which all mixtures were found completely fused, was 1900°C.

The heating time for the material to reach the maximum temperature is 2 h. The soaking period at maximum temperature was 15 min to effect complete melt homogenization. The melt was cooled at a rate of 100°C/h to, respectively, 1800°, 1700°, 1600°, 1500°, 1400°, 1300°, and 1200°C. On attaining one of the above temperatures, fast cooling (quenching) to 600-800°C was performed at a rate of 800 deg/min. The clinker was then cooled to room temperature over a period of 30-40 min. This produced three series of clinkers, with eight clinkers in each series.

The uncombined CaO content in the fused clinkers was determined by the ethyl glycerate and petrographic methods. The clinker phase-composition was estimated by x-ray and petrographic analysis methods.

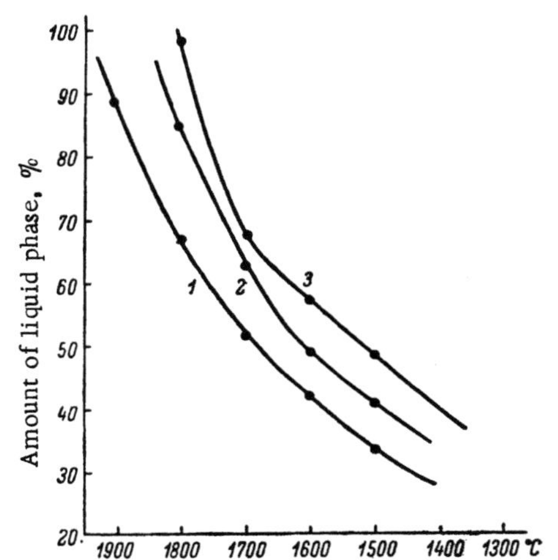

Fig. 1. Temperature dependence of the amount of liquid phase forming in clinkers. Clinker of 1) composition I; 2) composition II; 3) composition III.

Table 1. Chemical Composition of Starting Mixtures

Comp. No.	Chemical comp., %				Saturation coeff.	Modulus		Computed mineralogical comp.			
	SiO_2	Al_2O_3	Fe_2O_3	CaO		silica (n)	alumina (p)	C_3S	C_2S	C_4AF	C_3A
I	19.11	9.36	3.07	68.46	0.97	1.55	3.12	65.98	5.15	9.12	19.75
II	21.95	7.32	5.61	65.12	0.83	1.70	1.30	42.58	30.86	17.85	8.71
III	21.50	9.37	6.13	63.00	0.75	1.40	1.53	21.50	45.45	18.64	14.41

Table 2. Characteristics of Melt Compositions

Comp. No.	Melting temp., °C		Chemical comp.,%				Value of alumina modulus, P
	theoretical	actual	CaO	Al_2O_3	Fe_2O_3	SiO_2	
II	1920	1700	65.12	7.32	5.62	21.95	1.3
IIa	1500	1500	61.44	16.55	12.73	9.28	1.3
IIб	1450	1400	60.72	17.72	13.70	7.86	1.3
IIв	1940	1840	65.12	9.70	3.23	21.95	3.0

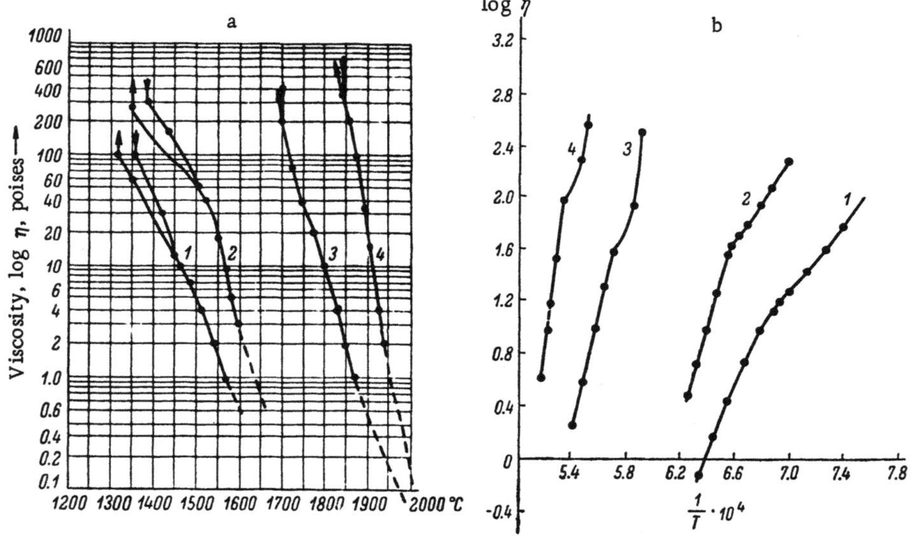

Fig. 2. a) Temperature dependence of melt viscosity; b) plot of the function $\log \eta = f(1/T)$. 1) Melt of composition IIb; 2) melt of composition IIa; 3) melt of composition II; 4) melt of composition IIc.

AMOUNT OF MELT AND ITS VISCOSITY

Figure 1 shows the change in amount of the liquid phase during equilibrium crystallization of the melts of compositions I, II, and III, computed on the basis of the $CaO-Al_2O_3-SiO_2$ phase-diagram. An increase in the value of the melt saturation coefficient from 0.75 to 0.97 is accompanied by a 150°C increase in the temperature at which the melts attain a truly liquid state. A change in amount of the melt also necessitates a change in its composition. Consequently, an interaction necessarily occurs between the melt and the precipitating crystal phases, evidenced by the

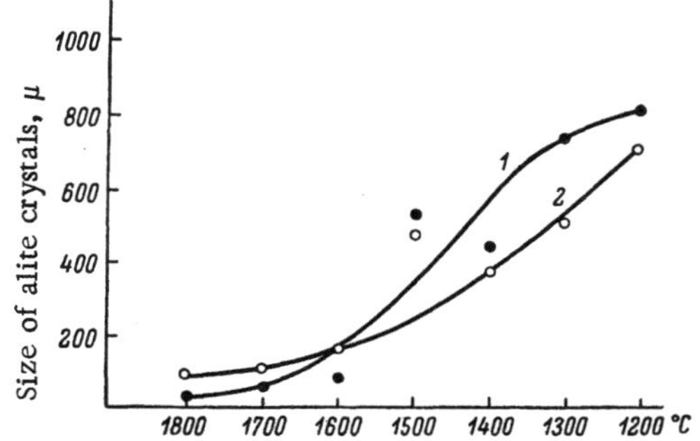

Fig. 3. Effect of melt cooling regime on size of alite crystals in clinkers. 1) Composition I; 2) composition III.

solution of some crystals and the growth of others [1]. The rate of these processes is determined by both the amount and the properties of the melt, especially its viscosity and the mobility of certain ions in it.

The viscosity of the following melts was determined (Table 2): (1) a completely fused melt of composition II; (2) the liquid phase in a sinter of this composition after cooling to, respectively, 1500°C (IIa) and 1455°C (IIb); and (3) the melt IIc, which differed from melt II by the value of its Al_2O_3 modulus. The characteristics of the melt compositions are given in Table 2. Calculation of the composition of the melts IIa and IIb was performed according to the phase diagram of the $CaO—Al_2O_3—SiO_2$ system. The Fe_2O_3 content in the starting mixtures was then calculated according to the value of the Al_2O_3 modulus.

In determining the viscosity, the raw materials were preannealed 3 h at 1500°C and then pulverized to give a specific surface of 3000 cm^2/g. The viscosity determinations were performed on an ORGRÉS viscometer, which operates on the principle of rotating, coaxial cylinders.

The temperature dependence of the viscosity of melts II and IIc (Fig. 2a, b) obeys an exponential law, since the function $\log \eta$ (T) has a linear character. On increase of the melt silicate modulus (i.e., SiO_2 content), not only does the temperature at which these melts attain a truly liquid state increase, but also the melt viscosity increases. Thus, melt IIb (7.86% SiO_2), for which T_{melt} (the temperature at which the melt attains a truly liquid state) is 1450°C, has a viscosity of 11 poises. The melt IIa (9.28% SiO_2) has a viscosity of 48 poises at a T_{melt} of 1510°C, while melt II (21.95% SiO_2) has a viscosity of 210 poises at T_{melt} = 1700°C. For such melts to have a viscosity of 1 poise, the temperature of melt II should be 1860°C, that of melt IIa, 1640°, and that of melt IIb, 1560°C. It is typical of the melts studied that those of higher SiO_2 contents have a narrower crystallization interval.

The viscosity of the melts studied depends to a somewhat lesser extent on the value of the alumina modulus. The temperature for attaining the truly liquid state rises to 1400°C on increase of the alumina modulus from 1.3 (II) to 3 (IIc). The effect of the alumina modulus on the melt viscosity decreases with rising temperature. A temperature difference of approximately 70°C is sufficient to cause these melts to attain a viscosity of 1 poise. This means that in the high-temperature region (1700-2000°C), melts of differing alumina moduli approach one another with respect to their technological properties.

As follows from the data of Fig. 3, the properties and composition of the resulting melts can be different up to temperatures where the truly liquid state is attained, depending on whether the system is heated or cooled. Thus, a mass of composition IIb has a viscosity of 40 poise while being heated to 1400°C, but a mass brought to this temperature while cooling has a viscosity of only 25 poises. The above phenomenon attests to the development of different structures for the melts and crystal phases depending on heating or cooling, and to a principal lack of correspondence in the crystallization processes for this and the other case. It is to be expected that the mineral formation processes will also differ, depending on which scheme is used in the heating process.

The mobility of the melt ions also changes as a function of the viscosity. The relationship between the coefficient of ionic diffusion and melt viscosity is expressed by D_η = const, i.e., the

Table 3. Effect of Cooling Regime on Size and Shape of Crystals in Fused Clinkers

Temp. range of slow-cooling, °C	Shape of alite crystals in clinker sections			Av. size of alite crystals in clinker, μ			Respective axes of alite crystals in clinkers			Shape of belite crystals in clinker sections			Av. size of belite crystals in clinkers, μ		
	I	II	III	I	II	III	I	II	III	I	II	III	I	II	III
1900	Platy crystals corroded by melt	Elongated plates, corroded by melt	Rough fragments of alite plates	300	180	50	12.5	12	—	Fine, round crystals	Round	Round and dendritic crystals	12	7	—
1900–1800	Fragments of plates	Cleaved plates corroded by melt, with inclusions of intermed.matter	Dissolved and cleaved plates	15	90	90	4	4.5	6	Split crystals	Dendritic and round	Dendritic and round	6	12	12
1900–1700	Dissolved plates of irreg. shape	Intergrown and cleaved plates, irreg. hexagons	Plates partially disintegrated, corroded by melt	50	300	100	3.5	—	7.5	Fine round grains, oriented	Dendritic, cleaved or irreg. shape	Dendritic, cleaved or irreg. shape	15	30	100
1900–1600	Plates corroded by melt around periphery	Platy	Plates corroded by melt	60	200	150	4	10	2	Round and resorbed	Dendritic	Round	12	30	100
1900–1500	Plates corroded by melt	Plates with well-developed faces	Elongated plates and irreg. hexagons	560	500	500	5	—	6	Round, re-dissolved	Dendritic and round	Round	45	70	120
1900–1400	Melted plates	Intergrown plates	Reg. hexagonal tablets, irreg. hexagons; occasional plates	400	640	350	3	8	8	Round and cleaved	Dendritic and round	Dendritic and round	70	20	120
1900–1300	Intergrown plates with well-developed faces	Intergrown, irregular hexagons	Plates greatly elongated, corroded by melt	800	1000	500	8	—	8	The same	The same	The same	60	13	100
1900–1200	Plates with well-develop. faces	Plates and elongated hexagonal tablets	Plates, irregular hexagons, hexagonal tablets	800	720	—	7	—	—	Cleaved grains	Poorly shaped masses of round and irreg. shape	Round and cleaved	30	—	100

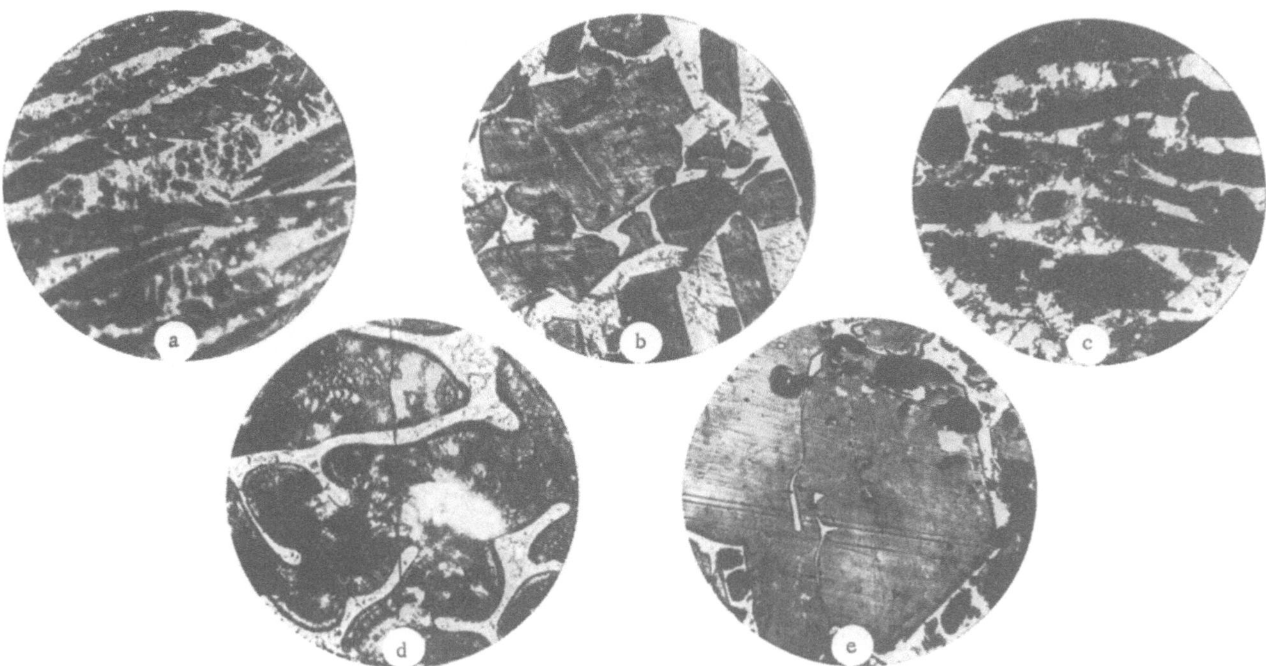

Fig. 4. Microstructure of clinkers; reflected light, 400×. a,b,c) Clinkers of composition I; d,e) clinkers of composition III. a) 1900–1700°C; b) 1900–1300°C; c) 1900–1200°C; d) 1900–1300°C; e) 1900–1200°C.

diffusion coefficient increases with a decrease in viscosity [2]. In turn, the ionic mobility is related to the diffusion coefficient by the Nernst—Einstein equation $D = KTB$, where K is the Boltzmann constant, T is the temperature, °K; and B is the ionic mobility, cm/sec · dyne.

The mobility of a Ca^{2+} ion in a melt having a viscosity of 1–2 poises at 1420–1480°C [3], obtained by our method of calculation, is 2.2–$2.9 \cdot 10^7$ cm/sec · dyne. Consequently, for melt clinkers of ordinary composition at 1800–2000°C, in cases where the melt viscosity is measured in ones or even fractions of a poise, the Ca^{2+} ion mobility is so large that the time necessary for an equilibrium distribution of Ca^{2+} throughout the liquid phase is negligibly small. However, the time for the equilibrium distribution of the silicon-oxygen and aluminum-oxygen radicals in the melt is two to three orders higher, and the resulting inadequate melt homogeneity with respect to the latter ions is reflected in the ensuing crystallization processes. However, the temperature regime during the cooling stage has a decisive effect on the crystallization process.

SIZE AND SHAPE OF CRYSTALS

Slow-cooling of the melts from 1900° to 1600–1500°C with a final quenching leads to a size increase of the alite crystals up to 560 μ for clinkers of composition I. In clinkers of composition II and III, the alite crystal size does not exceed 300 μ. In the sinters produced by slow-cooling the melts to 1500–1300°C, 1000-μ alite crystals are found (Table 3, Fig. 3).

For all clinker compositions studied, the maximum alite crystal size (350–1000 μ) is observed in the sinters produced by slow cooling the melts from 1900° to 1500–1400°C, with a final quench. The maximum alite crystal size decreases with slow-cooling the fused clinkers in the 1300–1200°C temperature range. However, the average size remains practically constant. The cleaved alite-crystals can be seen in the photomicrographs (Fig. 4c, e). X-ray and chemical analysis methods did not reveal the formation of any new phases in the above clinkers.

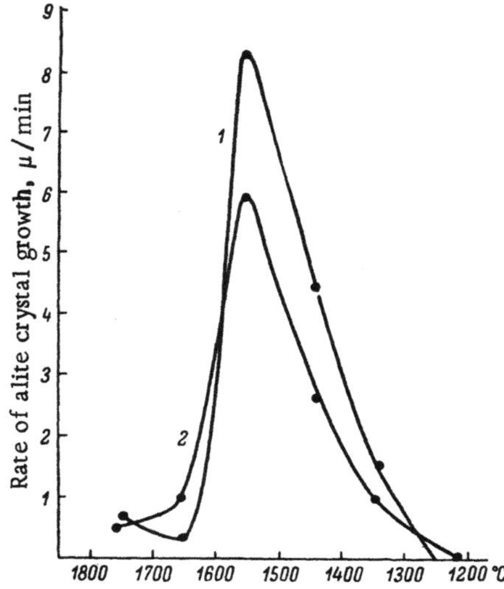

Fig. 5. Effect of melt cooling regime on alite crystal growth. 1) Clinker of composition I; 2) clinker of composition II.

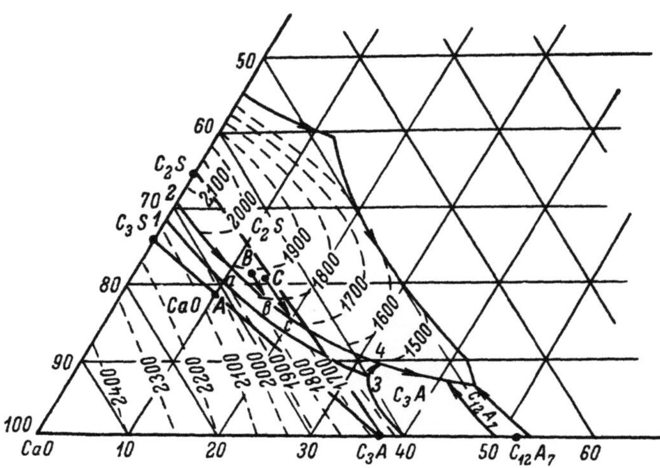

Fig. 6. Location of clinker compositions in the phase diagram of the CaO—Al₂O₃—SiO₂ system.

The size of the belite crystals in clinkers of compositions I and II varies independently of the cooling regime, within the limits of 4 to 70 μ. In clinkers of composition III, the belite crystals reach 160 μ in size. The belite crystal size increases negligibly on more prolonged soaking of the sinters in the 1600–1200°C temperature range.

Alite crystals of platy shape with an l:b ratio varying from 2 to 12.5 can be observed in all the clinkers studied. In this, of course, there is some deviation of crystal shape from the regular prismatic. Thus, alite crystals considerably redissolved by the melt, with inclusions of intermediate material (Fig. 4), are observed on slow cooling melts in the temperature range 1900–1700°C. The dissolution of the alite crystals is explained by the large amount of the reactive melt and an excess of $3CaO \cdot SiO_2$. The belite crystals in these clinkers have a round or dendritic shape.

Decreasing the temperature from which quenching was performed led to an ordering of the clinker structure. In clinkers rapidly cooled from 1500 to 1400°C, the alite crystals have distinct outlines. Together with this, alite crystals are also observed as hexagonal tablets and regular hexagons.

For clinkers with a saturation coefficient of 0.75 and 0.83 (compositions II and III), the belite crystals are typically of round and dendritic shape (Fig. 4), while for clinkers with a saturation coefficient of 0.97 (composition I), they are in the form of cleaved and only rarely round grains.

Consequently, both the size and shape of belite crystals change with the change in cooling regime.

GROWTH RATE OF ALITE CRYSTALS

The growth rates of alite crystals were calculated on the basis of the crystal-structure analysis of clinkers obtained on cooling melts of compositions I and II, and of III using a different regime. This corresponds to considering the difference in alite crystal size for clinkers of two different cooling regimes. It was noted for all compositions studied that the maximum growth-rate of

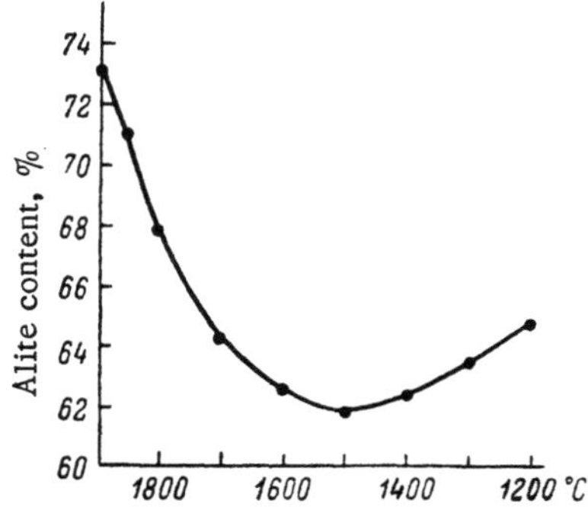

Fig. 7. Effect of melt cooling regime on alite content of clinkers of composition I.

alite crystals (4-8 μ/min) is observed in the temperature range 1400-1600°C (Fig. 5). The amount of melt in the sinter at these temperatures is 30-50%, and its viscosity is 1-40 poises (depending on composition). The rapid growth of crystals in this temperature range can depend on two reasons: first, melt supersaturation and, second, the involvement at these temperatures not only of fine structural elements (ions, etc.), but also of crystal nuclei or even coarser particles, which provide conditions for intermittent crystal growth.

The material being cooled over the temperature range 1800-1600°C is 60-100% composed of a melt in which all or a considerable portion of the silica is dissolved. Melts of such composition are distinguished by an increased tendency to crystallize resulting in a commensurate rate of crystal nuclei formation and crystal growth. As a result of this, the alite crystal size in clinkers obtained on slow-cooling melts in the above temperature range changes somewhat (from 15 to 60 μ for clinkers of composition I). The growth of crystals in the cooled material in the above temperature range occurs by the classical scheme. The crystal growth rate varies within the limits of 0.3 to 3 μ/min (Fig. 5).

Cooling the material over the temperature range 1400–1200°C does not result in a significant increase in alite crystal size, since there is practically no nutrient material in the melt for crystal growth. Under these conditions, crystal growth occurs as a result of the dissolution of metastable crystals and the relationship of the material to thermodynamically stable crystals.

Thus, fast-cooling of the melt in the temperature range to 1600°C is to be recommended for producing clinkers having a coarsely crystalline structure. After 1600° the melt should be slow-cooled at 100 deg/min to 1400°C. Then rapid cooling is possible from 1400°C to room temperature. Finely crystalline clinker is produced by rapid cooling of the material over the entire temperature range, and especially between 1600 and 1400°C.

PHASE COMPOSITION OF FUSED CLINKERS

The composition of the clinkers of series I is denoted by point A, of those of series II by point B, and of those of series III by point C on the phase diagram of the CaO—Al_2O_3—SiO_2 ternary system with conditional interchange of the sum of the oxides for Al_2O_3 (Fig. 6). According to the phase diagram, when a melt of composition I is cooled from 1900 to 1800°C, CaO is the primary crystalline phase; at 1800° the first crystals of 3CaO · SiO_2 appear. However, practically no free CaO (<1.5%) was observed in sinters obtained on quenching a melt of composition I from 1900° and on slow-cooling such a melt from 1900 to 1800°C, with a final quench.

Simultaneous crystallization of CaO and CaO · SiO_2 occurs on equilibrium crystallization of a melt of composition I in the temperature range 1800-1600°C. At temperatures from 1600-1455°C, dissolution of the CaO crystals and combination of the oxide with silica in the melt to form 3CaO · SiO_2 begin. The process occurs along the boundary curve between points 1 and 3. The x-ray analysis of sinters produced as a result of cooling this melt by a favorable regime to 1800°, 1700°, 1600°, and 1500°C with a final quench indicate that the sinters contain large amounts of the minerals 3CaO · SiO_2, 2CaO · SiO_2, 3CaO · Al_2O_3, as well as the calcium aluminoferrites, but, as will be shown below, in ratios not corresponding to equilibrium.

The primary crystalline phase during equilibrium crystallization of melts of compositions I and II is 2CaO · SiO_2. The first 2CaO · SiO_2 crystals appear at 1800° (melt of composition II) and at 1700° (melt of composition III). However, three crystalline phases are observed in the actual sinters obtained on quenching a melt of composition II from 1900° and on slow cooling a melt of composition II over the temperature range 1900–1700°C. These are $3CaO \cdot SiO_2$, $2CaO \cdot SiO_2$, and $3CaO \cdot Al_2O_3$. Only β-$2CaO \cdot SiO_2$ is observed in a sinter in composition III quenched from 1900°C. The x-ray patterns of this sinter do not display lines characteristic of $3CaO \cdot SiO_2$. The $3CaO \cdot Al_2O_3$ lines have a negligible intensity.

Consequently, the mineralogical composition of sinters produced with melt crystallization at different rates cannot be computed on the basis of phase-diagram data. Since the mineralogical composition of the clinkers formed during nonequilibrium crystallization depends strongly on the cooling rate, each empirical correction to the known computational formulas will have an approximate character.

A maximum amount of alite (~73%) is observed in the clinker obtained on quenching a melt of composition I from 1900°C. Here virtually all the SiO_2 goes into formation of alite, and only two phases occur in the clinker, alite, and intermediate material. Figure 7 shows the change in alite content of clinkers produced from a melt of composition I as the result of cooling according to differing regimes. Slow-cooling the melt to 1800°C with final quenching is accompanied by a decrease to 68% in the amount of the alite phase in the clinker. During slow-cooling the melt in the temperature range 1700–1300°C, the alite content in the clinker changes insignificantly and varies within the limits of 62–65%, decreasing with lower temperatures.

The alite content also changes a little with the final slow-cooling of the clinker to 1200°C. During this an increase in alite content is possible due to separation from the eutectic melt crystallizing under supersaturated conditions. Consequently, the amount of alite in the clinker can be controlled within definite limits by varying the cooling regime.

CONCLUSIONS

1. As a result of the study of crystallization processes in melts of different composition, the temperature regions were established within the limits of which the growth rate of alite crystals is different and dependent on the amount of melt present. By cooling the melts according to a chosen regime, one can effect controlled crystallization and obtain clinkers with a desired crystalline structure.

2. The alite content in clinker depends on the melt-cooling regime. Therefore, by changing the melt-cooling regime in the temperature range 1900–1200°C, one can effect controlled mineral formation, i.e., one can obtain, using a starting mixture of the same composition, Portland cement clinkers having differing alite contents.

LITERATURE CITED

1. N. A. Toropov and P. F. Rumyantsev, Tsement, No. 6, p. 3 (1964).
2. D. A. Vysotskii, Yu. M. Butt, and V. V. Timashev, Tr. Moskovsk. Khim.–Tekhnol. Inst., No. 45, p. 36 (1964).
3. Z. B. Entin, Abstract of Candidate's Dissertation, Moscow (1961).
4. L. Ya. Gol'dshtein and S. D. Okorokov, Tr. Giprotsementa, No. 26, p. 56 (1964).

EFFECT OF BARIUM AND STRONTIUM ON THE STRUCTURAL FORMATION OF CLINKERS PRODUCED FROM THE SLUDGE OF THE CHIMKENTSKII CEMENT FACTORY

M. R. Ramankulov, M. Sh. Sharafiev, and V. A. Popov

Sludge from the Chimkentskii Cement Factory containing KN-90 was used in the synthesis of barium- and strontium-containing clinkers. Barium and strontium oxides were added to the raw starting mixture in the amounts of 0.3, 0.5, 1, and 5%, computed as $2BaO \cdot SiO_2$ and $2SrO \cdot SiO_2$. Firing was performed at 1300-1400°C. Soaking at the maximum temperature was 3 h. The sinters were cooled in air. The free CaO content was determined in the resulting sinters by the ethyl glycerate method.

The presence in the cement clinkers of less basic calcium silicates, in addition to the highly basic ones, is evidently the reason free CaO was absent in our experiments after addition of up to 0.5% BaO. Consequently, when up to 0.5% BaO is added to the sludge composition, it enters the composition chiefly as a solid solution. The negligible free CaO precipitation on addition of BaO to the sludge (computed to give 1 and 5% $2BaO \cdot SiO_2$) explains the formation of the barium compounds. The somewhat higher indices of refraction of the clinker minerals show that on heating the oxides added to the charge predominantly dissolve in the calcium silicates.

A petrographic study of clinkers fired at 1300°C showed the following: the clinker without additives is characterized by the imperfect, poorly developed crystallization of clinker minerals and has a highly porous structure. The addition of 0.3-0.5% BaO noticeably improves the crystallization of the clinker minerals. On increase of the BaO additive to 1% (computed as $2BaO \cdot SiO_2$), the clinker porosity decreases considerably, and the clinker structure noticeably improves. The clinker is then characterized by well-developed crystallization.

Addition of BaO (computed to give 5% $2BaO \cdot SiO_2$) results in the disintegration of clinker minerals such as alite and belite. The decomposition products form in a finely dispersed state.

On addition of 0.3-0.5% SrO, the SrO enters the crystal lattice of the clinker minerals almost completely. An increase of the SrO additive (computed as $2SrO \cdot SiO_2$) from 1 to 5% results in some decomposition of the tricalcium silicate. Evidently, in this case solid solutions form which are thermodynamically unstable. On cooling, these solid solutions decompose. This displaces some of the $3CaO \cdot SiO_2$ from the crystal lattice. It is interesting to note that on increase of temperature to 1400°C, all the SrO added completely dissolves in the crystal lattice of the clinker minerals. The absence of free CaO attests to the fact that all the added SrO reacted with the clinker minerals to form a solid solution.

The petrographic studies show that 0.5% SrO additive noticeably improves clinker crystallization and gives a uniformly grained structure with well-developed mineral crystallization.

The favorable effect of SrO is only observed at 1300°C after addition of 0.3% SrO. Increase in the amount of SrO has a negative effect on the strength of cements.

CONCLUSIONS

1. The structure of clinkers improves noticeably with small BaO and SrO contents, with a corresponding increase in the hydration activity of the cements.

2. The optimum content of the latter oxides was established for different clinker annealing temperatures.

WORK IN THE FIELD OF REFRACTORIES

INTERACTION OF REFRACTORY METALS
WITH ALUMINA AND SILICA

T. P. Markholiya, B. F. Yudin, and N. I. Voronin

Problems concerning the interaction between refractory ceramics and refractory metals are of great importance with respect to the development of high-temperature technology. The present work considers the interaction between the basic constituents of the aluminosilicate refractories (alumina and silica) and the refractory metals.

The interaction in the SiO_2—Me system was studied in the work of Bacon et al. [1], which was devoted to the selection of refractory metals (Mo, W, Zr, Ir, and Ta) for the melting of quartz. These authors found that tungsten has the greatest stability with respect to SiO_2 in a vacuum or inert environment.

Armstrong et al. [2] studied the interaction of the refractory metals with SiO_2 at 1900°K under a vacuum of 10^{-7} atm. These authors discovered the presence of Me_2O_5-type oxides in the SiO_2—Ta and SiO_2—Nb systems, and of MoO_3 oxide in the SiO_2—Mo system. DeMaria et al. [3] and Porter and Inghram [4] used the mass-spectrometric method to determine the vapor compositions and reaction product partial pressures in the Al_2O_3—Mo and Al_2O_3—W systems at 2000-2500°K.

Of other work, an article deserving attention is the study of the surface interaction between Al_2O_3 and Mo, Nb, and Zr in a helium atmosphere, performed by Economos and Kingery [5]. No interaction between Al_2O_3 and Mo was observed up to 2100°K. However, noticeable interaction was observed between Al_2O_3 and Zr at 2100°K (ZrO_2 was observed in the solid phase) and between Al_2O_3 and Nb, also at 2100°K (formation of Nb_2O_5).

As we see, the investigations of the metal—oxide systems were performed under different conditions. A large number of them are of a purely qualitative nature and cannot provide a wholly precise description of the interaction.

The chief goal of the present work is to perform the thermodynamic analysis of the interaction processes of Al_2O_3 and SiO_2 with the refractory metals Mo, W, Nb, Ta, and Zr, and to compare experimental data we obtain with the results of such thermodynamic analysis.

As a rule, the interaction processes between ceramics and metals belong to the category of "complex processes," i.e., processes determined by the parallel occurrence of several reactions. In order to define the equilibrium composition in the systems under consideration, we constructed the following model of interaction.

Table 1. Interaction Processes between Al_2O_3 and Mo

Reaction No.	Principal reaction	Value of K_p
1	$Al_2O_3(sol) = 2AlO\,(gas) + 1/2\,O_2(gas)$	$K_{p1} = p_{AlO}^2\,p_{O_2}^{1/2}.$
2	$Al_2O_3(sol) = Al_2O\,(gas) + O_2\,(gas)$	$K_{p2} = p_{Al_2O}\,p_{O_2}.$
3	$Al_2O_3(sol) = 2Al\,(gas) + 3/2\,O_2\,(gas)$	$K_{p3} = p_{Al}^2\,p_{O_2}^{3/2}.$
4	$O_2\,(gas) = 2O\,(gas).$	$K_{p4} = p_O^2/p_{O_2}.$
5	$Mo\,(sol) + 1/2\,O_2\,(gas) = MoO\,(gas).$	$K_{p5} = p_{MoO}/p_{O_2}^{1/2}.$
6	$Mo\,(sol) + O_2\,(gas) = MoO_2(gas).$	$K_{p6} = p_{MoO_2}/p_{O_2}.$
7	$Mo\,(sol) + 3/2\,O_2\,(gas) = MoO_3\,(gas).$	$K_{p7} = p_{MoO_3}/p_{O_2}^{3/2}.$

N o t e. Expressions are given in the last column for the equilibrium constant (K_p) of the reaction being considered.

The first stage involves the dissociative vaporization of the refractory oxide, during which molecular oxygen is evolved. In the second stage of the process, the evolving oxygen oxidizes the refractory metal, and dissociates into atomic oxygen [6].

We note that the model assumed for the interaction is quite conditional, and may far from correspond to the actual mechanism of the process. However, from the thermodynamic standpoint, this is not important, since the latter interaction model encompasses all possible equilibrium components.

As an example, Table 1 records the plan of the interaction process between Al_2O_3 and Mo. The total molecular oxygen pressure generated in the first stage of the process (reactions 1-3) can be expressed by the pressure of the products of the Al_2O_3 reduction

$$p_{O_2}' = 1/4\,p_{AlO} + p_{Al_2O} + 3/4\,p_{Al}. \tag{1}$$

On the other hand, the amount of oxygen necessary for occurrence of the second stage of the process (reactions 4-7) is made up of (1) the partial pressures of the metal oxidation products, (2) the partial pressure of the atomic oxygen, and (3) the residual (i.e., equilibrium) pressure of the molecular oxygen:

$$p_{O_2}' = 1/2\,p_{MoO} + p_{MoO_2} + 3/2\,p_{MoO_3} + 1/2\,p_O + p_{O_2}. \tag{2}$$

According to the material balance, the right sides of Eqs. (1) and (2) are equal. Computing the values of the equilibrium constants of the reactions 1-7 (Table 1) and performing the corresponding algebraic conversions, we obtain the equation for the material balance

$$x\left[0.5K_{p5} + 0.5\sqrt{K_{p4}} + x(1 + K_{p6}) + 1.5K_{p7} + x^2\right] - \left[0.25\sqrt{\frac{K_{p1}}{x}} + \frac{K_{p2}}{x^2} + 0.75\sqrt{\frac{K_{p3}}{x^3}}\right] = 0. \tag{3}$$

Since $x = \sqrt{P_{O_2}}$,

$$p_{O_2} = x^2, \quad P_O = x\sqrt{K_{p4}}, \quad p_{AlO} = \sqrt{\frac{K_{p1}}{x}}, \quad p_{Al_2O} = \frac{K_{p2}}{x^2}, \quad p_{Al} = \sqrt{\frac{K_{p3}}{x^3}},$$

$$p_{MoO} = K_{p5} \cdot x, \quad p_{MoO_2} = K_{p6} \cdot x^2, \quad p_{MoO_3} = K_{p7} \cdot x^3.$$

Table 2. Total Vapor Pressure over the Al_2O_3—Me
and SiO_2—Me Systems at $2000\,°K$

Metal	Vapor pressure, atm	
	SiO_2—Me	Al_2O_3—Me
W	$1.13 \cdot 10^{-4}$	$1.79 \cdot 10^{-8}$
Mo	$2.01 \cdot 10^{-4}$	$2.38 \cdot 10^{-8}$
Nb	$2.44 \cdot 10^{-1}$	$2.22 \cdot 10^{-6}$
Ta	$1.11 \cdot 10^{-1}$	$1.05 \cdot 10^{-5}$
Zr	$2.23 \cdot 10^{2}$	$2.31 \cdot 10^{-1}$

Consequently, by solving Eq. (3) with respect to x, the equilibrium pressures can be determined for all the gaseous components of the reaction between Al_2O_3 and Mo. The equilibrium composition can also be determined for the other systems under consideration in an analogous manner.

The results of the thermodynamic analysis revealed the following.* Aluminum oxide is reduced on reaction with the refractory metals, chiefly to Al and Al_2O, during which the Al content in the vapor phase is approximately ten times larger than that of Al_2O. This well agrees with the work of Porter and Inghram [4].

Silicon dioxide is reduced to gaseous SiO, except during the reaction with Zr. In the latter case, the equilibrium pressure of silicon is above the saturation pressure. This can result in the precipitation of Si as a condensed phase.

Molybdenum and tungsten are oxidized by the refractory oxides to the higher oxides MeO_2 and MeO_3, while zirconium is oxidized to ZrO_2. Tantalum is oxidized by Al_2O_3 to TaO and by SiO_2 to Ta_2O_5. Since the dissociation pressures of all the niobium oxides are near one another, the probability of each forming is about the same.

Table 2 gives the total pressures of the products of the gaseous reaction between Al_2O_3 or SiO_2 and the refractory metals above $2000\,°K$. The activities of the refractory metals, relative to SiO_2 and Al_2O_3, were defined by comparison of these pressures as increasing in the sequence W, Mo, Nb, Ta, and Zr. The greater stability of tungsten is also confirmed in the work of Bacon [1]. It is also seen from Table 2 that SiO_2 is the most reactive component in the aluminosilicate refractories when in contact with refractory metals. The most mutually stable compositions are Al_2O_3—Mo and Al_2O_3—W.

The equilibrium pressures of the gaseous equilibrium products enable not only a definition of the predominant direction of the process, but also of the starting temperature of interaction. The latter is taken to mean the temperature at which the sum of the partial pressures equals the pressure of the surrounding medium. Beginning with this determination, we determined the starting temperatures of interaction for the systems considered as a function of pressure in the surrounding medium (Table 3). As seen from Table 3, the starting temperature of reaction decreases with a decrease of pressure in the surrounding medium.

The authors of the present work investigated the interaction (on contact) of the metals Mo, W, and Nb with Al_2O_3 and mixtures of Al_2O_3 and SiO_2. Metal rods were pressed into the specimens which were fired at $1800-2200\,°K$ under a vacuum of $10^{-7}-10^{-8}$ atm. In contrast to the earlier studies, the processes were studied for this work over lengthy soaking periods (5-50 h).

---

*The values of the equilibrium constant (K_p) were calculated using the thermodynamic functions for the reaction components given in handbooks [7, 8].

Table 3. Dependence of Starting Temperature of Interaction (°K) on the Ambient Pressure in the SiO_2—Me and Al_2O_3—Me Systems

System	Ambient pressure, atm								
	10^{-8}	10^{-7}	10^{-6}	10^{-5}	10^{-4}	10^{-3}	10^{-2}	10^{-1}	10^{-1}—1
SiO_2—W	1450	1600	1730	1840	2200	>2200	—	—	—
SiO_2—Mo	1400	1530	1680	1830	1950	>2000	—	—	—
SiO_2—Nb	<1200	<1200	1200	1260	1380	1510	1690	1900	>2000
SiO_2—Ta	<1200	1200	1300	1400	1500	1640	1800	2000	>2000
SiO_2—Zr	—	—	—	—	<1200	—	—	—	—
Al_2O_3—W	1980	2100	2250	>2300	—	—	—	—	—
Al_2O_3—Mo	1950	2100	2250	>2300	—	—	—	—	—
Al_2O_3—Nb	1670	1800	1950	2100	2300	—	—	—	—
Al_2O_3—Ta	1660	1760	1870	2000	2130	2300	—	—	—
Al_2O_3—Zr	<1200	<1200	1250	1360	1470	1600	1750	1920	2140

Table 4. Comparison of Calculated and Experimental
Values for the Starting Temperature of Interaction
under a Vacuum of 10^{-7} atm

System	Starting temperature of inter-action, °K	
	calculated	experimental
Al_2O_3—Mo	2100	>2200
Al_2O_3—W	2100	>2200
Al_2O_3—Nb	1800	2000
SiO_2—Mo	1530	1550
SiO_2—W	1600	1550

The phase composition and intensity of specimen disintegration were determined by x-ray and microscopic analysis. The results revealed the following.

Corundum ceramic does not react with Mo and W at temperatures up to 2200°K, even after a 50-h soaking. The x-ray structure analysis of specimens of powdered mixtures of the oxides and metals fired 2 h at 2100°K also did not reveal any new phases. The weight losses of the samples of pure Al_2O_3 and of mixtures of Al_2O_3 with Mo and W practically agree, attesting to the absence of discernible chemical reaction.

Corundum ceramic in contact with niobium rods does not react over short soaking periods at up to 2100°K; however, prolonged soakings result in the complete disintegration, already at 2000°K, of both the metal and the ceramic. The sample porosity increases with the increase of SiO_2 in the ceramic. Sample swelling is observed (the increase of sample dimensions reaches 4% for a 50-h soaking at 1800°K), and the weight losses increase (up to 10 wt.%).

The microscopic and x-ray structure analyses only reveal the presence of corundum, mullite, and a glassy phase in the samples. For the case of Mo and W, no new formations were discovered at the metal—oxide interface.

For the Nb case, the reaction zone around the Nb rod increases with increased SiO_2 content in the mixtures with Al_2O_3. This attests to the high activity of SiO_2. Unfortunately, it was impossible to identify the reaction-zone composition due to the absence of necessary literature data.

When the SiO_2 content is above 23% for a prolonged anneal at elevated temperatures, the ceramic becomes loose and disintegrates after a 50-h soaking at 2000°K.

The temperature dependence of the weight loss of SiO_2 and its mixtures was determined under a 10^{-7}–10^{-8} atm vacuum in order to establish the beginning temperature of interaction between pure SiO_2 and Mo and W. The temperature at which the weight losses per unit of SiO_2 volatilization surface equalled the weight losses of a SiO_2—Me mixture was taken as the beginning temperature of interaction. It turned out to be 1500°K, in good agreement with the results of thermodynamic analysis.

In conclusion, we offer Table 4, which compares the results of the thermodynamically computed starting temperature of interaction with the results we experimentally determined. For the data of this table it follows that the starting temperatures of interaction calculated for the case of the Al_2O_3—Me system is lower than that determined experimentally. This is explained by the kinetic peculiarities of the process. The calculated values for the SiO_2—Me system agree well with the experimental.

CONCLUSIONS

1. A technique for the thermodynamic analysis of the interaction processes in metal—metal oxide systems was described.

2. The starting temperatures of interaction in the Al_2O_3—Me and SiO_2—Me systems were calculated on the basis of the thermodynamic computations performed as a function of pressure in the surrounding medium. The activities of the refractory metals were determined relative to the refractory oxides Al_2O_3 and SiO_2.

3. The comparison of the results of the experimental investigation with the thermodynamic analysis shows them to be in good mutual agreement.

LITERATURE CITED

1. J. F. Bacon, A. Hasaris, and J. Wholley, Phys. and Chem. of Glasses, Vol. 1, No. 3, p. 90 (1960).
2. W. Armstrong, A. Chaklander, and M. Cleene, J. Am. Ceram. Soc., Vol. 45, No. 9, p. 407 (1962).
3. G. deMaria, R. P. Burns, J. Drowart, and M. G. Inghram, J. Chem. Phys., Vol. 32, No. 5, p. 1373 (1960).
4. R. F. Porter and M. G. Inghram, Met. Soc. Roy. Sci., Liege, Ser. 4, Vol. 18, p. 513 (1957).
5. G. Economos and W. D. Kingery, J. Am. Ceram. Soc., Vol. 36, No. 12, p. 403 (1953).
6. B. F. Yudin, T. P. Markholiya, and N. I. Voronin, Trudy Vses. Inst. Ogneuporov, Vol. 37, p. 204 (1965).
7. Thermodynamic Properties of Individual Materials [in Russian], Izd. Akad. Nauk SSSR, Moscow (1962).
8. A. N. Krestovnikov, L. P. Vladimirov, B. S. Gulyanitskii, and A. Ya. Fisher, Handbook on Equilibrium Calculations for Metallurgical Reactions [in Russian], Metallurgizdat, Moscow (1963).

INTERACTION OF METAL
AND REFRACTORY MATERIALS
DURING THE CONTINUOUS CASTING OF STEEL

A. K. Karklit and E. S. Borisovskii

One of the principal areas of engineering progress in ferrous metallurgy is the change-over to the continuous casting of steel, which permits automation of the casting process, reduction in losses, and an improvement in metal quality.

A most indispensable refractory component in the continuous casting of steel is the feeder to the intermediate ladle through which the steel enters the crystallizer. The channel section of the feeder should undergo little change during the casting process, so that the rate of metal arrival at the crystallizer is constant.

The refractory should be resistant to disintegration in the presence of liquid steel. On the other hand, the passage should not be too constricted. Stable refractory materials have now been developed for several conditions of continuous casting [1]. The general theory of the problem is encumbered by the absence of an adequate theory describing the interaction between steel and refractories.

Until recently, primarily the interaction between steel and fireclay nozzles has been studied. Most authors [2, 3, etc.] believe that disintegration of the aluminosilicate refractories occurs as a result of the interaction between manganese and silica by the reaction

$$2Mn + SiO_2 \rightarrow 2MnO + Si.$$

This theory is confirmed (insofar as is known) by the high MnO content in the layer of contact between the refractory and the steel, and also by the reduced $SiO_2 : Al_2O_3$ ratio in this layer as compared to the original refractory. However, the latter effect is not always observed. For example, studies conducted at the Institute of Refractories revealed that the $SiO_2 : Al_2O_3$ ratio increases when SiCa steel is being cast. No direct thermodynamic computations were found in the literature to determine the possibility that Mn reduces the SiO_2 under actual steel-casting conditions and for the typical concentrations of Mn in the steel and of SiO_2 in the refractory. However, there is a greater chemical affinity between silicon and oxygen than between manganese and oxygen [4].

Several authors indicate that SiO_2 is reduced to Si by molten Fe [5], and that the iron oxide contained in the refractory is reduced by Mn and Fe to the suboxides, while others say the reduction of SiO_2 barely occurs and is of no practical importance [6]. According to other data, a molten glassy slag layer forms at the metal—refractory interface at high temperatures. Initially, this has the composition of the refractory. As a result of the interaction of Mn, Si, and FeO, the FeO and MnO contents in steels having the molten slag tend to a metallurgically known equilibrium ratio between slag containing SiO_2, MnO, FeO, and Al_2O_3 and liquid steel containing Fe, Mn, Si, and FeO. The slag tends to be low-melting and is eroded [7].

It is also believed that the chief reason for the corrosion of refractories is the erosion of the resulting low-melting glass composed of the refractory components and the MnO from the steel [8]. The FeO plays a primary role in the chemical corrosion of refractories by steel containing less than 0.75% Mn [9]. Disintegration of the refractory (a steel-casting nozzle) in contact with the steel occurs chiefly by abrasion of the steel on the rough exterior layer of the refractory (forming during the casting process and composed of the Al_2O_3-containing glass [10]).

According to the electrochemical theory, a molten layer forms on the surface of the refractory, creating an electrolyte between the liquid metal and solid refractory. Cations such as Mn^{2+} are injected into the refractory by action of an electrical field. They form low-melting compounds, and the refractory disintegrates [11].

The following comments must be made relative to the above assumptions.

The thermodynamic computations performed in [12] show the improbability of a reaction involving the reduction of SiO_2 by Fe. The reaction between the Mn and Fe of the steel and the Fe_2O_3 of the refractory can occur, but is not necessarily the principal reason for disintegration of the aluminosilicate refractories (since their Fe_2O_3 content is small). The theory of electro-chemical disintegration of a refractory is not based on any computational or experimental data on the existence of any particularly significant potential differences among the components of the systems considered. It only predicts some softening of the external layer at 1500–1600°C temperatures. The most correct assumption takes into account the slag-like character of the eroded contact-layer where the low-melting, FeO-, and MnO-containing melts form.

Besides fireclay, magnesite, and (recently) high-Al_2O_3 refractories, zirconia, graphite-fireclay, and (rarely) chrome-magnesite have also been used for steel-casting nozzles. The mechanism of their decomposition by steel has so far been almost completely neglected. However, these refractories in some way react with the steel, leading to a constriction in the nozzle channel. The formation of nonmetallic deposits is often observed on the channel wall. The latter occurrence is known as "retardation" [1], and is the result of the interaction between the steel and the refractory.

It is known that the nonmetallic deposits on the walls of an operating channel displaying retardation result from Al_2O_3 or its compounds, which separate from the steel [11, et al.]. However, the literature contains no data on the chemical interaction in the contact layer between the refractory and the metal. The character of this interaction may be the reason why steels of the same composition cause a different degree of nozzle retardation with different refractories, and why refractories are observed as both retarded and corroded.

The interaction of bubble-free and bubbling carbon transformer steel was studied at the All-Union Institute of Refractories with high-Al_2O_3 fireclays of varying Al_2O_3 content, alumina–silicon carbide, zirconia, magnesite, and chrome-magnesite refractories. Several laws were observed during this study.

In all cases, the contact layer contains the very same oxides. Quantitative differences are at first observed with transition to another type of refractory causing an increase in the layer components (Fig. 1). Secondly, the composition of the contact layer is quantitatively different for the case of refractory retardation as opposed to the case of refractory corrosion. Figure 2 shows how the MnO content in the contact layer decreases with use of high-alumina rather than of fireclay refractories. This tends toward more retardation. This also occurs to

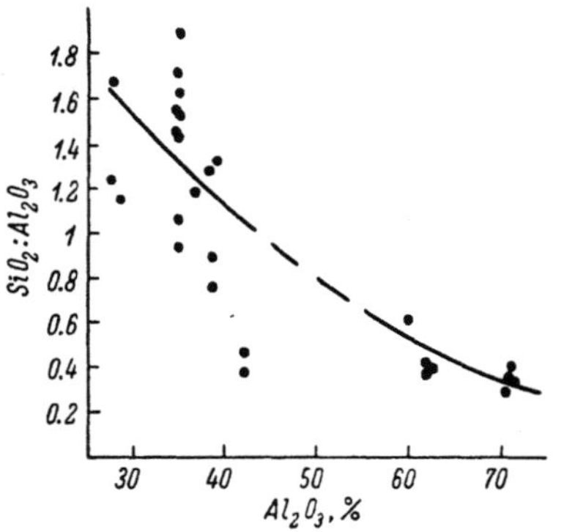

Fig. 1. Dependence of $SiO_2 : Al_2O_3$ ratio in the contact layer on Al_2O_3 content in the refractory.

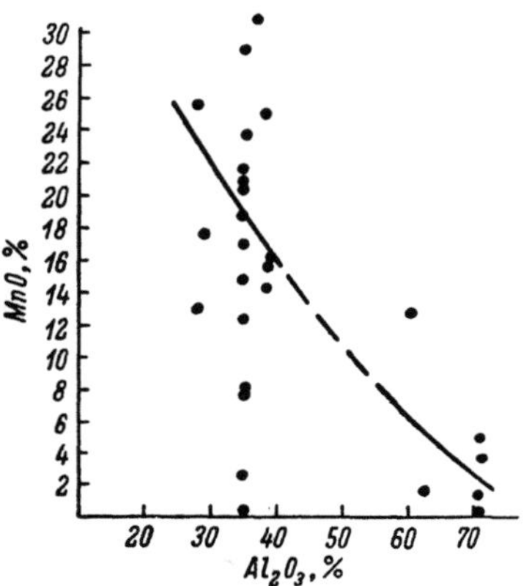

Fig. 2. Dependence of MnO content in the contact layer on Al_2O_3 content in the refractory.

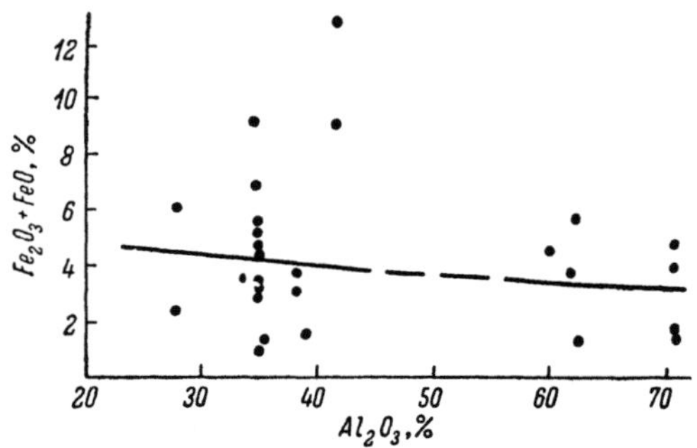

Fig. 3. Dependence of Fe_2O_3 + FeO content in the contact layer on Al_2O_3 content.

a lesser extent with the iron oxides (Fig. 3). However, if samples of the same refractories are chosen during the casting of different steels or of the same steel with different deoxidizers, the composition of the contact layer will be different. Here it is typical to observe some increase of Al_2O_3 and decrease of Mn content in the contact layer of steels causing channel retardation.

The generalization of microscopic data shows that the contact layer of all the refractories consists of a glass which generally has a greater or lesser amount of crystalline phases as inclusions. With semioxide refractories a relatively thick, almost homogeneous, glassy shell results. The contact layer becomes heterogeneous and the amount of the crystalline phases increases in accordance with the Al_2O_3 content in the refractory. For fireclay refractories, the crystalline phases are mullite and spinel. In the high-Al_2O_3 refractories corundum is formed while the mullite gradually disappears. At an Al_2O_3 content in the refractory exceeding 50%, the only crystal phases resulting are corundum and spinel. The contact layer is more homogeneous when Al is absent from the steel. According to the microstructure of the contact layers, the fireclay-graphite refractories differ from the fireclay ones only by the small amount of spinel. For the alumina-carborundum refractories, the crystalline phase constitutes a significant fraction of the layer. It contains mullite, spinel, and some corundum.

The microscopic study of the magnesite and chrome-magnesite refractories attests to the absence of substantial interaction between the refractory proper and the contact layer. The boundary between the refractory and the contact layer is abrupt; the contact layer (in presence of Al in the steel) is composed of spinels of the hertzenite-type, hematite, glass, and metal tricklings. It does not contain components which might contain a more or less significant amount of MgO. With the zirconia refractories, the layer is composed chiefly of ZrO_2 and SiO_2. The crystal phases are baddeleyite and spinel, the composition of which depends on the content of Al, Mn, Fe, and Mg in the steel. Here, the baddeleyite crystallizes chiefly near the refractory surface, while the spinel crystallizes in that portion of the slag in immediate contact with the steel. As examples, Figs. 4 and 5 present photographic microsections of the samples.

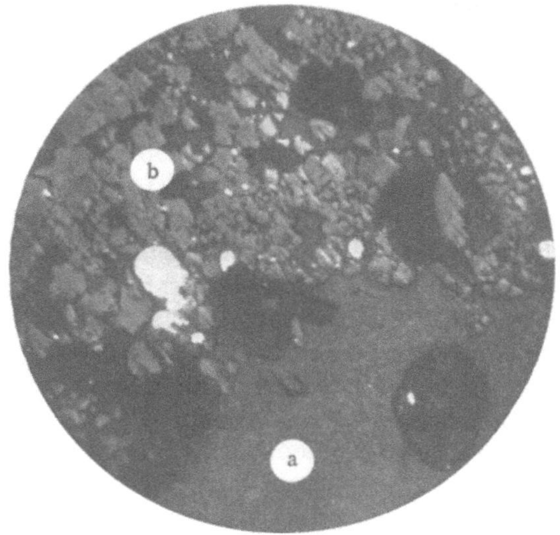

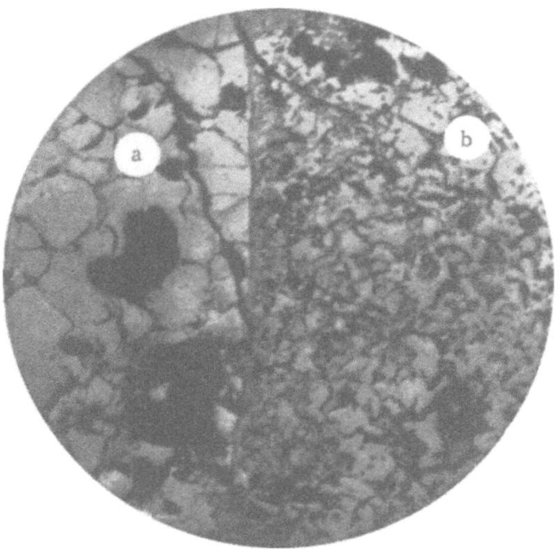

Fig. 4. Microsection of a fireclay nozzle
after casting transformer steel. Reflected
light, 200×. a) Unaltered layer of the refractory composed of glass with mullite
needles; b) layer of contact composed of
crystals of spinel, hematite, glass, and
metal tricklings.

Fig. 5. Microsection of a magnesite insert
after casting bubble-free carbon steel. Reflected light, 110×. a) Unaltered refractory
layer of periclase and glass; b) contact layer
with crystals of spinel, hematite, metal
tricklings, glass, and mullite.

The crystalline phases observed in solidified contact layers have melting points typically
above the temperature of the steel during casting. In most cases, the contact layer is therefore
heterogeneous, even when the metal is being cast. An increase in amount of the crystalline phase
generally coincides with an increase in refractory stability and with a tendency toward refractory retardation. The properties of the phases constituting the layer and their quantitative ratios
determine the nature of the interaction between the metal and the refractory.

The layer begins to form with the trickling of the oxide components of the steel (FeO,
MnO, MgO, etc.) at the heated refractory surface and at the interface of the two phases. Promot-
ing this process are (1) the presence of air oxygen adsorbed at the surface and pores of the refractory; (2) precipitation of oxygen dissolved in the steel due to the decrease in the temperature
of the steel when it contacts the refractory; and, (3) the presence of iron oxides and of other
nonmetallic inclusions in the steel.

The slag layer reacts with both the refractory and the steel. The second process, already
described in [7], obeys the known laws of interaction between metal slags and steel. The composition of the contact layer is generally determined by the composition of the steel, and that of
the refractory. The layer may be viscous or fluid, homogeneous or heterogeneous, depending
on the location of its composition in the phase diagram and depending on the viscosities in the
complex multicomponent system which corresponds to the slag composition. However, the interaction rate is determined by (a) the amount of oxygen in the steel (this explains the corrosive-
ness of bubbling steels as well as the considerable scatter in measuring the wear of refractories
by the same steel under approximately the same conditions [10 et al.]); and by (b) the diffusion
rate in portions of the contact layer adjoining the atmosphere.

The final result of the interaction (degree of retardation or corrosion) depends entirely
on the slag viscosity and rate of interaction. Here the viscosity is chiefly determined by the

Table 1. Effect of Steel and Refractory Components on Viscosity of the Contact Layer, and the Result of Interaction

Effect of steel components	Result	Effect of refractory components
0	N	0
0	N	+
0	Corrosion	−
−	N	0
+	Retardation	0
−	Corrosion	−
+	Corrosion and retardation	−
−	The same	+
+	Retardation	+

Note. Conventional symbols: 0, no effect; +, increases viscosity; −, lowers viscosity; N, no change.

nature and rate of interaction, a quantitative result. Such treatment permits determination of the possible combinations giving the final interaction results, as shown in Table 1.

Of the nine cases in Table 1, the first three are unrealistic, since unavoidable oxidized components in steel always form a slag layer at the solid—liquid interface on a rough refractory surface. We do not include materials which would not be wetted by these components. The best case is shown in the fourth line. In practice it is approximately achieved with refractories of high chemical stability in the presence of aluminum in the steel, corresponding to a viscosity increase in the contact layer. The second and third lines from the bottom are also of interest. With the proper combination, the action of the surrounding atmosphere can be held to a level of slight corrosion satisfactory for casting requirements.

The aluminosilicate refractories exemplify application of the above considerations. With an increase in Al_2O_3 content, the corrosion of nozzles by molten steel also increases the tendency toward retardation, associated with the increase of Al_2O_3 in the adjacent slag. Mixtures with sufficiently high Al_2O_3 content are not liquid at molten-steel temperatures (1500-1550°C). Crystals of various compounds accordingly separate from them when the composition of the mixture is situated in their field of crystallization on the phase diagram. This is confirmed by microscopic studies. The viscosity of the adjacent slag increases on formation and growth of the crystalline phases. The retardation becomes more intense if the Al_2O_3 enters the contact layer between the refractory and steel (e.g., on reoxidation of bubble-free carbon steel by aluminum).

In the same way, refractories are corroded by the manganese in bubbling steel. The latter contains considerable manganese not in equilibrium with silicon. The direct introduction of manganese into the adjacent slag lowers the slag viscosity and makes it more amenable to wetting. Like the bubble-free steel, bubbling steel contains small amounts of manganese, but has almost no silicon whatsoever. However, it was considerable oxygen, which promotes transfer of the steel components to the slag. This leads to an increase of the Fe and Mn content in the adjacent slag, and to a decrease in its viscosity.

These concepts on the interaction mechanism of steel with refractories led to the discovery of several ways of perfecting the feeding arrangements for the continuous casting of steel. For casting bubbling, low-carbon steel this means using materials containing 72, 82, 92, 95, and 97% Al_2O_3. For casting bubble-free carbon steel (deoxidized by aluminum to 300 g/ton), this means using zirconia and alumina—silicon carbide refractories. A subsequent, deeper study on the process of contact layer formation between the steel and the refractory (to be continued) should help correctly resolve any questions regarding the choice of refractories for other conditions.

LITERATURE CITED

1. S. S. Kazakevich and E. S. Borisovskii, Tr. Vses. Inst. Ogneuporov, Vol. 31, p. 98 (1961).
2. J. R. Rait, Trans. Brit. Ceram. Soc., Vol. 4, No. 42, p. 57 (1943).
3. A. M. Cance, Iron and Steel Inst. Spec. Report, Vol. 22, p. 331 (1938).

4. S. T. Rostovtsev, Theory of Metallurgical Processes [in Russian], Metallurgizdat, Moscow (1950).
5. Maekava and Nakagava, J. Iron and Steel Inst., Japan, Vol. 41, No. 12, p. 1237 (1955).
6. J. Natkaniec and L. Majenesk, Hutnik, Vol. 25, No. 4, p. 172 (1958).
7. C. B. Post and G. V. Inerson, J. Metals, Vol. 1, No. 1, p. 15 (1948).
8. Midsuno and Fukuda, J. Iron and Steel Inst., Japan, Vol. 41, No. 9, p. 913 (1955).
9. L. H. von Vlack and R. A. Flinn, Mag. Cast., Vol. 37, No. 3, p. 136 (1960).
10. R. B. Snow and J. A. Shea, Steel, Vol. 123, No. 4, p. 88 (1948).
11. F. Visseria, Silicates Industr., Vol. 20, No. 5 (1955).
12. A. D. Kramarov, Physicochemical Processes in Steel Production [in Russian], Metallurgizdat, Moscow (1954).

PHYSICOCHEMICAL PROCESSES OF REFRACTORY DISINTEGRATION IN THE METALLURGY OF COPPER

V. A. Ragozinnikov, I. L. Shchetnikova,
and K. V. Vorob'eva

Two of the principal processes in the pyrometallurgy of copper are the converter treatment of copper matte and the fire refining of crude copper.

Basic refractories have been used in the production of copper since 1910. The history of development of the converter method of crude copper production is linked with constant research on highly stable refractory materials. In this connection, it should be stated that the literature data are generally limited to a description of the changes in refractories after service.

The low stability of refractories in use (particularly for laying the tuyère zone of copper-melting converters [1–5]) and the difficulties in studying their disintegration mechanism have made special studies necessary.

The tuyere refractory of the converter is subjected to the action of molten sulfides (of Fe, Cu, Zn, and Pb), low-melting slags of fayalite composition, and metallic copper and its oxides. The principal reagents interacting with the tuyères in copper refining furnaces are metallic copper and its oxides, as well as the slags forming during the refining process.

The operating temperatures at the surfaces of various refractory tuyère components are different, and range within the limits of 1200–1500°C. During operation, the melt components, consisting chiefly of SiO_2 and iron and copper oxides, migrate to the refractory surface, giving it a banded structure. The resulting bands differ in phase and chemical composition and in physico-chemical properties.

Spent magnesia refractories have deep cracks progressing parallel to the tuyère operating surface, where spalling occurs most. The following characteristics of spent refractories in a copper converter were determined by a petrographic study:

1. The phase composition of the refractory binder changes. The forsterite content (in the transition band) and the fayalite content (in the operational band) increase with transition to a more intensely heated contour in the binder.

2. The peripheral grains of the periclase aggregate are saturated with iron oxides, resulting in point depositions of magnesioferrite. The degree of periclase saturation by iron is higher in the hotter zones. The periclase grains contacting the iron oxides or the copper have a corroded appearance, attesting to chemical interaction.

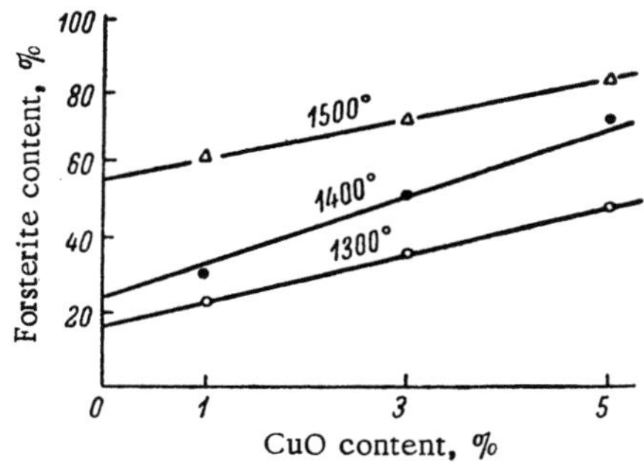

Fig. 1. Effect of copper on the rate of forsterite formation.

3. It was established that a metastable chemical compound, huggenite, forms in the transition band. This compound was not earlier observed in magnesia refractories used in copper-melting converters.

Chemical reaction of the copper oxides with both periclase and chromite is quite evident in magnesia refractories after service in copper refining furnaces.

The most interesting results were obtained with the study of slag-resistant refractories by a new technique [6]. This enabled description of the influence of copper compounds on the mechanism of refractory disintegration.

It was determined that the disintegration of magnesia refractories by converter slag depends on the content of copper compounds in the slag. The extent of refractory disintegration becomes larger with increase of copper content in the slag.

An analogous dependence was also obtained during tests with different types of refractories (forsterite, chrome-magnesite, magnesite-chromite, magnesite with spinel binder, and periclase-spinel) used for converter slags of different copper content. The periclase-spinel components have the best resistance to the action of converter slag. The forsterite refractories suffer the greatest disintegration.

The experiments described below were proposed with the aim of clarifying the chemical processes transpiring in the refractories of copper melting furnaces.

Stoichiometric mixtures were prepared of chemically pure, finely ground (passing through 10,000 openings/cm^2) MgO and SiO$_2$, to produce CuO-doped forsterite. Study of the ionization curves and the chemical composition of the samples showed that the synthesis of forsterite is considerably accelerated in the presence of the copper oxides (Fig. 1).

The MgO—CuO (Cu$_2$O) system [7] was studied in order to explain the catalyzing mechanism of CuO on the process of forsterite formation. The investigation was conducted using the anneal-quench method, and by chemical and high-temperature x-ray analysis. The melting points of the mixtures, prepared from chemically pure MgO and CuO or Cu$_2$O, were previously determined under a hot-stage microscope up to a temperature of 1500°C.

It was determined that one metastable chemical compound, corresponding to a 1:1 composition (i.e., CuMgO$_2$), forms in the MgO—CuO system. It is known as huggenite, and has a melting temperature of 950°C. It decomposes above 1210°C, forming a solid solution and precipitating MgO and copper oxides. The indices of refraction and crystal lattice parameters of huggenite were determined.

According to analysis of the study, the following describes the chemical interaction and disintegration of magnesia refractories on their exposure to the products of the converter process.

Fayalite forms on oxidation and slag formation of iron sulfide:

$$2FeS + 3O_2 + SiO_2 = Fe_2SiO_4 + 2SO_2.$$

In the absence of free oxygen (neutral or reducing environment), fayalite reacts with the MgO of the refractory by the reaction

$$Fe_2SiO_4 + 2MgO = Mg_2SiO_4 + 2FeO$$

Fig. 2. Thermogram of a 1:1 oxide mixture of Cu_2O and Cr_2O_3.

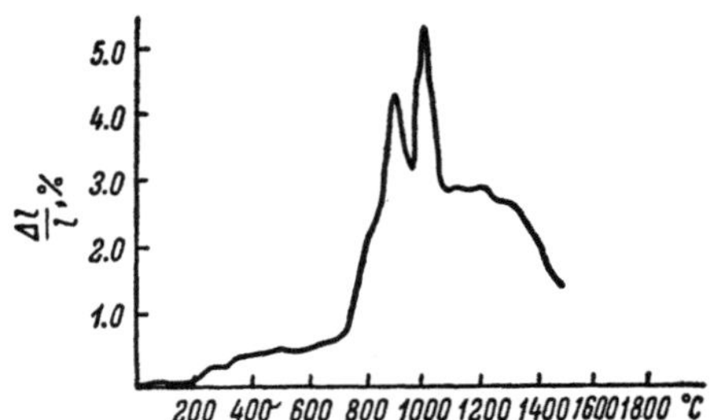

Fig. 3. Linear thermal expansion curve of a 1:1 oxide mixture of Cu_2O and Cr_2O_3.

with eventual formation of two different types of solid solutions. These are $2(Mg, Fe)O \cdot SiO_2$ and magnesiowustite $(Mg, Fe)O$.

In the presence of free oxygen, the $(Mg, Fe)O$ solid solution oxidizes to magnesioferrite, forming $(Mg, Fe)SiO_3$ and $(Mg, Fe)O \cdot Fe_2O_3$ solid solutions with the fayalite melt. The periclase refractory transforms to magnesioferrite and silicates. The high-temperature silicate phase (Mg_2SiO_4) transforms to the low-temperature phase in accordance with concentration of SiO_2 and iron in the refractory. (The low-temperature phase consists of olivines enriched by fayalite. These have a melting temperature under 1330°C.)

The unfavorable catalytic effect of the copper compounds consists of the formation and ultimate decomposition of the metastable compound huggenite. This process depends on the precipitation of free MgO from the reaction with the SiO_2 melt, resulting in formation of forsterite and later of a low-temperature silicate phase.

The good stability of periclase-spinel components in presence of melt products is explained by the crystal structure of the binder [8, 9]. The physicochemical transformations in periclase-spinel components on interaction with melt reagents stem chiefly from the substitution of Fe^{2+} for the Mg ions in the crystal lattice. This results in a continuous series of solid solutions which are to a small extent reflected by the properties of the chrome spinels.

The interaction processes between periclase and the melt reagents in crude-copper refining are analogous to those described above.

The chemical interaction of the chrome spinellide and the copper oxides was first established in the study of spent chromium-containing refractories. This essentially consists of formation of a compound $Cu_2O \cdot Cr_2O_3$, by the chrome spinellide interacting with the copper oxides. The formation of this new chemical compound is confirmed by the x-ray analysis data.

The copper oxide-chromite, $Cu_2O \cdot Cr_2O_3$, forms prismatic crystals (often polysynthetic twins) around the periphery of the chrome spinellide grains and finally completely displaces the original chrome-spinellide. $Cu_2O \cdot Cr_2O_3$ is characterized by intense double-refraction, the absence of internal reflections, and a negative reflectance averaging between cuprite and the chrome spinellide.

It was established by high-temperature x-ray analysis methods that the compound $Cu_2O \cdot Cr_2O_3$ forms from the oxides at 800°C. It does not decompose on heating to 1400°C with a final cooling.

The formation of $Cu_2O \cdot Cr_2O_3$ from the oxides is accompanied by an exothermic effect (Fig. 2) and occurs with an increase in volume (Fig. 3).

Formation of the new compound $Cu_2O \cdot Cr_2O_3$ on interaction of the copper oxides and chrome spinellide results in destruction and disintegration of the initial refractory structure.

CONCLUSIONS

1. Fundamental changes were found to occur in the mineralogical composition and physicochemical properties of magnesia refractories used for converter treatment of copper mattes.

2. The physicochemical processes in the disintegration of magnesia refractories by the converter melt products were studied for the first time. Fayalite is the principal agent causing disintegration and forms during the oxidation and slag formation of the iron sulfides. As a result of the interaction between the periclase refractory and the slag, low-melting compounds form which are of the olivine-type [2 (Mg, Fe)O $\cdot$ SiO$_2$], concentrated in fayalite. These also influence wear of the refractory during service.

It was established that copper compounds in the melt accelerate the chemical disintegration process of the refractories. The unfavorable effect of copper compounds is evident from the acceleration of the transition of the periclase refractory to forsterite via a metastable chemical compound, huggenite, and then to a low-temperature silicate phase. The conditions of huggenite formation were determined, as well as its stability field and chemical formula. The indices of refraction and crystal lattice parameters of huggenite were determined.

3. It was shown that the chemical compound $Cu_2O \cdot Cr_2O_3$ forms during the interaction of chromium-containing refractories with the copper oxides during fire refining, causing intensification of wear.

LITERATURE CITED

1. N. I. Voronin, Ogneupory, No. 5, p. 329 (1939).
2. C. B. Clark and J. S. McDowell, J. Metals, No. 2, p. 119 (1959).
3. W. F. Rochow and C. A. Bracheres, J. Metals, No. 9 (1956).
4. T. V. Demikhova et al., Trudy Inst. Metallurgii i Obogashchenia, Akad. Nauk KazSSR, No. 4, p. 109 (1962).
5. T. V. Demikhova and Yu. V. Vorontsov, Tsvetnye Metally, No. 9, p. 37 (1963).
6. V. A. Ragozinnikov, I. L. Shchetnikova, P. S. Mamykin, et al., Tsvetnye Metally (in press).
7. V. A. Ragozinnikov, I. L. Shchetnikova, V. M. Ust'yantsev, Trudy Vost. Inst. Ogneuporov, No. 6, p. 122 (1966).
8. S. M. Zubakov, Mineral Formation in Chrome-Magnesite Refractories [in Russian], Izd. Akad. Nauk KazSSR, Alma-Ata (1960).
9. A. S. Berezhnoi, in: Scientific Transactions of the Ukrainian Scientific Research Institute for Refractories, No. 6 (1962), p. 5.